CliffsNotes®

AP® Biology 2021 Exam

by
Phillip E. Pack, Ph.D.

Houghton Mifflin Harcourt
Boston • New York

About the Author

Phillip E. Pack, Ph.D., taught AP Biology for 11 years. He is currently Professor of Biology at Woodbury University in Burbank, California. He teaches courses in biology, human biology, botany, field botany, environmental studies, and evolution, and co-teaches various interdisciplinary courses, including energy and society (with architecture faculty) and natural history of California and nature writing (with English faculty).

Editorial

Executive Editor: Greg Tubach

Senior Editor: Christina Stambaugh

Production Editor: Jennifer Freilach

Technical Editor: Kellie Ploeger Cox, Ph.D.

Proofreader: Elizabeth Kuball

CliffsNotes® AP® Biology 2021 Exam

Library of Congress Control Number: 2020934009
ISBN: 978-0-3583-5352-2 (pbk)

Printed in the United States of America
DOO 10 9 8 7 6 5 4 3 2 1

www.hmhbooks.com

To Beckett, age 3
—an AP Bio student on the horizon

Acknowledgments

I am the author of this book, but there are many who have contributed to its final form.

I am indebted to students past and present at Woodbury University for their challenging questions and their stimulating discussions. It is no exaggeration to say that I am a student in my own classroom.

I am grateful for the support and encouragement of the editorial team: to Kellie Ploeger Cox, technical editor, for her attention to accuracy; and to Christina Stambaugh, senior editor, for her scrupulous editing, penetrating questions, and insightful suggestions. A special thanks to Greg Tubach, executive editor, for his persistent encouragement and unending prodding, without which I'd still be writing a chapter somewhere in the middle of the book.

Finally, I would like to thank my wife, Mary McGinnis, and daughter, Megan McGinnis-Pack, for their patience, love, and understanding through the course of this project.

Table of Contents

Chapter 1

An Overview of the AP Biology Exam

This chapter introduces you to the AP Biology exam. You'll learn how this book can aid you in your preparation and what to bring on exam day. It also details the format of the exam and how it is scored. The "What's on the Exam" section outlines the topics that are covered on the exam and offers strategies for answering the two question types: multiple-choice and free-response. The last section of this chapter details how to best use the practice exams in this book to enhance your study.

How You Should Use This Book

The Advanced Placement program is designed to encourage students to take challenging courses in high school and receive college credit for their efforts. Many high schools offer classes especially designed for the AP program, but any course or program of study, whatever it is called, is appropriate preparation for taking the AP exam if the content is at the college level. This book helps you to prepare for the Advanced Placement Examination in Biology. It does this in three ways:

- It reviews the important material that you need to know for the actual AP exam. These reviews are detailed but written in an organized and condensed format, making them especially useful for studying.
- It provides you with review questions that reinforce the review. Many of the review questions, like those on the AP exam, require considerable thought to determine the correct answer. In addition, some of the review questions ask you to apply the reviewed material to new situations and, as a result, increase your breadth of understanding. Answers with complete explanations provide additional opportunities to understand the material.
- It offers two complete practice exams, giving you the opportunity to evaluate your knowledge and your test-taking skills. Taking these practice exams helps to improve your AP exam score because they are similar in content and format to the actual AP exam. Answers with complete explanations are given for each question, and a scoring worksheet is provided to help you determine your score.

When preparing for a test, have you ever wished that you had a copy of your teacher's lecture notes? The review sections in this book are very much like lecture notes. Each section contains all of the important terminology with brief descriptions. All of the important biological processes are outlined with a key word or phrase, listed in an easy-to-remember sequence. After each key word or phrase, a short explanation is given. When you study the material the first time, you can read the key words and the short explanations. When you review, you can just study the key words, rereading the explanations only as needed.

You should consider this book, however, as a supplement to your textbook, your laboratory exercises, and your teacher's lectures. Much of the excitement and adventure of biology can be obtained only through hands-on activities and discussions with teachers. In addition, textbooks provide background information, extensive examples, and thought-provoking questions that add depth to your study of biology. Each time you study a topic in class, after listening to the lectures and reading the textbook, use this book to review. Underline or highlight material to help you remember it. Write in the margins, noting any additional material that you heard in lectures or read in your textbook that you or your teacher thinks is important. Then answer the review questions and read the answer explanations at the end of each chapter. This will reinforce your learning.

At the end of your biology course, this book will be a single, condensed source of material to review before the AP exam. Begin your final preparation several weeks before the AP exam by reviewing the material in each chapter. Then take the two practice exams at the end of this book.

What to Bring to the Exam

1. A No. 2 pencil and an eraser are required for the multiple-choice section.
2. A pen with black or dark-blue ink is required for the free-response section.
3. A calculator with a 4-function (+, –, ×, ÷) and square-root capability is allowed for the entire exam. Programmable, graphing, and cell-phone calculators are not permitted. A calculator will be especially useful for the free-response questions but may be used for multiple-choice questions as well. Buy this calculator as soon as possible so that you can begin using it early in your biology course. Practice with the calculator frequently so that you are as familiar with it as you are with your cell phone. You don't want to spend time figuring out how to take a square root for the first time during the AP exam.
4. You are not allowed to bring your own scratch paper. For the multiple-choice section, you can use the margins of your exam booklet. For the free-response section, scratch paper is provided.
5. Obviously, you are not allowed to use any prepared notes. However, you will be provided with a list of equations and formulas. The Equations and Formulas pages are provided in this book at the beginning of "Part 3: AP Biology Practice Exams."

Exam Format

The AP Biology exam consists of two sections. Section I consists of 60 multiple-choice questions. These questions are often presented in sets of three to five related questions that refer to a descriptive paragraph, a set of data, or a figure. Each multiple-choice question provides four answer choices. Section II consists of six free-response questions. Two of these questions are long and evaluate your understanding and ability to interpret experimental data. One of these two questions may require graphing. The other four are short-answer questions. All questions, long and short, have four parts, each of which provides focus for your answer. Before you begin a free-response question, read the question carefully to organize your thoughts—underline or circle key words, record notes, or create an outline on provided paper. You have 90 minutes to complete each section of the exam, and each section counts toward 50% of your exam score. The exam is administered in May of each year, along with AP exams in other subjects.

	Type of Question	Number of Questions	Question Weight	Section Weight	Approximate Time Recommended per Question	Time Allowed
Section I	Multiple-Choice	60	1 point each	50%	1½ minutes	90 minutes
Section II	Long Free-Response	2	8–10 points each	50%	20 minutes	90 minutes
	Short Free-Response	4	4 points each		12 minutes	

Exam Grading

Exams are graded on a scale of 1 to 5, with 5 being the best. Most colleges accept a score of 3 or better as a passing score. If you receive a passing score, colleges give you college credit (applied toward your bachelor's degree), advanced placement (you can skip the college's introductory course in biology and take an advanced course), or both. You should check with the biology department at the colleges you're interested in to determine how they award credit for the exam.

The distribution of student scores for some recent AP Biology exams is shown here.

	Exam Grade	College Grade Equivalent	Student Performance*			
			2016	2017	2018	2019
Extremely well qualified	5	A+, A	6.6%	6.4%	7.2%	7.2%
Well qualified	4	A–, B+, B	21.0%	21.0%	21.6%	22.2%
Qualified	3	B–, C+, C	33.6%	36.7%	32.8%	35.3%
Possibly qualified	2	n/a	28.8%	27.5%	28.3%	26.6%
No recommendation	1	n/a	10.1%	8.4%	10.2%	8.8%
Mean score (1 to 5)			2.85	2.90	2.87	2.92

**Percentages may not total 100 due to rounding.*

The multiple-choice section is designed with a balance of easy and difficult questions to produce a mean score of 50%. Scores for free-response questions vary significantly with individual questions and from year to year. On recent exams, mean scores ranged from 2.01 to 5.20 for the 10-point free-response questions and 0.96 to 2.66 for the 4-point free-response questions. Clearly, both sections of the exam are difficult. They are deliberately written that way so that the full range of students' abilities can be measured. In spite of the exam difficulty, however, more than 63% of the students taking recent exams received a score of 3 or better. Therefore, the AP exam is difficult, but most (prepared) students do well.

What's on the Exam

Theodosius Dobzhansky, a famous geneticist, once wrote an essay entitled "Nothing in Biology Makes Sense Except in the Light of Evolution." Similarly, many diverse topics in biology cannot be fully appreciated without studying them through the multiple lenses of other biology topics. Biology is not just a set of individual concepts or processes to be studied in isolation. Biology is a web of interconnecting themes. To fully grasp a theme, you must understand how it is shaped and influenced by other themes.

To address this web of interconnecting themes, the College Board has organized the course around four broad principles, called "Big Ideas," each of which encompasses a variety of unifying concepts. These are the four Big Ideas:

Big Ideas	Topics
Big Idea 1: Evolution	Evolution, Heredity, Gene Expression, Ecology
Big Idea 2: Energetics	Chemistry, Cells, Respiration, Photosynthesis, Cell Communication, Ecology
Big Idea 3 Information Storage and Transmission	Chemistry, Cells, Cell Communication, Heredity, Gene Expression, Ecology
Big Idea 4: Systems Interactions	Chemistry, Cells, Respiration, Photosynthesis, Heredity, Evolution, Ecology

As you can see, each Big Idea has multiple overlapping topics. That's why they are called *Big* Ideas. Although each chapter of this book focuses on an individual topic, as you progress through your study of the topics, each Big Idea will begin to take shape.

This book reviews every concept included in the AP Biology curriculum framework and its accompanying lab manual (*AP Biology Investigated Labs, An Inquiry-Based Approach*). It carefully *excludes* those concepts that are omitted from the framework and lab manual (but are often included in college biology textbooks). This book is what you need to know—no more, no less. Keep in mind, however, that the AP exam varies from year to year and will have questions that only cover a select portion of all the framework concepts.

The table that follows outlines the major topics that are covered on the exam and how much of that topic is represented. Depending on the exam, the weight of each topic varies. But realistically, except for evolution, they are all fairly equally represented. Evolution gets a slightly higher weight. The take-home message here is you need to know all of the topics.

Chapter	Topic	Exam Weighting
2	Chemistry of Life	8–11%
3	Cell Structure and Function	10–13%
4	Cellular Respiration	12–16%
5	Photosynthesis	
6	Cell Communication	10–15%
7	Cell Cycle	
8	Heredity	8–11%
9	Gene Expression and Regulation	12–16%
10	Evolution	13–20%
11	Ecology	10–15%

As you probably have already discovered, biology consists of a lot of technical words, concepts, and processes. It is often much easier to study a topic in detail, when the connections among the words, concepts, and processes are presented together, before going on to the next topic. As you read your textbook and review with this book, it is important to remember that the AP exam will test not just your knowledge of individual topics, but how various topics contribute to overlapping themes. Multiple-choice questions, and especially free-response questions, will evaluate how well you understand this big picture of biology.

Keep in mind that the big picture is supported by content. For free-response questions, the quality of your content is often determined by the detail that you provide. That detail is in this book. But the AP curriculum also indicates certain material that you do not need to know for the exam. If that material appears in this book, it does so to help you understand a concept or to connect the material with your textbook. That information, however, what you are *not* expected to know for the exam, is clearly identified.

Strategies for Multiple-Choice Questions

On the AP exam, questions for the multiple-choice section are provided in a booklet. While reading the questions in the booklet, feel free to cross out answer choices you know are wrong or underline important words. After you've selected the answer from the various choices, you carefully fill in bubbles labeled A, B, C, or D on an answer sheet. Mark only your answers on the answer sheet. Since unnecessary marks can produce machine-scoring errors, be sure to fill in the bubbles carefully and erase errors and stray marks thoroughly.

Some specific strategies for answering the multiple-choice questions follow:

1. **Don't leave any answers blank.** There is no penalty for guessing. You get 1 point for each correct answer. If you leave it blank or if you get it wrong, you get 0 points. If you're not sure of the answer to a question, eliminate any answer choices you think are wrong and then select one of the remaining answer choices. If you can't eliminate any wrong answers, you still have a 25% probability (1 chance in 4) of choosing the correct answer by guessing. If you can eliminate one or more wrong answers, your probability of getting it right increases.
2. **Don't let easy questions mislead you.** The multiple-choice questions range from easy to difficult. On one exam, 92% of the candidates got the easiest question right, but only 23% got the hardest question right. Don't let the easy questions mislead you. If you come across what you think is an easy question, it probably is. Don't suspect that it's a trick question.

3. **Budget your time.** You have 90 minutes to answer 60 multiple-choice questions; that's about 1½ minutes per question. Read the question and consult any diagrams or graphs. Read all the answer choices, crossing out any you think are wrong. Then choose or, if necessary, guess the correct answer and mark your answer sheet. Remember, there's no penalty for wrong answers! It's better to move on to the next question so that you will have the opportunity to try all of the questions.
4. **Skip hard questions.** If you come across a hard question that you can't answer quickly, skip it, and mark the question to remind you to return to it if time permits. If you can eliminate some of the answer choices, mark those also so that you can save time when you return. It's important to skip a difficult question, even if you think you can eventually figure it out, because for each difficult question you spend 3 minutes on, you could have answered two easy questions. If you have time at the end of the exam, you can always go back. If you don't have time, at least you will have had the opportunity to try all of the questions. Also, if you don't finish the section, don't be overly concerned. Since the exam is designed to obtain a mean score of 50%, it is not unusual for a student to run out of time before reaching the end of the section. But don't leave any answers blank (see #8, below).
5. **Judge time requirements for questions.** Some multiple-choice questions begin with a long description followed by three to five associated questions. These questions make good use of your time because once you've read the introduction, you're ready to answer all of the associated questions. On the other hand, some multiple-choice questions with long introductory descriptions are followed by a single question. These questions require proportionately more time than if followed by multiple questions. Skip these questions with long introductions and a single question if you think you're running low on time. Return to them if time is available.
6. **Carefully answer reverse multiple-choice questions.** In a typical multiple-choice question, you need to select the answer choice that is true. On the AP exam, you may find a *reverse* multiple-choice question where you need to select the *false* answer choice. These questions usually use the word "EXCEPT" in sentences such as "All of the following are true EXCEPT . . ." or "All of the following occur EXCEPT. . . ." A reverse multiple-choice question is more difficult to answer than regular multiple-choice questions because it requires you to know three true pieces of information about a topic before you can select the false answer choice. It is equivalent to correctly answering four true-false questions to get 1 point; and if you get one of the four wrong, you get them all wrong. Reverse multiple-choice questions are also difficult because halfway through the question you can forget that you're looking for the false answer choice. To avoid confusion, do the following: After reading the opening part of the question, *read each answer choice and mark a T or an F next to each one to identify whether it is true or false.* If you're able to mark a T or an F for each one, then the correct answer is the choice marked with an F. Sometimes you won't be sure about one or more choices, or sometimes you'll have two choices marked F. In these cases, concentrate on the uncertain choices until you can make a decision.
7. **Return to difficult questions *only* if you have time.** Here's one thing to consider when skipping a question: If you return to a question, you will need to read the question, read the answer choices, and consult the diagrams. This is a costly strategy because you already spent time doing that once. Only do this if you've already tried to answer all of the other questions in Section I.
8. **Save the last minute to mark all unanswered questions.** Because the exam is designed to obtain a mean score of 50%, some students may not have enough time to read all of the questions. Should this happen to you, be sure to mark answers for all of your remaining unanswered questions. Remember, there's no penalty for wrong answers!

Strategies for Free-Response Questions

The free-response questions are provided in a separate booklet. Read each of the four parts of a question thoroughly, circling key words. Next, write a brief outline using key words to organize your thoughts. If you choose not to write an outline, go back and reread the question halfway through writing your answer. It's easy to get carried away, and by the end of your response you might be answering a different question. Also, make sure that you haven't written the same answer for more than one part; if so, choose the part most appropriate for your answer and write something different for the other part.

Each of the two long free-response questions is worth 8–10 points. Each of the four short free-response questions is worth 4 points.

Strategies for answering the free-response questions follow:

1. **Don't approach the free-response section with apprehension.** Most students approach the free-response section of the exam with more anxiety than they have when approaching the multiple-choice section. However, in terms of the amount of detail in the knowledge required, the free-response section is easier. On these questions, *you* get to choose what to write. You can get an excellent score without writing every relevant piece of information. Besides, you don't have time to write an entire book on the subject. A general answer that addresses the question with a limited number of specifics will get a good score. Additional details may (or *may not*) improve your score, but the basic principles are the most important elements for a good score. In contrast, a multiple-choice question focuses on a very narrow and specific body of knowledge, which you'll either know or you won't; it doesn't let you select from a range of correct information.
2. **Keep your answers brief for the short free-response questions.** Some short questions will be very general and seem to be asking for a whole lot of information, as if it were a long question. If you get a short question like this, don't freak out because you think that it will take 20 minutes to write down everything you know. Instead, try to answer each part of the question as concisely and briefly as possible. Come back and add more details if you have time. Other short questions will have parts that will require more specific answers and won't require a lengthy explanation.
3. **Give specific information in your answer.** You need to give specific information for each free-response question. Don't be so general that you don't really say anything. Give more than just terminology with definitions. You need to use the terminology to explain biological processes. The combination of using the proper terminology and explaining processes will convince an AP exam reader that you understand the answer. Give some detail when you know it—names of processes, names of structures, names of molecules—and then tell how they're related. The exam reader is looking for specific information. If you say it, you get the points. You don't have to say everything, however, to get the maximum number of points.
4. **Answer each part of a free-response question separately.** You should answer each part of the question in a separate paragraph, which helps the exam reader recognize each part of your answer. Questions are formally divided into parts, such as a, b, c, and d, so label your paragraphs a, b, c, and d.
5. **Answer all parts of a free-response question.** It is extremely important that you give a response for each part of the question. Don't overload the detail on one part at the expense of saying nothing in another part because you ran out of time. Each part of the question is apportioned a specific number of points. If you give abundant information in one part and nothing in the remaining parts, you receive only the maximum number of points allotted to the part you completed. Each part of a long question may earn between 1 and 4 points; in a short question, each part is worth 1 point. You won't get any extra points above the established maximum, even if what you write is Nobel Prize–quality.
6. **Don't answer more parts than required.** Some free-response questions may give you a choice of parts to answer. Choose the parts that you know the most about and answer only those parts. Do not answer extra parts. *There is no extra credit on this exam!* In general, an exam reader will not read beyond the required number of answers. In cases where the exam reader does read the extra parts, you may lose points if you contradict something you said correctly in an earlier required part.
7. **Budget your time.** You have 90 minutes to answer six free-response questions. Allow 20 minutes for each of the two long questions (40 minutes total) and about 12 minutes for each of the four short questions (48 minutes total). That's 5 minutes per part for the long questions and 3 minutes per part for the short questions. When you begin this section, identify the questions you think you can answer the best and answer those questions first. However, just as it's most important to answer all parts of a question, it's best to respond to ALL of the free-response questions and all of their parts instead of leaving blank parts. You'll probably know *something* about every part of each question, so be sure you get that information written. If you are nearing the end of the 90-minute period and you still have several questions to answer, use that time to write something for each of the remaining questions. One point, especially on a short free-response question, is a lot better than zero.

8. **Don't worry if you make a factual error.** What if you write something that is incorrect? The AP exam readers look for correct information. They search for key words and phrases and award points when they find them. If you use the wrong word to describe a process, or identify a structure with the wrong name, no formal penalty is assessed. If you're going to get any points, however, you need to write correct information. Also, you'll lose points if you contradict something you said correctly earlier.
9. **Don't be overly concerned about grammar, spelling, punctuation, or penmanship.** The AP exam readers don't penalize for incorrect grammar, spelling, or punctuation or for poor penmanship. They are interested in *content.* However, if your grammar, spelling, or penmanship impairs your ability to communicate, then the exam readers cannot recognize the content, and your score will suffer.
10. **Don't write a standard essay.** Don't spend your time writing a standard essay with an introduction, supporting paragraphs, and a conclusion. Just dive right in and answer the question directly. On the other hand, your response cannot be an outline; it must have complete sentences and be written in paragraph form.
11. **Don't repeat the question in your answer.** Or do so only briefly. The exam reader knows the question.
12. **Improve your score by incorporating drawings.** Drawings and diagrams may sometimes add as much as 1 point to your free-response score. But the drawings must be explained in your response, and the drawings must be labeled with supporting information. If not, the AP exam reader will consider them doodles, and you will get no additional points.
13. **Pay attention to direction words.** A direction word is the first word in a free-response question that tells you how to answer the question. The direction word tells you what you need to say about the subject matter that follows. Here are the most common direction words found on the AP exam:
    - *Describe* means to characterize or give an account in words.
    - *Explain* means to clarify or make understandable.
    - *Identify* means to name a structure, process, or concept (no explanation is required).
    - *Justify* means to provide evidence or reasoning to support a claim.
    - *Represent* means to provide information in a graph, drawing, table, or mathematical expression.
    - *Compare* means to discuss two or more items with an emphasis on their *similarities.*
    - *Contrast* means to discuss two or more items with an emphasis on their *differences.*

    Specialized direction words are used for questions that evaluate your quantitative skills. These words include *design* (an experiment), *calculate* or *determine* (a value), and *construct* and *label* (a graph). These words have specific meanings for laboratory analyses and are discussed in Chapter 12, "Review of Laboratory Investigations."

Taking the Practice Exams

For each of the practice exams, a scoring template is provided. Each exam is followed by an answer key for the multiple-choice questions, explanations for these questions, and scoring standards for the free-response questions (often called a rubric).

To get the full benefit of simulating a real AP exam, set aside at least 3 hours for each practice exam. Begin with the multiple-choice section (Section I), and after 90 minutes, stop and move on to the free-response section (Section II). Allow yourself 90 minutes to write out your full answers. By using the actual times that the real AP exam allows, you will learn whether the time you spend on each multiple-choice and free-response question is appropriate. When you're done taking a practice exam, score your exam using the multiple-choice answer key and the free-response scoring standards. Then go back and answer any multiple-choice questions that you were unable to complete in the allotted 90 minutes. When you are done, read all of the multiple-choice explanations, even those for questions you got right. The explanations are thorough and provide you with information and suggestions. Even if you know the answers, reading the provided explanations is good review.

Although you've heard it so many times, practice *will* improve your test performance (although it's unlikely to make you perfect). So be sure to complete both practice exams and review all of the answer explanations. Good luck!

PART 1

SUBJECT AREA REVIEWS WITH REVIEW QUESTIONS AND ANSWERS

Chapter 2

Chemistry of Life

Review

A major difference between an AP biology course and a regular high school biology course is the emphasis on detail. In many cases, that detail derives from a description of the molecular structure of molecules and the chemistry of metabolic reactions. It is the understanding of biological processes at the molecular level that provides you with a more thorough understanding of biology. The AP examiners want to know whether you have this kind of understanding. With that in mind, your studying should begin with a brief review of chemistry and the characteristics of major groups of biological molecules.

Although chemistry deepens your understanding of biology, the AP exam will not ask you to draw molecular structures, to distinguish fine differences between molecules, or to identify specific names of molecules. You will only need to be able to recognize very major distinctions between groups of similar types of molecules.

Atoms, Molecules, Ions, and Bonds

An **atom** consists of a nucleus of positively charged protons and neutrally charged neutrons. Negatively charged electrons are arranged outside the nucleus. **Molecules** are groups of two or more atoms held together by chemical bonds. Chemical bonds between atoms form because of the interaction of their electrons. The **electronegativity** of an atom, or the ability of an atom to attract electrons, plays a large part in determining the kind of bond that forms. There are three kinds of bonds, as follows:

1. **Ionic** bonds form between two atoms when one or more electrons are *transferred* from one atom to the other. This bond occurs when the electronegativities of the atoms are very different and one atom has a much stronger pull on the electrons (high electronegativity) than the other atom in the bond. The atom that gains electrons has an overall negative charge, and the atom that loses electrons has an overall positive charge. Because of their positive or negative charges, these atoms are **ions.** The attraction of the positive ion to the negative ion constitutes the ionic bond. Sodium and chlorine form ions (Na^+ and Cl^-), and the bond formed in a molecule of sodium chloride (NaCl) is an ionic bond.
2. **Covalent** bonds form when electrons between atoms are *shared,* which means that neither atom completely retains possession of the electrons (as happens with atoms that form strong ionic bonds). Covalent bonds occur when the electronegativities of the atoms are similar.
   - **Nonpolar covalent** bonds form when electrons are *shared equally.* When the two atoms sharing electrons are identical, such as in oxygen gas (O_2), the electronegativities are identical, and both atoms pull equally on the electrons.
   - **Polar covalent** bonds form when electrons are *shared unequally.* Atoms in this kind of bond have electronegativities that are different, and an unequal distribution of the electrons results. The electrons forming the bond are closer to the atom with the greater electronegativity and produce a negative charge, or **pole,** near that atom. The area around the atom with the weaker pull on the electrons produces a positive pole. In a molecule of water (H_2O), for example, electrons are shared between the oxygen atom and each hydrogen atom. Oxygen, with a greater electronegativity, exerts a stronger pull on the shared electrons than does each hydrogen atom. This unequal distribution of electrons creates a negative pole near the oxygen atom and positive poles near each hydrogen atom (Figure 2-1).
   - **Single covalent, double covalent,** and **triple covalent** bonds form when two, four, and six electrons are shared, respectively.
3. **Hydrogen** bonds are weak bonds between *molecules.* They form when a positively charged *hydrogen* atom in one covalently bonded molecule is attracted to a negatively charged area of another covalently bonded molecule. In water, the positive pole around a hydrogen atom forms a hydrogen bond to the negative pole around the oxygen atom of *another* water molecule (Figure 2-1).

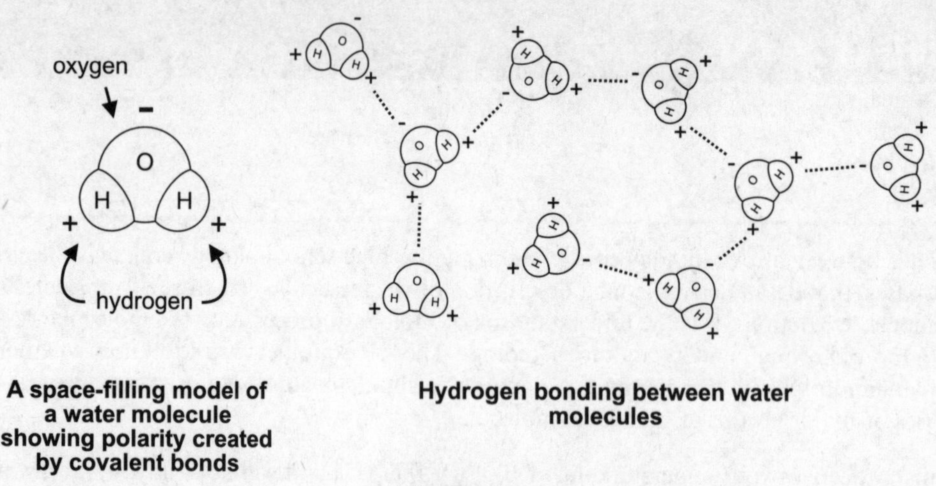

Figure 2-1

When you think of chemical bonds, imagine a continuum based on the differences of electronegativities (Figure 2-2). The left end represents bonds that form when no differences exist in the electronegativities of the atoms. Electrons are shared equally, and nonpolar covalent bonds form. The right end represents bonds that form when very large differences in electronegativities exist. Electrons are transferred from one atom to another, and ionic bonds form. When the electronegativities of the atoms are different, but not strongly so, the electrons are shared unequally, and polar covalent bonds form; this activity is represented by the center of Figure 2-2. The kind of bond that forms between two atoms and the strength of that bond depend upon the difference of electronegativities of the atoms and might occur any place along the line shown in Figure 2-2.

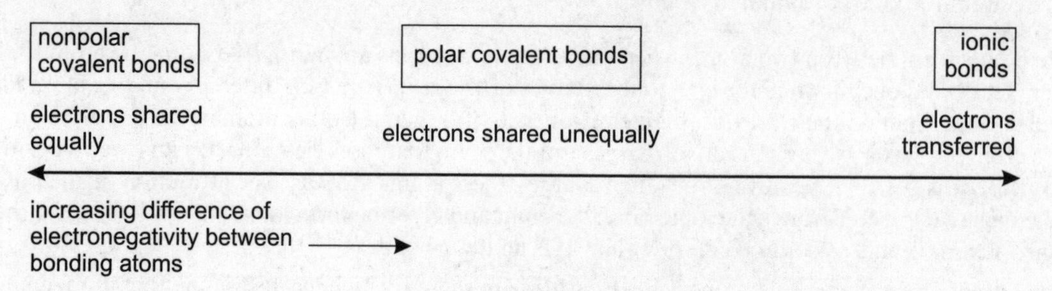

Types of Chemical Bonds

Figure 2-2

Properties of Water

The polarity of water molecules and the hydrogen bonds that result from that polarity contribute to some very special properties for water:

1. *Water is an excellent* **solvent.** Ionic substances are soluble (they dissolve) in water because the poles of the polar water molecules interact with the ionic substances and separate them into ions. Substances with polar covalent bonds are similarly soluble because of the interaction of their poles with those of water. Substances that dissolve in water are called **hydrophilic** ("water loving"). Because they lack charged poles, nonpolar covalent substances do not dissolve in water and are called **hydrophobic** ("water fearing"). Because so many kinds of molecules are soluble in water, water is often referred to as a universal solvent. In general, a **solute** is a substance that dissolves in a solvent. When that solvent is water, the solution is called an **aqueous** solution.
2. *Water has a high* **specific heat capacity.** Specific heat is the degree to which a substance changes temperature in response to a gain or loss of heat. Water has a high specific heat, changing temperature very slowly with changes in its heat content. You must add a relatively large amount of energy to warm (and boil) water or

remove a relatively large amount of energy to cool (and freeze) water. When sweat evaporates from your skin, a large amount of heat is taken with it and you are cooled ("evaporative cooling"). When water changes physical states, from solid to liquid or liquid to gas, energy is absorbed but the water temperature remains constant. The absorbed energy is used only to change the physical state of the water by breaking the hydrogen bonds that tether the water molecules together. In the reverse reactions, from gas to liquid or liquid to solid, the energy released reestablishes the hydrogen bonds. In Figure 2-3, the horizontal lines between the physical states of water indicate the absorption of energy without a rise in temperature. The energy associated with each of these transitions has a special name:

- **Heat of fusion** is the energy required to change water from a solid to a liquid.
- **Heat of vaporization** is the energy required to change water from a liquid to a gas.

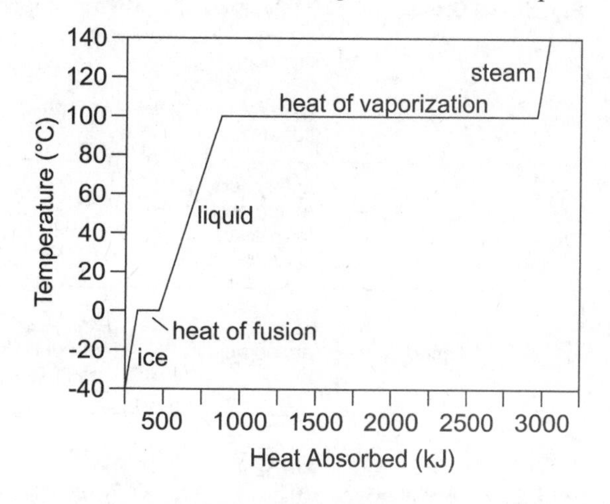

Temperature of Water as a Function of Heat Absorbed

Figure 2-3

3. *Ice floats.* Unlike most substances that contract and become more dense when they freeze, water *expands* as it freezes, becomes less dense than its liquid form, and, as a result, floats in liquid water. Hydrogen bonds are typically weak, constantly breaking and reforming, allowing molecules to periodically approach one another. In the solid state of water, the weak hydrogen bonds between water molecules become rigid and form a crystal that keeps the molecules separated and less dense than its liquid form. If ice did not float, it would sink and remain frozen due to the insulating protection of the overlaying water. This would have a profound effect on the survival of organisms inhabiting the bottom of bodies of water.
4. *Water has strong* **cohesion** *and high* **surface tension.** Cohesion, or the attraction between *like* substances, occurs in water because of the hydrogen bonding between water molecules. The strong cohesion between water molecules produces a high surface tension, creating a water surface that is firm enough to allow many insects to walk upon it without sinking.
5. *Water has strong* **adhesion.** Adhesion is the attraction of *unlike* substances. This results from the attraction of the poles of water molecules to other polar substances. If you wet your finger, you can easily pick up a straight pin by touching it because the water on your finger adheres to both your skin and the pin. Similarly, some people wet their fingers to help them turn pages. When water adheres to the walls of narrow tubing or to absorbent solids like paper, it demonstrates **capillary action** by rising up the tubing or creeping through the paper.

Organic Molecules

Organic molecules are those that have carbon atoms. In living systems, large organic molecules, called **macromolecules,** may consist of hundreds or thousands of atoms. Most macromolecules are **polymers,** molecules that consist of a single group of atoms **(monomer)** repeated many times.

Four of carbon's six electrons are available to form bonds with other atoms. Thus, you will always see four lines connecting a carbon atom to other atoms, each line representing a pair of shared electrons (one electron from carbon and one from another atom). Complex molecules can be formed by stringing carbon atoms together in a straight line or by connecting carbons together to form rings. The presence of nitrogen, oxygen, and other atoms adds additional variety to these carbon molecules.

Many organic molecules share similar properties because they have similar clusters of atoms, called **functional groups.** Each functional group gives the molecule a particular property, such as acidity or polarity. The more common functional groups with their properties are listed in Figure 2-4.

Functional Group		Examples	Characteristics
— OH	hydroxyl	alcohols (e.g., ethanol), glycerol, sugars	polar, hydrophilic
—C(=O)OH	carboxyl	acetic acid, amino acids, fatty acids, sugars	polar, hydrophilic, weak acid
—NH₂ (—N with two H)	amino	amino acids	polar, hydrophilic, weak base
—P(=O)(O⁻)—O⁻	phosphate	DNA, ATP, phospholipids	polar, hydrophilic, acid
—CH₃ (—C with three H)	methyl	fatty acids, oils, waxes	nonpolar, hydrophobic

Functional Groups

Figure 2-4

Four important classes of organic molecules—carbohydrates, lipids, proteins, and nucleic acids—are discussed in the following sections. Although specific examples are provided, the AP exam does not require that you know their formulas. However, you should know the general characteristics that distinguish one group of molecules from another.

Carbohydrates

Carbohydrates are classified into three groups according to the number of sugar (or saccharide) molecules present:

1. A **monosaccharide** is the simplest kind of carbohydrate. It consists of a single sugar molecule, such as fructose or glucose (Figure 2-5). (Note that the symbol C for carbon may be omitted in ring structures; a carbon exists wherever four bond lines meet.) Sugar molecules have the formula $(CH_2O)_n$, where *n* is any number from 3 to 8. For **glucose,** *n* is 6, and its formula is $C_6H_{12}O_6$. The formula for **fructose** is also $C_6H_{12}O_6$, but as you can see in Figure 2-5, the placement of the carbon atoms is different. Two forms of glucose, **α-glucose** and **β-glucose,** differ simply by a reversal of the H and OH on the first carbon (clockwise, after the oxygen). As you will see below, even very small changes in the position of certain atoms may dramatically change the chemistry of a molecule.

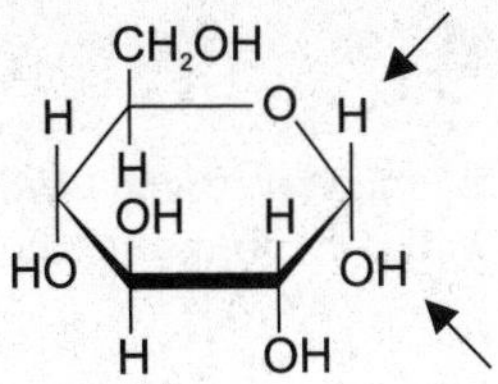

Alpha Glucose

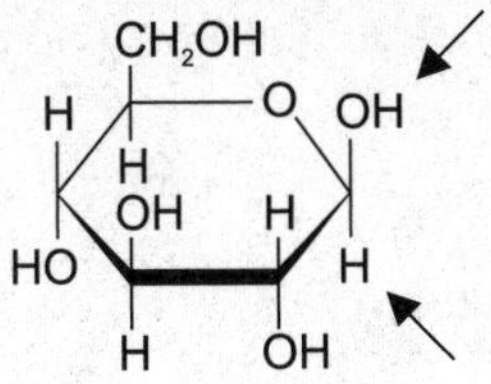

Beta Glucose

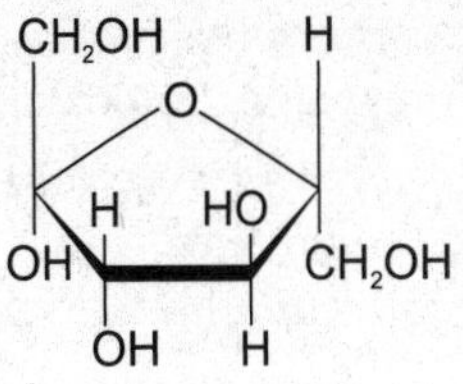

Fructose

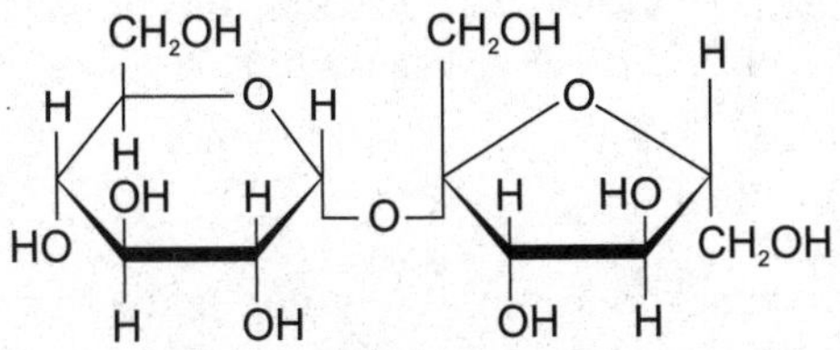

Sucrose

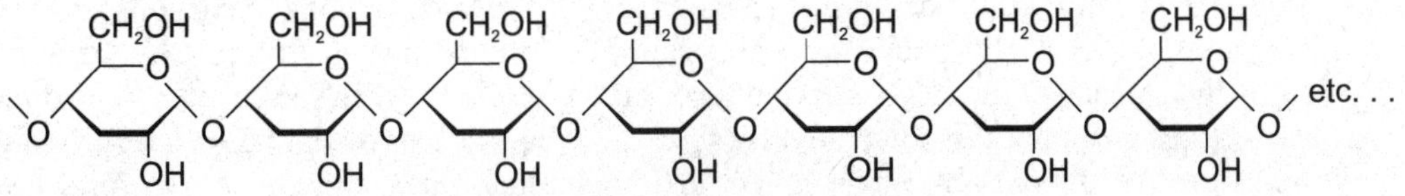

Starch

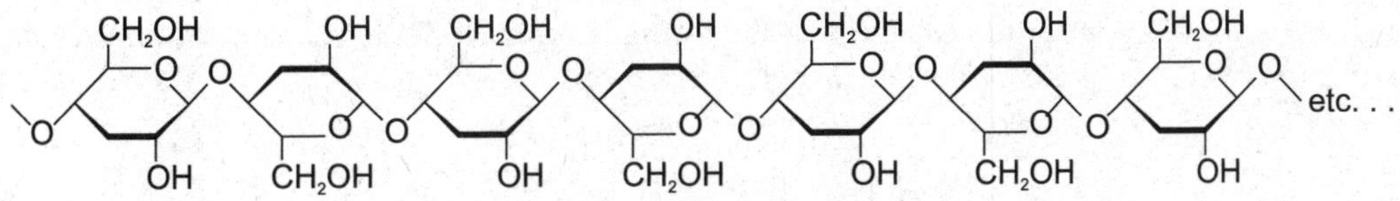

Cellulose

Carbohydrates

Figure 2-5

2. A **disaccharide** consists of two sugar molecules joined by a **glycosidic linkage.** During the process of joining, a water molecule is lost. Thus, when glucose and fructose link to form sucrose, the formula is $C_{12}H_{22}O_{11}$ (not $C_{12}H_{24}O_{12}$) (Figure 2-5). This type of chemical reaction, where a small molecule is lost, is generally called a **condensation reaction** (or specifically, a **dehydration reaction,** if the lost molecule is water). The reverse reaction, where one molecule is *split* to form two molecules by the *addition* of water, is called **hydrolysis.** The formation of some common disaccharides with their dehydration reactions follows:
   - glucose + fructose = H_2O + **sucrose** (common table sugar)
   - glucose + galactose = H_2O + **lactose** (the sugar in milk)
   - glucose + glucose = H_2O + **maltose** (a product of the breakdown of starch)

3. A **polysaccharide** consists of a series of connected monosaccharides. Thus, a polysaccharide is a polymer because it consists of repeating units of a monosaccharide. The following examples of polysaccharides may contain thousands of glucose monomers (Figure 2-5):
   - **Starch** is a polymer of α-glucose molecules. It is the principal *energy storage* molecule in plant cells.
   - **Glycogen** is a polymer of α-glucose molecules. It differs from starch by its pattern of polymer branching. It is a major *energy storage* molecule in animal cells.

- **Cellulose** is a polymer of β-glucose molecules. It serves as a *structural* molecule in the walls of plant cells and is the major component of wood.
- **Chitin** is a polymer similar to cellulose, but each β-glucose molecule has a nitrogen-containing group attached to the ring. Chitin serves as a *structural* molecule in the walls of fungus cells and in the exoskeletons of insects, other arthropods, and mollusks.

The α-glucose in starch and the β-glucose in cellulose illustrate the dramatic chemical changes that can arise from subtle molecular changes: The bonds in starch (specifically, the α-glycosidic linkages) can easily be broken down (digested) by humans and other animals, but only specialized organisms, like the bacteria in the guts of termites, can break the bonds in cellulose (specifically, the β-glycosidic linkages).

Lipids

Lipids are a class of substances that are nearly insoluble in water (and other polar solvents) but are highly soluble in nonpolar substances (like ether or chloroform). There are three major groups of lipids:

1. **Triglycerides (triacylglycerols)** include fats and oils. They consist of three **fatty acids** attached to a **glycerol** molecule (Figure 2-6). Fatty acids are hydrocarbons (chains of covalently bonded carbons and hydrogens) with a carboxyl group (–COOH) at one end of the chain. Fatty acids vary in structure by the number of carbons and by the placement of single and double covalent bonds between the carbons. The double bond in a fatty acid creates a bend at the bond, slightly spreading the triglyceride apart. As a result, saturated fatty acids pack together more tightly, have higher melting temperatures, and are usually solid at room temperature (fats); in contrast, unsaturated fatty acids pack together more loosely, have lower melting temperatures, and are usually liquid at room temperature (oils).

   During the formation of a triglyceride, the carboxyl group of each fatty acid bonds to a hydroxyl group of the glycerol, releasing a water molecule. Thus, the formation of each bond is a dehydration reaction, and a total of three molecules is released when a triglyceride forms.

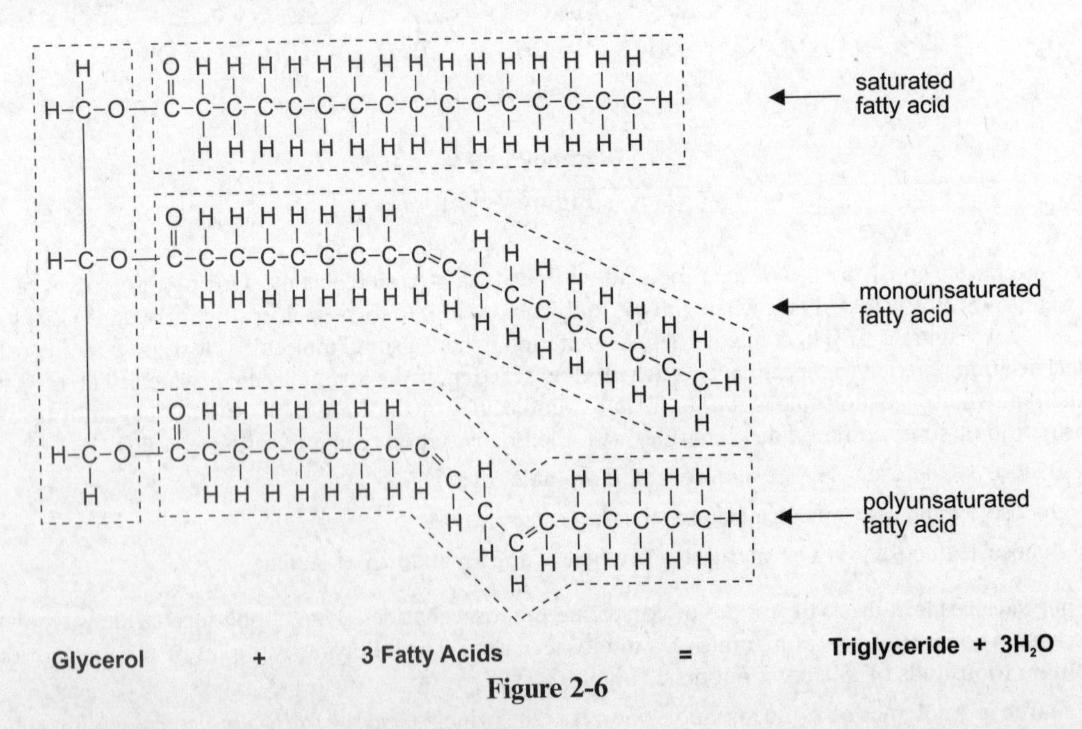

Figure 2-6

- A **saturated** fatty acid has a single covalent bond between each pair of carbon atoms, and each carbon has two hydrogens bonded to it (three hydrogens bonded to the last carbon). You can remember this by thinking that each carbon is "saturated" with hydrogen.

- A **monounsaturated** fatty acid has *one double* covalent bond, and each of the two carbons in this bond has only one hydrogen atom bonded to it.
- A **polyunsaturated** fatty acid is like a monounsaturated fatty acid except that there are *two or more double* covalent bonds.

2. A **phospholipid** looks just like a triglyceride except that one of the fatty acid chains is replaced by a phosphate group ($-PO_3^{2-}$) (Figure 2-7). An additional and variable group of atoms (indicated by R, for radical, in Figure 2-7) is covalently attached to the phosphate group. The two fatty-acid "tails" of the phospholipid are nonpolar and hydrophobic, and the phosphate "head" is polar and hydrophilic. Phospholipids are often found oriented in sandwich-like formations, with the hydrophobic tails grouped together on the inside of the sandwich and the hydrophilic heads oriented toward the outside and facing an aqueous environment. Such formations of phospholipids provide the structural foundation of cell membranes.

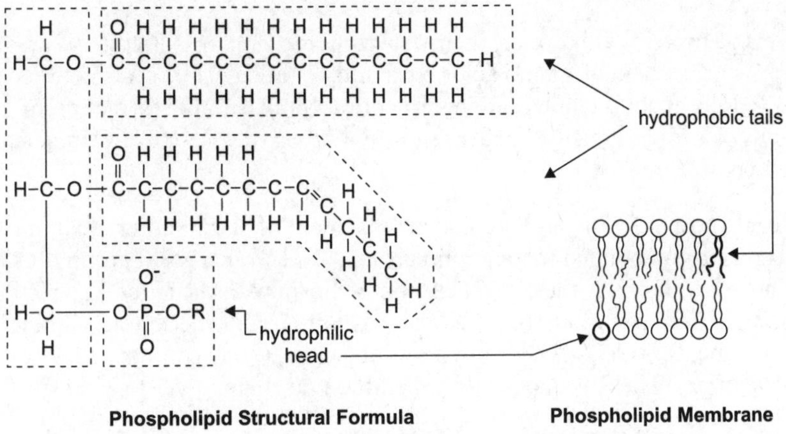

Phospholipid Structural Formula

Phospholipid Membrane

Figure 2-7

3. **Steroids** are characterized by a backbone of four linked carbon rings (Figure 2-8). Examples of steroids include cholesterol (a component of cell membranes) and certain hormones, including testosterone and estrogen.

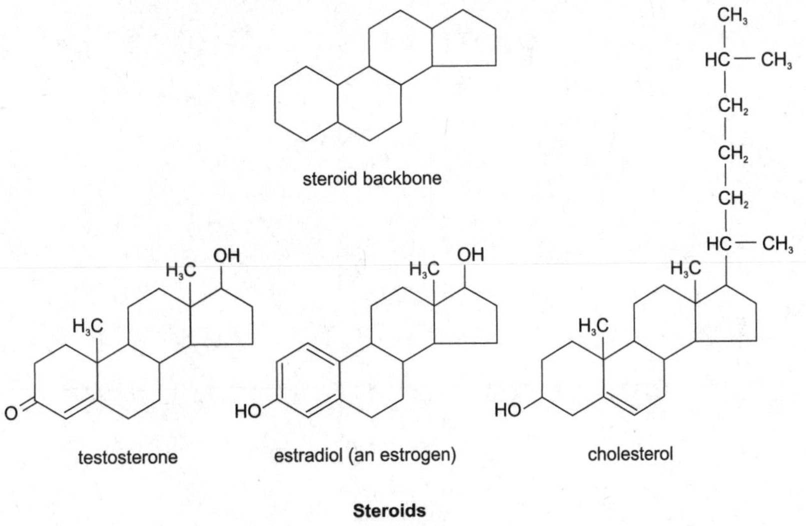

Steroids

Figure 2-8

Proteins

Proteins can be grouped according to their functions. Some major categories follow:

1. **Structural proteins,** such as keratin in the hair and horns of animals, collagen in connective tissues, and silk in spider webs
2. **Storage proteins,** such as casein in milk, ovalbumin in egg whites, and zein in corn seeds
3. **Transport proteins,** such as those in the membranes of cells that transport materials into and out of cells and as oxygen-carrying hemoglobin in red blood cells
4. **Defensive proteins,** such as the antibodies that provide protection against foreign substances that enter the bodies of animals
5. **Enzymes** that regulate the rate of chemical reactions

Although the functions of proteins are diverse, their structures are similar. All proteins are polymers of **amino acids;** that is, they consist of a chain of amino acids covalently bonded. The bonds between the amino acids are called **peptide bonds,** and the chain is a **polypeptide,** or peptide. Similar to the formation of carbohydrate polymers, the peptide bonds of proteins form by dehydration synthesis; that is, one H_2O molecule is released during the formation of each peptide bond.

One protein differs from another by the number and arrangement of the 20 different amino acids. Each amino acid consists of a central carbon bonded to an amino group ($-NH_2$), a carboxyl group ($-COOH$), and a hydrogen atom (Figure 2-9). The fourth bond of the central carbon is shown with the letter R, which indicates an atom or group of atoms that varies from one kind of amino acid to another. For the simplest amino acid, glycine, the R is a hydrogen atom. For serine, R is CH_2OH. For other amino acids, R may contain sulfur (as in cysteine) or a carbon ring (as in phenylalanine). The R group is also called the side chain.

The protein, then, is a polymer of different amino acids with $-NH_2$ at one end and $-COOH$ at the other end. Each amino acid carries an R group with specific properties. For example, the R group may make the amino acid hydrophobic or hydrophilic, polar or nonpolar, or acidic or basic. As a result, proteins can differ widely in structure (size and shape) and chemistry, which, in turn, makes for dramatic differences in protein function.

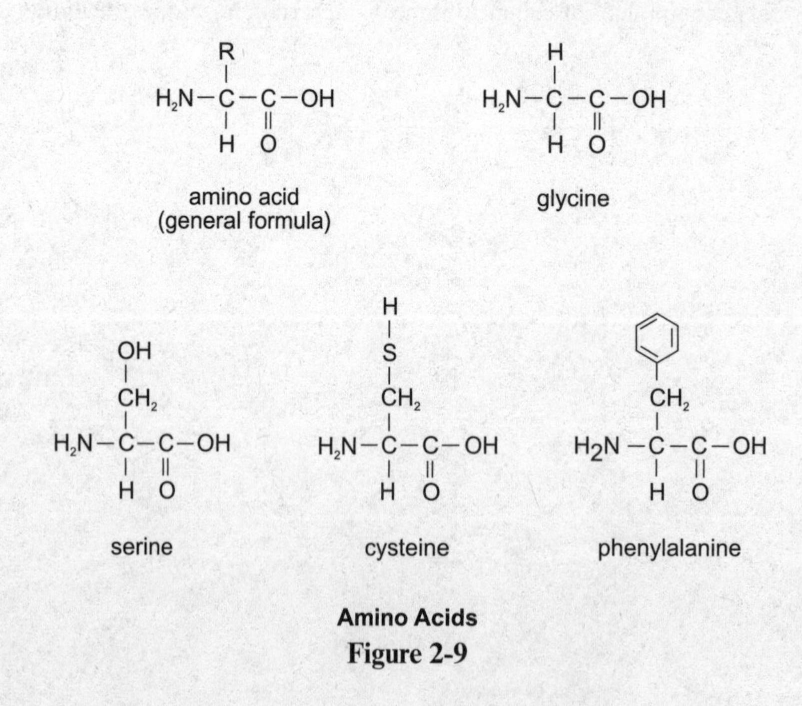

Amino Acids
Figure 2-9

Four levels describe the structure of a protein:

1. The **primary structure** of a protein describes the order of amino acids. Using three letters to represent each amino acid, the primary structure for the protein antidiuretic hormone (ADH) can be written as Cys-Tyr-Phe-Gln-Asn-Cys-Pro-Arg-Gly. Using the one letter abbreviations, ADH is written as C-Y-F-Q-N-C-P-R-G. You do not need to know the abbreviations for each amino acid for the exam; rather, just be aware that they represent amino acids.
2. The **secondary structure** of a protein is a three-dimensional shape that results from hydrogen bonding between the amino and carboxyl groups of nearby amino acids. The bonding produces a spiral **(alpha helix)** or a folded plane that looks much like the pleats on a skirt **(beta pleated sheet).** Proteins whose shapes are dominated by these two patterns often form **fibrous proteins.**
3. The **tertiary structure** of a protein includes additional three-dimensional shaping and often dominates the structure of **globular proteins.** The following factors contribute to the tertiary structure.
   - **Hydrogen bonding** between R groups of amino acids.
   - **Ionic bonding** between R groups of amino acids.
   - The **hydrophobic effect** that occurs when hydrophobic R groups move toward the center of the protein (away from the water in which the protein is usually immersed).
   - The formation of **disulfide bonds** that occurs when the sulfur atom in the amino acid cysteine bonds to the sulfur atom in another cysteine (forming cystine, a kind of "double" amino acid). This disulfide bridge helps maintain the folds of the amino acid chain (Figure 2-10).

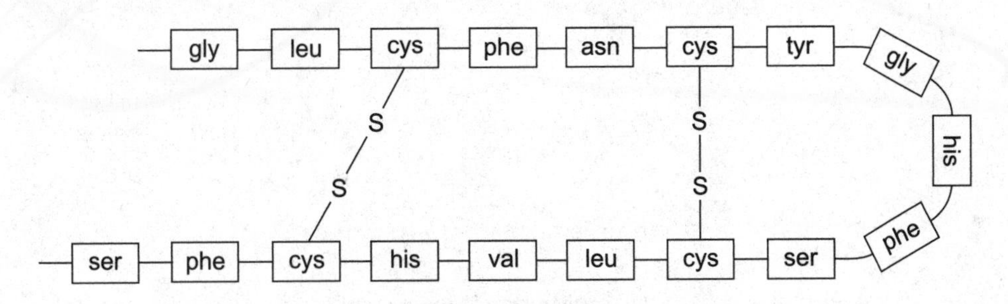

Disulfide Bridges in a Polypeptide

Figure 2-10

4. The **quaternary structure** describes a protein that is assembled from two or more separate peptide chains. The globular protein hemoglobin, for example, consists of four peptide chains that are held together by hydrogen bonding and interactions among R groups.

Nucleic Acids

The genetic information of a cell is stored in molecules of deoxyribonucleic acid (DNA). The DNA, in turn, passes its genetic instructions to ribonucleic acid (RNA) for directing various metabolic activities of the cell.

DNA is a polymer of **nucleotides.** A DNA nucleotide consists of three parts—a **nitrogen base,** a five-carbon sugar called **deoxyribose,** and a **phosphate group.** There are four DNA nucleotides, each with one of the four nitrogen bases, as follows (see Figure 2-11):

1. Adenine—a double-ring base (purine)
2. Thymine—a single-ring base (pyrimidine)
3. Cytosine—a single-ring base (pyrimidine)
4. Guanine—a double-ring base (purine)

Pyrimidines are single-ring nitrogen bases, and purines are double-ring bases. You can remember which of these bases are purines by the words "pure silver," where pure suggests purine and Ag (*a*denine and *g*uanine) is the chemical symbol for silver. The first letter of each of these four bases is often used to symbolize the respective nucleotide (A for the adenine nucleotide, for example).

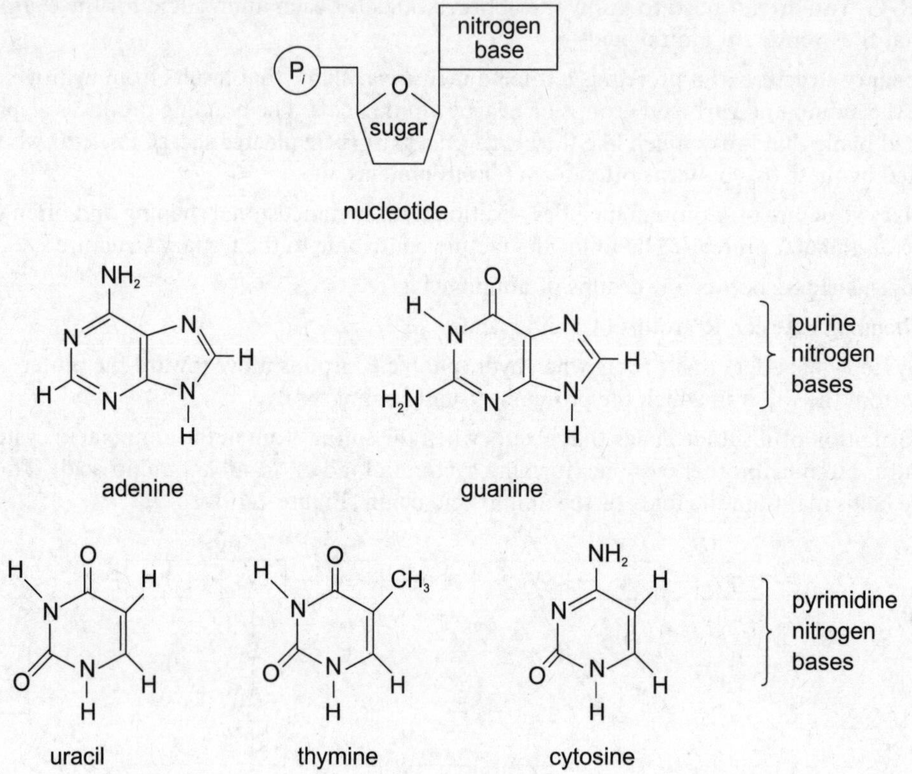

Nitrogen Bases

Figure 2-11

DNA nucleotides form a single-stranded DNA molecule when the phosphate group of one nucleotide joins to the sugar of the adjacent nucleotide (Figure 2-12). Two of these single strands of DNA, paired by weak hydrogen bonds between the bases, form a double-stranded DNA. When bonded in this way, DNA forms a two-stranded spiral, or double helix. *Note that adenine always bonds with thymine and guanine always bonds with cytosine (always a purine with a pyrimidine)* (Figure 2-12).

The two strands of a DNA helix are antiparallel, that is, oriented in opposite directions. One strand is arranged in the $5' \rightarrow 3'$ direction; that is, it begins with a phosphate group attached to the *fifth* carbon of the deoxyribose (5′ end) and ends where the phosphate of the next nucleotide would attach, at the *third* deoxyribose carbon (3′). The adjacent strand is oriented in the opposite, or $3' \rightarrow 5'$, direction.

RNA differs from DNA in the following ways:

1. The sugar in the nucleotides that make an RNA molecule is ribose, not deoxyribose as it is in DNA.
2. The thymine nucleotide does not occur in RNA. It is replaced by uracil. When pairing of bases occurs in RNA, uracil (instead of thymine) pairs with adenine.
3. RNA is usually single-stranded and does not form a double helix as it does in DNA.

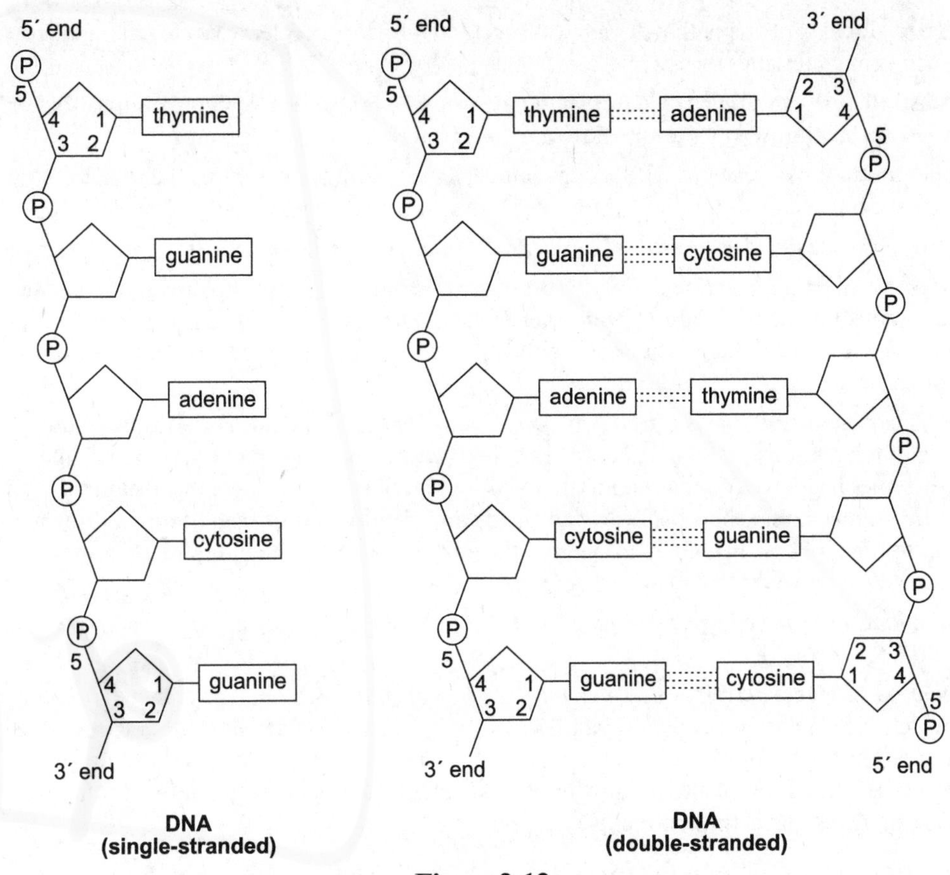

Figure 2-12

For clarification purposes, the structural formulas of the nucleotide bases are given in Figure 2-11. But the AP exam requires only that you know the general structure of a nucleotide (nitrogen base + five-carbon sugar + phosphate), the names of the nucleotides, which are double-ring purines and which are single-ring pyrimidines, and the differences between DNA and RNA.

Chemical Reactions in Metabolic Processes

In order for a chemical reaction to take place, the reacting molecules (or atoms) must first collide and then have sufficient energy (**activation energy**) to trigger the formation of new bonds. Although many reactions can occur spontaneously, the presence of a **catalyst** accelerates the rate of the reaction because it lowers the activation energy required for the reaction to take place. A catalyst is any substance that accelerates a reaction but does not undergo a chemical change itself. Since the catalyst is not changed by the reaction, it can be used over and over again.

Chemical reactions that occur in biological systems are referred to as **metabolism**. Metabolism includes the breakdown of substances (**catabolism**), the formation of new products (**synthesis** or **anabolism**), or the transferring of energy from one substance to another. Metabolic processes have the following characteristics in common:

1. The net direction of metabolic reactions, that is, whether the overall reaction proceeds in the forward direction or in the reverse direction, is determined by the concentrations of the reactants and the end products. Chemical **equilibrium** describes the condition where the rate of reaction in the forward direction equals the rate in the reverse direction, and, as a result, there is no net production of reactants or products. When the reaction shown below is in chemical equilibrium, the rate at which molecules of C and D are formed (in the forward direction) is the same as the rate at which molecules of A and B are formed (in the reverse direction).

$$A + B \rightleftarrows C + D$$

2. **Enzymes** are globular proteins that act as catalysts (activators or accelerators) for metabolic reactions. Note the following characteristics of enzymes:
   - *The* **substrate** *is the substance or substances upon which the enzyme acts.* For example, the enzyme amylase catalyzes the breakdown of the substrate amylose (starch).
   - *Enzymes are substrate specific.* The enzyme amylase, for example, catalyzes the reaction that breaks the α-glycosidic linkage in starch but cannot break the β-glycosidic linkage in cellulose.
   - *An enzyme is unchanged as a result of a reaction.* It can perform its enzymatic function repeatedly.
   - *An enzyme catalyzes a reaction in both forward and reverse directions.* The direction of *net* activity is determined by substrate concentrations and other factors. The net direction of an enzyme reaction can be driven in the forward direction by keeping the product concentration low (by its removal, or conversion to another product).
   - *The efficiency of an enzyme is affected by temperature and pH.* The human body, for example, is maintained at a temperature of 98.6°F, near the optimal temperature for most human enzymes. Above 104°F, these enzymes begin to lose their ability to catalyze reactions as they become **denatured**, that is, they lose their three-dimensional shape as hydrogen bonds and peptide bonds begin to break down. Most enzymes have an optimal pH of around 7.0; however, the enzyme pepsin, which digests proteins in the stomach, becomes active only at a low pH (very acidic).
   - *The standard suffix for enzymes is "-ase,"* so it is easy to identify enzymes that use this ending, although some do not.
   - *The* **induced-fit model** *describes how enzymes work.* Within the protein (the enzyme), there is an active site with which the reactants readily interact because of the shape, polarity, or other characteristics of the active site. The interaction of the reactants (substrate) and the enzyme causes the enzyme to change shape. The new position places the substrate molecules into a position favorable to their reaction. Once the reaction takes place, the product is released.

3. **Cofactors** are nonprotein molecules that assist enzymes.
   - **Coenzymes** are *organic* cofactors that usually function to donate or accept some component of a reaction, often electrons. Some vitamins are coenzymes or components of coenzymes.
   - **Inorganic cofactors** are often metal ions, like Fe^{2+} and Mg^{2+}.

4. **ATP** (adenosine triphosphate) is a common source of activation energy for metabolic reactions (Figure 2-13). ATP is essentially an RNA adenine nucleotide with two additional phosphate groups. When ATP releases its energy, a **hydrolysis reaction** breaks the last phosphate bond of the ATP molecule to form ADP (adenosine *di*phosphate) and an inorganic phosphate group (P_i), like this:

$$ATP + H_2O \rightarrow ADP + P_i + energy$$

In the reverse dehydration reaction, new ATP molecules are assembled by **phosphorylation** when ADP combines with a phosphate group using energy obtained from some energy-rich molecule (like glucose).

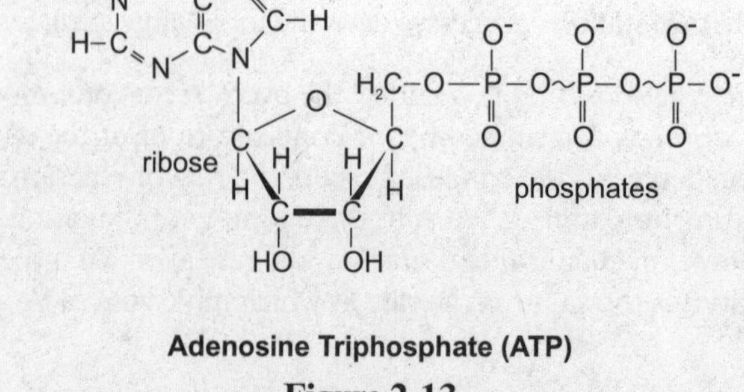

Adenosine Triphosphate (ATP)

Figure 2-13

How do living systems regulate chemical reactions? How do they know when to start a reaction and when to shut it off? One way of regulating a reaction is by regulating its enzyme. Here are five common ways in which this is done:

1. **Enzymes** have two kinds of binding sites—one an active site for the substrate and one or more possible allosteric sites for an **allosteric effector.** There are two kinds of allosteric effectors:
   - An **allosteric activator** binds to the enzyme and induces the enzyme's *active* form.
   - An **allosteric inhibitor** binds to the enzyme and induces the enzyme's *inactive* form.

   Some inhibitors bind irreversibly, permanently changing the structure of the enzyme by modifying an amino acid. Other inhibitors are weakly bonded to the enzyme by ionic or hydrogen bonds, and their effects are reversible.
2. In **feedback inhibition,** an end product of a series of reactions acts as an allosteric inhibitor, shutting down one of the enzymes catalyzing the reaction series.
3. In **competitive inhibition,** a substance that mimics the substrate inhibits an enzyme by occupying the active site. The mimic displaces the substrate and prevents the enzyme from catalyzing the substrate.
4. In **noncompetitive inhibition,** a substance inhibits the action of an enzyme by binding to the enzyme at a location other than the active site (i.e., allosteric site). The inhibitor changes the shape of the enzyme, which disables its enzymatic activity. Many toxins and antibiotics are noncompetitive inhibitors.
5. In **cooperativity,** an enzyme becomes more receptive to additional substrate molecules after one substrate molecule attaches to an active site. This occurs, for example, in enzymes that consist of two or more subunits (quaternary structure), each with its own active site. A common example of this process (though not an enzyme) is hemoglobin, whose binding capacity to additional oxygen molecules increases after the first oxygen binds to an active site.

Review Questions

Multiple-Choice Questions

The questions that follow provide a review of the material presented in this chapter. Use them to evaluate how well you understand the terms, concepts, and processes presented. Actual AP multiple-choice questions are often more general, covering a broad range of concepts, and often more lengthy. For multiple-choice questions typical of the exam, take the two practice exams in this book.

Directions: Each of the following questions or statements is followed by four possible answers or sentence completions. Choose the one best answer or sentence completion.

1. Which of the following molecules orient themselves into sandwich-like membranes because of hydrophobic components within the molecule?

A. glycogen molecules
B. cellulose molecules
C. phospholipid molecules
D. protein molecules

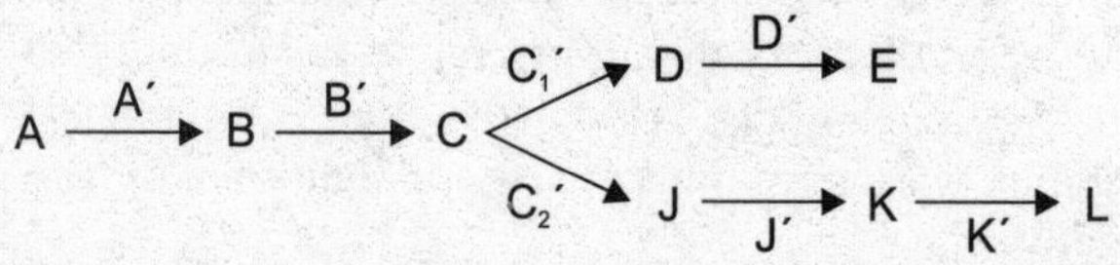

$A \xrightarrow{A'} B \xrightarrow{B'} C$; $C \xrightarrow{C_1'} D \xrightarrow{D'} E$; $C \xrightarrow{C_2'} J \xrightarrow{J'} K \xrightarrow{K'} L$

2. In the series of metabolic reactions shown above, C_1' catalyzes the conversion of C to D, and C_2' catalyzes the conversion of C to J. Assume that product E is an allosteric effector that inhibits enzyme D′. Normally, products E and L are consumed by other reactions. Which of the following would likely happen if product E were not consumed by other reactions?

A. The net rate of production of product B would decrease.
B. The net rate of production of product C would decrease.
C. The net rate of production of product D would decrease.
D. The net rate of production of product J would decrease.

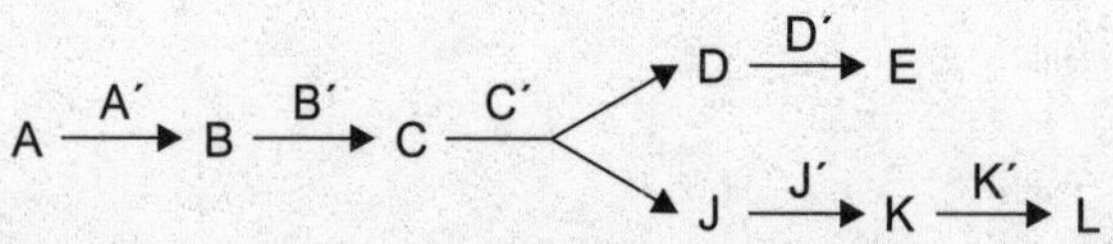

$A \xrightarrow{A'} B \xrightarrow{B'} C \xrightarrow{C'}$ D and J; $D \xrightarrow{D'} E$; $J \xrightarrow{J'} K \xrightarrow{K'} L$

3. In the series of metabolic reactions shown above, C′ catalyzes the splitting of C into D and J. Assume that product E is an allosteric effector that inhibits enzyme C′. If product E were not consumed in a subsequent reaction, which of the following would likely happen?

A. The rate of production of product D would increase.
B. The rate of production of product E would increase.
C. The rate of production of product J would increase.
D. The rate of production of all products—D, E, J, K, and L—would decrease.

4. Each of the following molecules is a polymer EXCEPT:

A. protein
B. glucose
C. cellulose
D. starch

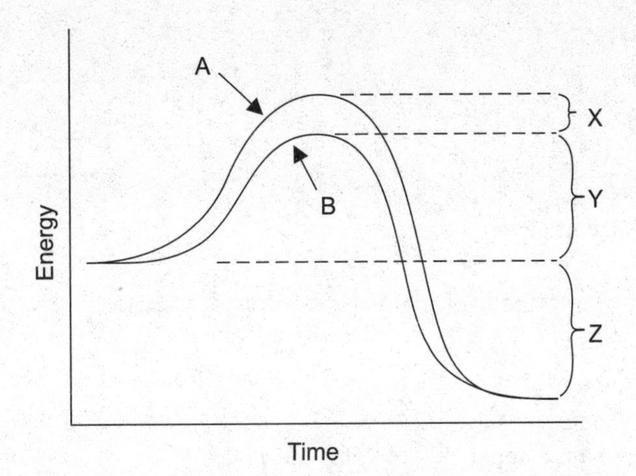

5. For the graph given above, the two curves describe the potential energy of substances during the progress of a chemical reaction. All of the following items could apply to this graph EXCEPT:

A. Curve B could be showing the influence of an enzyme.
B. The sum of energy in the products of the reaction is less than the sum of energy in the reactants.
C. The activation energy of this reaction could be given by X + Y + Z.
D. This reaction graph could describe the reaction ATP → ADP + P_i.

Questions 6–9 refer to the molecules below. For each question, indicate the molecule that is being described.

A.

B.

C.

D.

6. a monosaccharide

7. a polysaccharide

8. a polypeptide

9. a major component of cell membranes

10. Hydrophilic properties are characteristic of all of the following EXCEPT:

A. polar molecules
B. molecules soluble in water
C. molecules that readily ionize in water
D. the long hydrocarbon chain components of some molecules

11. All of the following are carbohydrates EXCEPT:

A. polypeptide
B. glycogen
C. glucose
D. polysaccharide

Free-Response Questions

The AP exam has long and short free-response questions. The long questions have considerable descriptive information that may include tables, graphs, or figures. The short questions are brief but may also include figures. Both kinds of questions have four parts and generally require that you bring together concepts from multiple areas of biology.

The questions that follow are designed to further your understanding of the concepts presented in this chapter. Unlike the free-response questions on the exam, they are narrowly focused on the material in this chapter. For free-response questions typical of the exam, take the two practice exams in this book.

Directions: The best way to prepare for the AP exam is to write out your answers as if you were taking the exam. Use complete sentences and do *not* use outline form or bullets. You may use diagrams to supplement your answers, but be sure to describe the importance or relevance of your diagrams.

---

1. The protein albumin that surrounds the yolk of an egg is a clear liquid when raw and a white solid when cooked. Explain why cooking causes this change.

2. Cells are made up mostly of water. Explain why the specific heat of water is important to a cell.

3. A nucleic acid molecule has a distinct 3′ end and a distinct 5′ end. Explain the significance of these ends during the assembly of a double-helix DNA molecule.

4. The rate of a reaction with and without the presence of an enzyme is provided in the table below. Explain the significance of temperature and the presence of an enzyme on the rate of a reaction.

Reaction Rates with and without an Enzyme at Different Temperatures				
Temperature (°C)	15°C	25°C	35°C	45°C
Reaction Rate – Enzyme Absent (mmol/m^3/sec)	0.0001	0.0002	0.0003	0.0004
Reaction Rate – Enzyme Present (mmol/m^3/sec)	1.0000	2.5000	4.0000	0.0004

5. Describe each of the following:

a. the structure of an enzyme
b. how enzymes function
c. how enzymes are regulated

Answers and Explanations

Multiple-Choice Questions

1. C. Phospholipids are composed of glycerol molecules bonded to two fatty acids and one phosphate group. The phosphate group is a hydrophilic "head," and the long hydrocarbon chains of fatty acids are hydrophobic "tails." In cell membranes, phospholipids orient themselves into two layers, with the hydrophobic tails pointing to the inside of the "sandwich."

2. C. When product E is no longer consumed by other reactions, it is available to inactivate enzyme D′. As quantities of product E accumulate, more and more of D′ would be inactivated. As a result, the rate of production of E would decrease and quantities of product D would accumulate. As product D accumulates, its rate of production decreases. At the same time, the rate of the reverse reaction, of D to C, increases. Now, more of C would be available for conversion to J (and then to K and L), and as C increases, the rate of production of J increases. Eventually, the rate of production of D would equal the rate of the reverse reaction (of D to C), and chemical equilibrium between C and D would be reached (the net rate of production of D would be zero).

3. D. The effect of the allosteric effector E is to inhibit enzyme C′. As quantities of product E accumulate, increasing amounts of C′ would become inactivated. As a result, fewer and fewer quantities of C would be converted to products D and J. Thus, the rate of production of D and J, as well as E, K, and L, would decrease. As quantities of C increase, the rate of the reverse reaction of C to B (and then to A) would increase. In the end, A, B, and C would be in chemical equilibrium, and the rate of production of products D, E, J, K, and L would be zero.

4. B. Glucose is a monomer consisting of a single glucose molecule. Starch and cellulose are polymers consisting of repeating units of glucose. Protein is a polymer of amino acids.

5. C. The activation energy is given by X + Y for curve A or Y for curve B. Curve B shows how the activation energy would be lowered if an enzyme were present. Since the products (right side of the curve) have less energy than the reactants, energy is released. This kind of reaction, where energy is released, is called an exergonic reaction. In contrast, if the products had more energy than the reactants, it would be an endergonic reaction. The reaction $ATP \rightarrow ADP + P_i$ is an exergonic reaction where the energy released is used as activation energy for other metabolic reactions.

6. A. This is the ring structure of glucose.

7. B. This is amylose, a starch found in plants.

8. D. This polypeptide contains five amino acids.

9. C. This is a phospholipid. Note that this phospholipid contains one saturated and one monounsaturated fatty acid.

10. D. Long hydrocarbon chains are nonpolar and, therefore, hydrophobic. Any polar molecule is hydrophilic. When a substance ionizes in water, it dissolves; thus, it is hydrophilic.

11. A. A polypeptide is a protein. Glucose is a monosaccharide carbohydrate, while glycogen is a polysaccharide carbohydrate.

Free-Response Questions

1. When a protein is heated above a critical temperature, it begins to lose its three-dimensional structure. When the secondary, tertiary, and quaternary structures of a protein break down, as they will when excessively heated, the structure of the protein is permanently destroyed.

2. Because water has a high specific heat capacity, its temperature changes very slowly in response to energy changes. As a result, metabolic activities occurring in the cell that release or absorb energy do not significantly change the temperature of the cell, allowing the internal temperature of the cell to remain fairly constant.

3. A double-helix DNA molecule consists of two single-strands of DNA. The base pairing between nucleotides requires that the two strands are arranged in opposite directions, or antiparallel.

4. Both temperature and the presence of an enzyme increase the rate of a reaction. Because molecules are moving faster at higher temperatures, there are more collisions and, therefore, more reactions. As a catalyst, an enzyme speeds up reactions by facilitating the coming together of the reactants (thus, lowering activation energy). At 45°C, the influence of the enzyme is eliminated because the high temperature denaturizes it, and the reaction rate falls back to the rate that occurs in the absence of an enzyme.

5. **a.** Enzymes are globular proteins. Proteins, in turn, are polymers of amino acids—chains of amino acids, bonded to each other by peptide bonds. The general formula for an amino acid is a central carbon atom bonded to an amino group ($-NH_2$), a carboxyl group ($-COOH$), and a hydrogen atom. A fourth bond is made with a group of atoms that varies with each of the 20 amino acids. This variable group can be a single hydrogen atom or a group of many atoms sometimes including sulfur, nitrogen, or carbon rings. The individual amino acids in a protein interact with one another, giving the protein special spatial and functional characteristics. These characteristics impart to an enzyme unique attributes that allow it to catalyze specific reactions of specific substrates. The characteristics of proteins (and, therefore, enzymes) are derived from four features of the protein's structure. The first, described by the primary structure, is the kind and arrangement of amino acids in the protein. A secondary structure originates from hydrogen bonding between amino and carboxyl groups of amino acids. This secondary structure is the three-dimensional shape of a helix or a pleated sheet. Further interactions between amino acids give proteins a tertiary structure. These interactions include hydrogen bonding and ionic bonding between R groups, the "hiding" of hydrophobic R groups into the interior of the protein, and a disulfide bridge between two cysteine amino acids. The summation of all of the interactions gives enzymes a globular shape.

   b. The function of an enzyme is to speed up the rate of, or catalyze, a reaction. The induced-fit model describes how enzymes work. In this model, there are specific active sites within the enzyme to which substrate molecules weakly bond. When substrate molecules bond to the active sites, the enzyme changes shape in such a way as to reduce the activation energy required for a bond to form between the substrate molecules. With less energy required, bonding proceeds at a faster rate.

   Many enzymes require a cofactor to catalyze a reaction. Cofactors include coenzymes (nonprotein, organic molecules) and metal ions (like Fe^{2+} or Mg^{2+}).

   c. There are several ways that enzymes are regulated. Allosteric enzymes are controlled by allosteric effectors, substances that bind to the enzyme and inhibit (or activate) the enzyme. Sometimes an allosteric inhibitor is a product of a series of reactions partly catalyzed by the allosteric enzyme. This is an example of feedback inhibition. Allosteric effectors bind to special sites in the enzyme. In competitive inhibition, however, an inhibitor binds to the active site, competing with substrate molecules. As a result, the activity of the enzyme is inhibited. Some toxins and drugs (including penicillin, an antibiotic) are examples. Environmental factors also contribute to the activity of enzymes. Enzymes operate best at specific temperatures and pH. Enzymes in the stomach, for example, are active only when the pH is low.

This answer provides quite a bit of detail on the structure of proteins. Although the material is relevant, you may be able to condense your answer to allow time to answer other questions if you were actually taking the AP exam. On the other hand, if time were available, you could give examples of some specific enzymes or coenzymes (both of which you'll learn about in subsequent chapters).

Chapter 3

Cell Structure and Function

Review

Structure and Function of the Cell

The **cell** is the basic functional unit of all living things. The **plasma membrane (cell membrane)** bounds the cell and encloses the **nucleus** and **cytoplasm.** The **cytoplasm** consists of specialized bodies called **organelles** suspended in a fluid matrix, the **cytosol,** which consists of water and dissolved substances such as proteins and nutrients.

The **plasma membrane** separates internal metabolic events from the external environment and controls the movement of materials into and out of the cell. The plasma membrane is a **double phospholipid membrane** (phospholipid **bilayer**) with the polar hydrophilic heads forming the two outer faces and the nonpolar hydrophobic tails pointing toward the inside of the membrane (Figure 3-1).

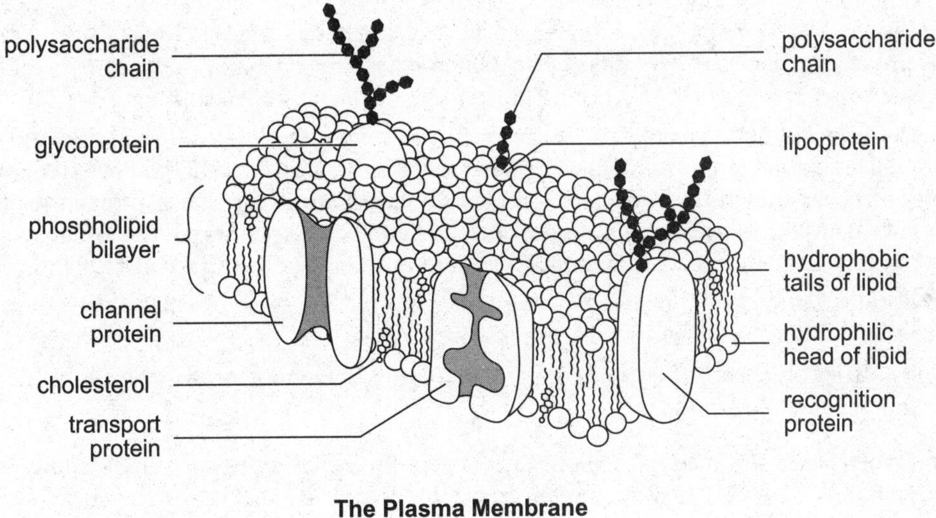

The Plasma Membrane

Figure 3-1

Proteins are scattered throughout the flexible phospholipid membrane. Proteins may attach loosely to the inner or outer surface of the membrane, or they may extend into the membrane. Similar to how phospholipids are oriented, proteins may span across the membrane with their hydrophobic regions embedded in the membrane and their hydrophilic regions exposed to the aqueous solutions bordering the membrane. The molecules that comprise the membrane are not fixed in a permanent pattern but move within the membrane. This mosaic nature of scattered proteins within a flexible matrix of phospholipid molecules describes the **fluid mosaic model** of the cell membrane. Variations in the fatty acid makeup of the phospholipids influence the fluidity of the membrane. Phospholipids with saturated fatty acids pack more tightly, leading to a more rigid membrane. In contrast, unsaturated fatty acids, which bend at their double-covalent bonds, limit packing and result in a more flexible membrane. Additional features of the plasma membrane follow:

1. The **phospholipid** membrane is selectively permeable. Only small, uncharged, polar molecules (such as H_2O) and hydrophobic molecules (nonpolar molecules like O_2, CO_2, and lipid-soluble molecules such as hydrocarbons) freely pass across the membrane. In contrast, large polar molecules (such as glucose) and all ions are impermeable.
   - **Glycolipids** are lipids to which a short polysaccharide chain is attached. The polysaccharide chain extends away from the outer surface of the membrane into the external environment. Glycolipids help establish cell identity—they help identify the cell as a *self* cell rather than a foreign cell or a virus-infected cell. There is no phosphate group in a glycolipid.
2. **Proteins** in the plasma membrane provide a wide range of functions and include the following:
   - **Channel proteins** provide open passageways through the membrane for certain hydrophilic (water-soluble) substances such as polar and charged molecules. **Aquaporins** are channel proteins of certain cells (such as those found in kidneys and plant roots) that dramatically increase the passage rate of H_2O molecules.
   - **Ion channels** allow the passage of ions across the membrane. In nerve and muscle cells, ion channels called **gated channels** open and close in response to specific chemical or electrical stimuli to allow the passage of specific ions (such as Na^+ and K^+).
   - **Carrier proteins** bind to specific molecules, which are then transferred across the membrane after the carrier protein undergoes a change of shape. The passage of glucose into a cell is by a carrier protein.
   - **Transport proteins** use energy (in the form of ATP, adenosine triphosphate) to transport materials across the membrane. When energy is used for this purpose, the materials are said to be actively transported, and the process is called **active transport.** The **sodium-potassium pump,** for example, uses ATP to maintain higher concentrations of Na^+ and K^+ on opposite sides of the plasma membrane.
   - **Recognition proteins**, like glycolipids, give each cell type a unique identification. This identification provides for a distinction between cell types, between *self* cells and foreign cells, and between normal cells and cells infected with viruses. Recognition proteins are actually **glycoproteins,** proteins with short polysaccharide chains that extend away from the outer surface of the membrane. The differences between blood types, for example, are the result of recognition proteins on the surface of red blood cells.
   - **Receptor proteins** provide binding sites for hormones or other trigger molecules. In response to the hormone or trigger molecule, a specific cell response is activated.
   - **Adhesion proteins** attach cells to neighboring cells or provide anchors for the internal filaments and tubules that give stability to the cell.
3. **Cholesterol** molecules distributed throughout the phospholipid bilayer provide some stability to the plasma membranes of *animal cells.* At higher temperatures, cholesterol helps maintain firmness, but at lower temperatures, it helps keep the membrane flexible.

Organelles are bodies within the cytoplasm that serve to physically separate the various metabolic reactions that occur within eukaryotic cells. Within these bodies, chemical reactions are isolated and can take place without interference or competition with other reactions that might be occurring nearby. Many of these bodies also provide large surface areas to maximize the space over which these chemical reactions can take place. Also, cells can be specialized for specific functions depending on the kinds and number of organelles in that cell. Descriptions of the important organelles, as well as other structures in the cell, follow (Figure 3-2):

1. The **nucleus** is bounded by the **nuclear envelope** consisting of *two* phospholipid bilayers, each similar to the plasma membrane. The nucleus contains DNA (deoxyribonucleic acid), the hereditary information of the cell. Normally, the DNA is spread out within the nucleus as a thread-like matrix called **chromatin.** When the cell begins to divide, the chromatin condenses into rod-shaped bodies called **chromosomes,** each of which, before dividing, is made up of two long DNA molecules and various histone (protein) molecules. The histones serve to organize the lengthy DNA, coiling it into bundles called **nucleosomes.** Also visible within the nucleus are one or more **nucleoli,** concentrations of DNA in the process of manufacturing the components of **ribosomes.** The nucleus also serves as the site for the separation of chromosomes during cell division. On the surface of the nuclear envelope are **nuclear pores,** which serve as passageways for proteins and RNA molecules.

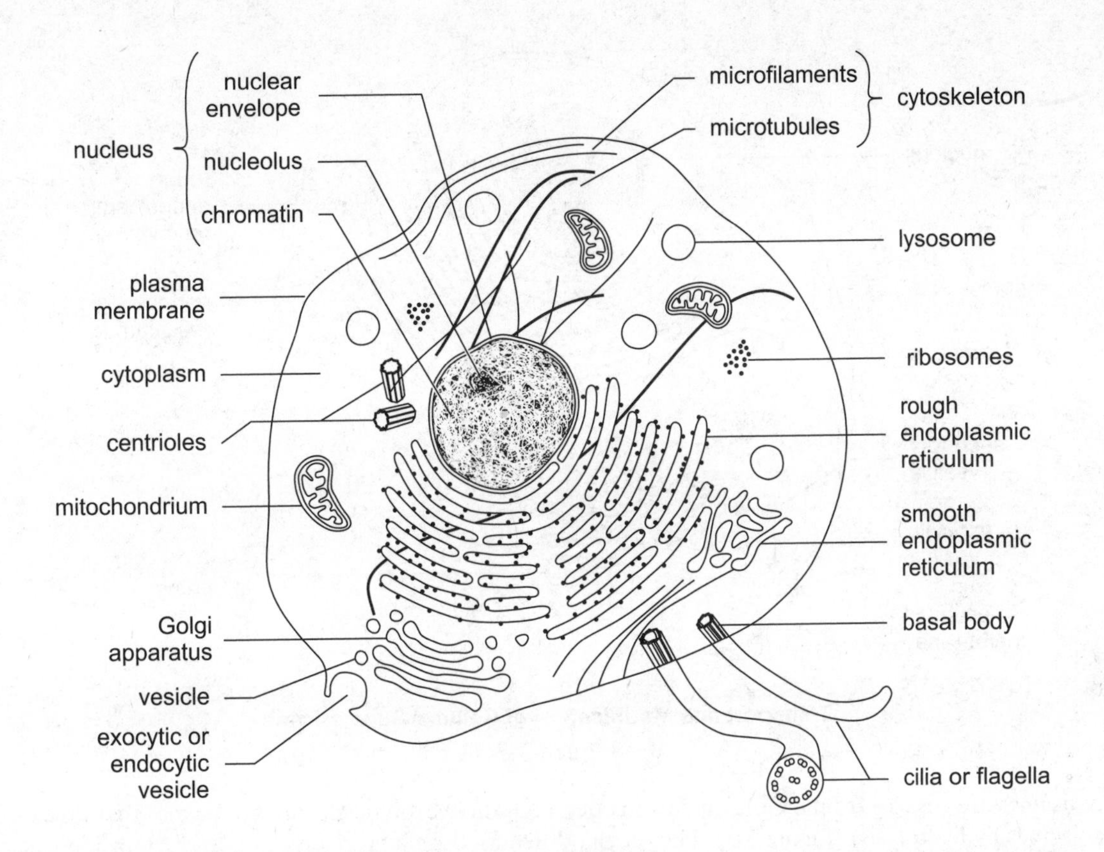

An Animal Cell
Figure 3-2

2. **Ribosome** subunits are manufactured in the nucleus and consist of RNA molecules and proteins. The two subunits, labeled 60S and 40S, move across the nuclear envelope through nuclear pores and into the cytoplasm, where they are assembled into a single 80S ribosome. (An S value expresses how readily a product forms sediment in a centrifuge, with larger values representing larger and heavier products.) In the cytoplasm, ribosomes assist in the assembly of amino acids into proteins.
3. The **endoplasmic reticulum,** or **ER,** consists of stacks of flattened sacs that begin as an extension of the outer bilayer of the nuclear envelope. In cross section, the ER appears as a series of maze-like channels, closely associated with the nucleus. When ribosomes are present, the ER (called **rough ER**) creates **glycoproteins** by attaching polysaccharide groups to polypeptides as they are assembled by the ribosomes. **Smooth ER,** without ribosomes, is responsible for various activities, including the synthesis of lipids and steroid hormones, especially in cells that produce these substances for export from the cell. In liver cells, smooth ER is involved in the breakdown of toxins, drugs, and toxic by-products from cellular reactions.
4. A **Golgi apparatus (Golgi body, Golgi complex)** is a group of flattened sacs **(cisternae)** arranged like a stack of bowls. They collect and modify proteins and lipids made in other areas of the cell and package them into **vesicles,** small, spherically shaped sacs that bud from the outside surface of the Golgi apparatus. For example, a glycoprotein made and packaged into a vesicle by the ER may be transported to the Golgi apparatus, where it is modified as it passes through its chambers (Figure 3-3). At the outer side of the Golgi apparatus, the modified protein can be packaged into a secretory vesicle, which migrates to and merges with the plasma membrane, releasing its contents to the outside of the cell. Other packaged substances may be retained within the cell for other purposes.

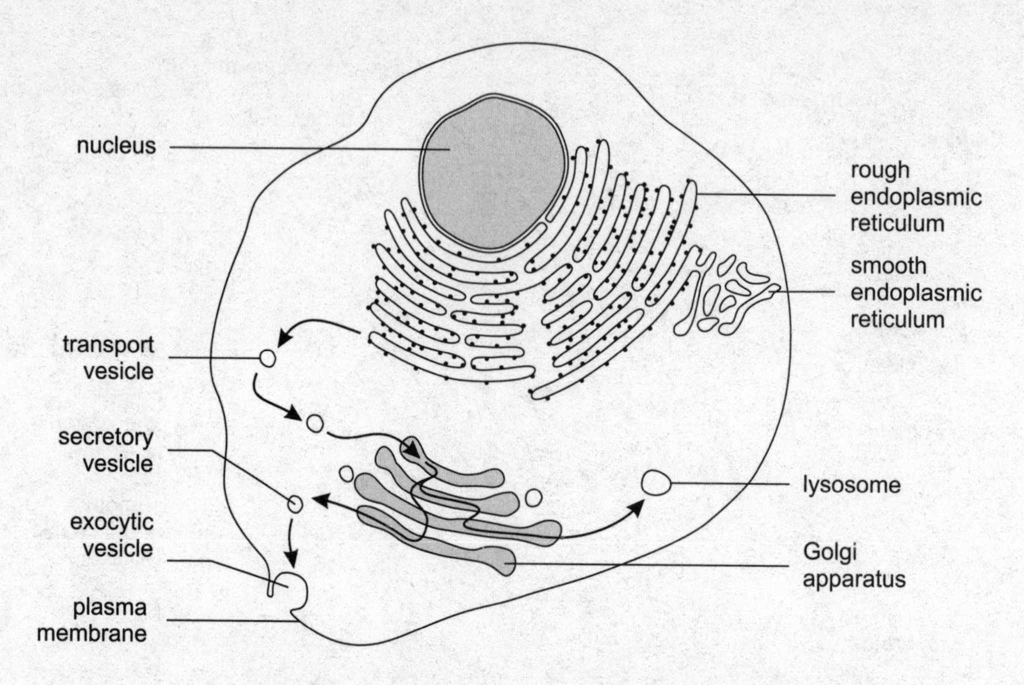

Transport and Modification of Cellular Substances

Figure 3-3

5. **Lysosomes** are vesicles from a Golgi apparatus that contain hydrolytic enzymes (enzymes that break down molecules by hydrolysis) (Figure 3-3). They break down food, cellular debris, and foreign invaders (such as bacteria) and generally contribute to a recycling of cellular nutrients. A low pH (acidic), favorable to the activity of the enzymes, is maintained inside the lysosome. As a result, any enzyme that might escape from the lysosome remains inactive in the neutral pH of the cytosol.
6. **Mitochondria** carry out aerobic respiration, a process in which energy (in the form of ATP) is obtained from carbohydrates, fats, and occasionally proteins. Mitochondria have two bilayer membranes, allowing the separation of metabolic processes that occur inside the inner membrane from those occurring in the intermembrane space.
7. **Chloroplasts** carry out **photosynthesis,** the plant process of incorporating energy from sunlight into carbohydrates. Chloroplasts, like mitochondria, also have two membranes.
8. **Microtubules, intermediate filaments,** and **microfilaments** are three protein fibers of decreasing diameter, respectively. All are involved in establishing the shape of or in coordinating movements of the **cytoskeleton,** the internal structure of the cytoplasm.
   - **Microtubules** are made of the protein **tubulin** and provide support and motility for cellular activities. They are found in the **spindle apparatus,** which guides the movement of chromosomes during cell division, and in flagella and cilia (described in item 10), structures that project from the plasma membrane to provide motility to the cell.
   - **Intermediate filaments** provide support for maintaining the shape of the cell.
   - **Microfilaments (actin filaments)** are made of the protein **actin** and are involved in cell motility. They are found in muscle cells and in cells that move by changing shape, such as phagocytes (white blood cells that wander throughout the body, attacking bacteria and other foreign invaders). In plants, microfilaments promote the movement of cytoplasmic materials around the cell **(cytoplasmic streaming).**

9. **Flagella** and **cilia** are structures that protrude from the cell membrane and make wave-like movements. Flagella and cilia are classified by their lengths, by their numbers per cell, and by their movement: Flagella are long, few, and move in a snake-like motion; cilia are short, many, and move with a back-and-forth, serpentine movement. A single flagellum propels sperm, while the numerous cilia that line the respiratory tract sweep away debris. Structurally, both flagella and cilia consist of microtubules arranged in a "9 + 2" array—nine pairs (doublets) of microtubules arranged in a circle surrounding a pair of microtubules (Figure 3-4).

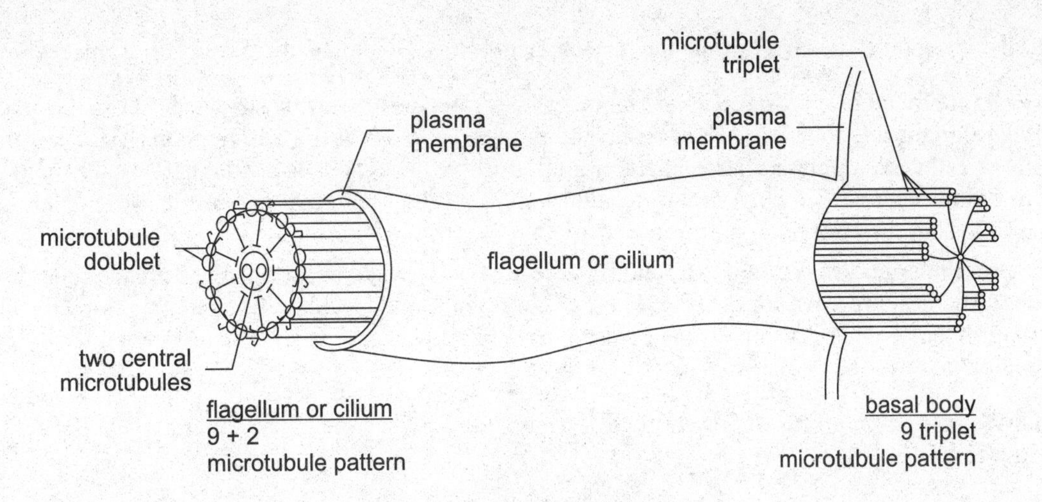

Flagella, Cilia, and Basal Bodies
Figure 3-4

10. **Centrioles** and **basal bodies** act as **microtubule organizing centers (MTOCs).** A pair of centrioles (enclosed in a **centrosome**) located outside the nuclear envelope gives rise to the microtubules that make up the spindle apparatus used during cell division. Basal bodies organize the development of flagella and cilia and anchor them to the cell surface. Both centrioles and basal bodies are made up of nine triplets of microtubules arranged in a circle (Figure 3-4). Plant cells lack centrioles and only *lower* plants (such as mosses and ferns) with motile sperm have flagella and basal bodies.
11. **Vacuoles** and **vesicles** are fluid-filled, membrane-bound bodies.
    - **Transport vesicles** move materials between organelles.
    - **Secretory vesicles** move materials from the Golgi apparatus to the plasma membrane.
    - **Food vacuoles** are temporary receptacles of nutrients. Food vacuoles often merge with lysosomes, whose digestive enzymes break down the food.
    - **Contractile vacuoles** are specialized organelles in single-celled organisms that collect and pump excess water out of the cell.
    - **Central vacuoles** are large bodies occupying most of the interior of many plant cells. When fully filled, they exert **turgor,** or pressure, on the cell walls, thus maintaining rigidity in the cell. The central vacuole provides other functions as well, some of which specialize the cell for specific functions. These include the following:
        - It may store starch, nutrients, pigments, cellular waste, or toxins (nicotine, for example).
        - It may carry out functions such as digestion that are otherwise assumed by lysosomes in animal cells.
        - It provides cell "growth" by simply absorbing water to allow expansion of the cell. In contrast, animal cells require nutrients to build macromolecules to generate growth.
        - It renders a large surface area-to-volume ratio of cytoplasm as it interfaces with the plasma membrane and the outside environment. This occurs because the central vacuole occupies so much of the cell that the organelles and cytoplasm are flattened into a narrow area between the central vacuole and the plasma membrane.

The **extracellular** region is the area outside the plasma membrane. The following may occur in this region:

- **Cell walls** are found in plants, fungi, and many protists (**protists** are mostly single-celled organisms). Cell walls develop outside the plasma membrane and provide support for the cell. In plants, the cell wall consists mainly of **cellulose,** a polysaccharide made from β-glucose. The cell walls of fungi are usually made of chitin. **Chitin** is a modified polysaccharide, differing from cellulose in that one of the hydroxyl groups is replaced by a group containing nitrogen.
- The **extracellular matrix** is found in animals, in the area between adjacent cells. The area is occupied by fibrous structural proteins, adhesion proteins, glycoproteins, and glycolipids secreted by the cells. The matrix provides mechanical support and helps bind adjacent cells together. The most common substance in this region is the protein collagen.

Cell junctions serve to *anchor* cells to one another or to provide a passageway for cellular *exchange.* They include the following (Figure 3-5):

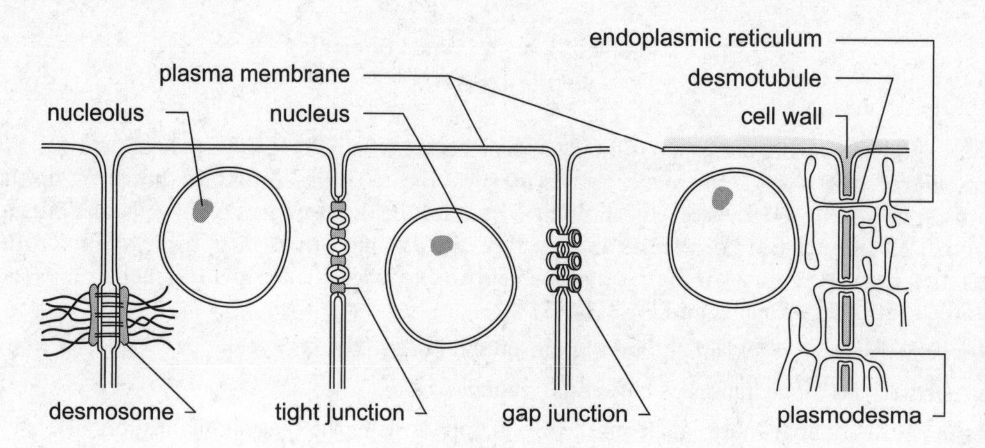

Cell Junctions
Figure 3-5

- **Anchoring junctions** are protein attachments between adjacent *animal* cells. One such junction, the **desmosome,** consists of proteins that bind adjacent cells together, providing mechanical stability to tissues. Desmosomes are also associated with intermediate filaments that extend into the interior of the cell and serve to hold cellular structures together.
- **Tight junctions** are tightly stitched seams between *animal* cells. The junction completely encircles each cell, producing a seal that prevents the passage of materials between the cells. Tight junctions are characteristic of cells lining the digestive tract, where materials are required to pass through cells (rather than intercellular spaces) to enter the bloodstream.
- **Communicating junctions** are passageways between cells that allow the transfer of chemical or electrical signals. Two kinds of communicating junctions occur, as follows:
  - **Gap junctions** are narrow tunnels between *animal* cells. The proteins that make up these junctions prevent cytoplasmic proteins and nucleic acids of each cell from mixing, but allow the passage of ions and small molecules. In this manner, gap junctions allow communication between cells through the exchange of materials or through the transmission of electrical impulses. Gap junctions are essentially channel proteins of two adjacent cells that are closely aligned. Because the proteins of each cell extend beyond the plasma membranes before they meet, a small *gap* occurs between the two plasma membranes.
  - **Plasmodesmata** (singular, **plasmodesma**) are narrow channels between *plant* cells. A narrow tube of endoplasmic reticulum, surrounded by cytoplasm and the plasma membrane, passes through the channel.

Note that plant cells can *generally* be distinguished from animal cells by the following:

1. The *presence* of cell walls, chloroplasts, and central vacuoles in *plant cells* and their absence in *animal* cells
2. The *presence* of centrioles and cholesterol in *animal cells* and their absence in *plant* cells

Prokaryotes and Eukaryotes

The cells described so far are those of **eukaryotic** organisms. Eukaryotes include all organisms except for bacteria and archaea. Bacteria and archaea are **prokaryotes** and lack most of the structures described above. They generally consist of only a plasma membrane, a DNA molecule, ribosomes, cytoplasm, and often a cell wall. In addition, they differ from eukaryotes in the following respects:

1. Prokaryotes do not have a nucleus.
2. The hereditary material in prokaryotes exists as a single "naked" DNA molecule without the proteins that are associated with the DNA in eukaryotic chromosomes.
3. Prokaryotic ribosomes are smaller than those of eukaryotes.
4. The cell walls of bacteria, when present, are constructed from peptidoglycans, a polysaccharide-protein molecule. The cell walls of archaea are chemically diverse and may contain proteins, glycoproteins, and/or polysaccharides, but *not* peptidoglycans, cellulose (as in plants), or chitin (as in fungi).
5. Flagella, when present in prokaryotes, are not constructed of microtubules and are not enclosed by the plasma membrane. The flagella deliver motion by twisting like a screw.

Movement of Substances

Various terms are used to describe the movement of substances between cells and into and out of a cell. These terms differ in the following respects:

1. The movement of substances may occur across a **selectively permeable membrane** (such as the plasma membrane). A selectively permeable membrane allows only specific substances to pass.
2. The substance whose movement is being described may be *water* (the *solvent*) or it may be the substance dissolved in the water (the *solute*).
3. Movement of substances may occur from higher to lower concentrations (*down* or *with* the concentration gradient) or the reverse (*up* or *against* the gradient).
4. Solute concentrations between two areas may be compared. A solute may be **hypertonic** (a higher concentration of *solutes*), **hypotonic** (a lower concentration of *solutes*), or **isotonic** (an equal concentration of *solutes*) relative to another region.
5. The movement of substances may be *passive* or *active.* Active movement requires the expenditure of energy and usually occurs up a gradient.

Bulk flow is the collective movement of substances (solvent and solutes) in the same direction in response to a force or pressure. Blood moving through a blood vessel is an example of bulk flow.

Passive transport processes describe the movement of substances from regions of higher concentrations to regions of lower concentrations (*down* a concentration gradient) and do not require expenditure of energy. Rates of passive transport increase with higher concentration gradients, higher temperatures, and smaller particle size. The different passive transport processes are as follows:

1. **Simple diffusion,** or **diffusion,** is the *net* movement of substances from an area of higher concentration to an area of lower concentration. This movement occurs as a result of the random and constant motion characteristic of all molecules (or atoms or ions), motion that is independent from the motion of other molecules. Since, at any one time, some molecules may be moving up the gradient and some molecules may be moving down the gradient (remember, the motion is random), the word "net" is used to indicate the overall, eventual result of the movement. Ultimately, a state of equilibrium is attained, where molecules are uniformly distributed but continue to move randomly.

2. **Osmosis** is the diffusion of *water* molecules across a selectively permeable membrane. When water moves into a body by osmosis, hydrostatic pressure (osmotic pressure) may build up inside the body. **Turgor pressure** is the hydrostatic pressure that develops when water enters the cells of plants and microorganisms.
3. **Plasmolysis** is the movement of water out of a *cell* (by osmosis) that results in the collapse of the cell (especially plant cells with central vacuoles). In contrast, when water moves into a cell (by osmosis), the cell volume increases and the cell expands. **Cell lysis** occurs when swelling causes the cell to burst (especially animal cells and other cells that lack a cell wall).
4. **Facilitated diffusion** is the diffusion of *solutes* or *water* through channel proteins or carrier proteins in the plasma membrane. Some *channel* proteins facilitate the movement of ions such as Na^+, K^+, Ca^{2+}, or Cl^- across the plasma membrane, while other channel proteins, the aquaporins, facilitate the movement of water across the plasma membrane. *Carrier* proteins can facilitate the movement of ions, as well as some larger organic molecules such as amino acids or glucose.
5. **Countercurrent exchange** describes the diffusion of substances between two regions in which substances are moving by bulk flow in opposite directions. For example, the direction of water flow through the gills of a fish is opposite to the flow of blood in the blood vessels. Diffusion of oxygen from water to blood is maximized because the relative motion of the molecules between the two regions is increased and because the concentration gradients between the two regions remain constant along their area of contact.

Active transport is the movement of *solutes against* a gradient and requires the expenditure of *energy* (usually in the form of ATP). Transport proteins in the plasma membrane transfer solutes such as small ions (Na^+, K^+, Cl^-, H^+), amino acids, and monosaccharides across the membrane. Active transport differs from facilitated diffusion in several ways. First, active transport does not result from random movements of molecules (as does any kind of diffusion). Rather, active transport moves *specific* solutes across a membrane from *lower to higher* concentrations (opposite direction of diffusion). The term "active" in active transport implies the use of energy, whereas the various processes of diffusion are passive.

Vesicular transport uses vesicles or other bodies in the cytoplasm to move macromolecules or large particles across the plasma membrane. Types of vesicular transport are described here:

- **Exocytosis** describes the process of vesicles fusing with the plasma membrane and releasing their contents to the outside of the cell. This is common when a cell produces substances for export.
- **Endocytosis** describes the capture of a substance outside the cell when the plasma membrane merges to engulf it. The substance subsequently enters the cytoplasm enclosed in a vesicle. There are three kinds of endocytosis.
  - **Phagocytosis** ("cellular eating") occurs when *undissolved* material enters the cell. The plasma membrane wraps around the solid material and engulfs it, forming a phagocytic vesicle. Phagocytic cells (such as certain white blood cells) attack and engulf bacteria in this manner.
  - **Pinocytosis** ("cellular drinking") occurs when *dissolved* substances enter the cell. The plasma membrane folds inward to form a channel, allowing the liquid to enter. Subsequently, the plasma membrane closes off the channel, encircling the liquid inside a vesicle.
  - **Receptor-mediated** endocytosis, a form of pinocytosis, occurs when *specific molecules* in the fluid surrounding the cell bind to specialized receptors that concentrate in coated pits in the plasma membrane. The membrane pits, the receptors, and their specific molecules (called **ligands**) fold inward, and the formation of a vesicle follows. Proteins that transport cholesterol in blood (low-density lipoproteins, or LDLs) and certain hormones target specific cells by receptor-mediated endocytosis.

Review Questions

Multiple-Choice Questions

The questions that follow provide a review of the material presented in this chapter. Use them to evaluate how well you understand the terms, concepts, and processes presented. Actual AP multiple-choice questions are often more general, covering a broad range of concepts, and often more lengthy. For multiple-choice questions typical of the exam, take the two practice exams in this book.

Directions: Each of the following questions or statements is followed by four possible answers or sentence completions. Choose the one best answer or sentence completion.

1. The cellular structure that is involved in producing ATP during aerobic respiration is the

A. nucleus
B. nucleolus
C. chloroplast
D. mitochondrion

2. Which of the following cellular structures are common to both prokaryotes and eukaryotes?

A. ribosomes
B. nucleoli
C. mitochondria
D. Golgi bodies

3. The plasma membrane consists principally of

A. proteins embedded in a carbohydrate bilayer
B. phospholipids embedded in a protein bilayer
C. proteins embedded in a phospholipid bilayer
D. proteins embedded in a nucleic acid bilayer

4. When the concentration of solutes differs on the two sides of a membrane permeable only to water,

A. water will move across the membrane by osmosis
B. water will move across the membrane by active transport
C. water will move across the membrane by plasmolysis
D. solutes will move across the membrane from the region of higher concentration to the region of lower concentration

5. All of the following characterize microtubules EXCEPT:

A. They are made of protein.
B. They are involved in providing motility.
C. They develop from the plasma membrane.
D. They make up the spindle apparatus observed during cell division.

6. Lysosomes are

A. involved in the production of fats
B. involved in the production of proteins
C. often found near areas requiring a great deal of energy (in the form of ATP)
D. involved in the degradation of cellular substances

7. Mitochondria

A. are found only in animal cells
B. produce energy (in the form of ATP) with the aid of sunlight
C. are often more numerous near areas of major cellular activity
D. are microtubule organizing centers

Questions 8–11 refer to the following key. Each answer in the key may be used once, more than once, or not at all.

A. active transport
B. bulk flow
C. osmosis
D. facilitated diffusion

8. Movement of solutes across a plasma membrane from a region of higher solute concentration to a region of lower solute concentration with the aid of proteins

9. Movement of water across a membrane from a region of higher concentration of water to a region of lower concentration of water without the aid of proteins

10. Movement of urine through the urinary tract

11. Movement of solutes across a plasma membrane requiring the addition of energy

12. The movement of water out of a cell resulting in the collapse of the plasma membrane is called

A. endocytosis
B. bulk flow
C. cell lysis
D. plasmolysis

13. The movement of molecules during diffusion can be described by all of the following EXCEPT:

A. Molecular movements are random.
B. Net movement of solute molecules is from a region of higher concentration to a region of lower concentration.
C. Each molecule moves independently of other molecules.
D. Solute molecules always move down the concentration gradient.

14. Plant and animal cells differ mostly in that

A. only animal cells have mitochondria
B. only plant cells have plasma membranes with cholesterol
C. only plant cells have cell walls
D. only plant cells have ribosomes attached to the endoplasmic reticulum

15. A smooth endoplasmic reticulum exhibits all of the following activities EXCEPT:

A. assembling amino acids to make proteins
B. manufacturing lipids
C. manufacturing hormones
D. breaking down toxins

16. All of the following are known to be components of cell walls EXCEPT:

- **A.** phospholipids
- **B.** chitin
- **C.** polysaccharides
- **D.** peptidoglycans

17. A saturated suspension of starch is enclosed in a bag formed from dialysis tubing, a material through which water can pass, but starch cannot. The bag with the starch is placed into a beaker of distilled water. All of the following are expected to occur EXCEPT:

- **A.** There will be a net movement of water from a hypotonic region to a hypertonic region.
- **B.** There will be a net movement of solute from a hypertonic region to a hypotonic region.
- **C.** The dialysis bag with its contents will gain weight.
- **D.** No starch will be detected outside the dialysis bag.

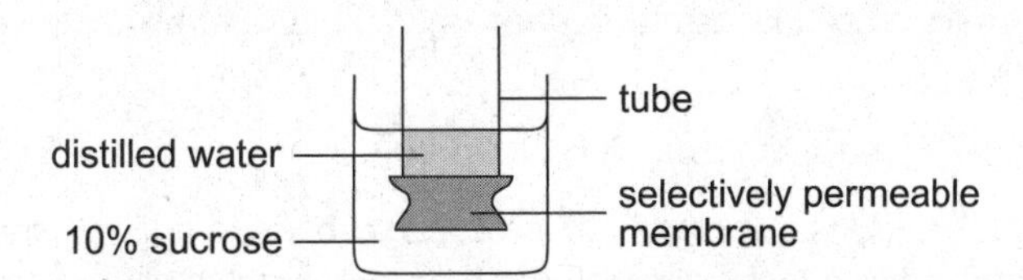

18. As shown above, a tube covered on one end by a membrane impermeable to sucrose is inverted and half filled with distilled water. It is then placed into a beaker of 10% sucrose to a depth equal to the midpoint of the tube. Which of the following statements is true?

- **A.** The water level in the tube will rise to a level above the water in the beaker.
- **B.** The water level in the tube will drop to a level below the water in the beaker.
- **C.** There will be no change in the water level of the tube, and the water in the tube will remain pure.
- **D.** The concentration of sucrose in the beaker will increase.

Free-Response Questions

The AP exam has long and short free-response questions. The long questions have considerable descriptive information that may include tables, graphs, or figures. The short questions are brief but may also include figures. Both kinds of questions have four parts and generally require that you bring together concepts from multiple areas of biology.

The questions that follow are designed to further your understanding of the concepts presented in this chapter. Unlike the free-response questions on the exam, they are narrowly focused on the material in this chapter. For free-response questions typical of the exam, take the two practice exams in this book.

Directions: The best way to prepare for the AP exam is to write out your answers as if you were taking the exam. Use complete sentences and do *not* use outline form or bullets. You may use diagrams to supplement your answers, but be sure to describe the importance or relevance of your diagrams.

---

1. The membranes of the rough endoplasmic reticulum have a very large surface area. Describe how a large surface area aids the activities of the structure.

2. The plasma membrane provides a flexible boundary to the cell. Describe the differences you would expect to find in the makeup of the plasma membrane for a plant cell in a leaf growing with full exposure to the sun compared to one growing in the shade.

3. Glycoproteins produced in the rough endoplasmic reticulum may ultimately be exported from the cell. Describe the pathway of the glycoprotein from the ER to the outside of the cell.

4. Describe each of the following:

   a. the structure of the plasma membrane
   b. the various ways in which the plasma membrane permits interactions with the outside environment

5. Compare and contrast the cellular characteristics of prokaryotes and eukaryotes.

6. Describe the various activities that occur within cells and the methods that cells use to separate these activities from one another.

Answers and Explanations

Multiple-Choice Questions

1. **D.** Aerobic respiration takes place in the mitochondrion. ATP is also produced in the chloroplast, but that is from photosynthesis.

2. **A.** Prokaryotes lack nucleoli, mitochondria, and Golgi bodies.

3. **C.** The plasma membrane consists principally of proteins embedded in a phospholipid bilayer. The phospholipid bilayer establishes a hydrophobic boundary across which solutes and solvents cannot penetrate. Many of the proteins in the membrane serve to facilitate the movement of these materials across the membrane, either by passive diffusion (channel proteins, ion channels, carrier proteins) or by active transport.

4. **A.** When there is a concentration gradient, water will move across a membrane unassisted by ATP or channel proteins. In contrast, solutes (the dissolved substances) cannot cross the membrane unassisted.

5. **C.** Microtubules originate from basal bodies or centrioles (microtubule organizing centers, or MTOCs), not from the plasma membrane.

6. **D.** Fats usually originate from smooth ER; proteins originate from ribosomes or rough ER; answer choice C would be appropriate for mitochondria.

7. **C.** Since mitochondria produce ATP (but not with the aid of sunlight, as stated in answer choice B), they are often found near areas of major cellular activity, areas that require large amounts of energy.

8. **D.** Note that this question asks about solutes moving down a concentration gradient across a plasma membrane and without ATP.

9. **C.** Note that this question asks about water moving down a gradient.

10. **B.** The movement of urine through the urinary tract is by bulk flow, a collective movement of substances moving in the same general direction. This is in contrast to diffusion, osmosis, and other molecular motions, in which the motion of particular molecules with respect to other molecules is being described.

11. **A.** The energy requirement indicates active transport.

12. **D.** If the solute concentration is higher outside than inside the cell, water moves out of the cell (by osmosis). This causes the cell volume to decrease, resulting in plasmolysis, the collapse of the cell. In contrast, cell lysis occurs when water *enters* the cell, causing the cell volume to increase to the point where the cell bursts.

13. **D.** Since the motion of the molecules is random, at any particular moment there are sure to be some molecules moving against the concentration gradient. It is only the *net* movement of molecules that moves down the gradient.

14. **C.** Animal cells, not plant cells, have plasma membranes that contain cholesterol. Both animals and plants have cells with mitochondria and have ribosomes attached to the ER.

15. **A.** Ribosomes assemble amino acids into proteins. Such activity would be associated with *rough* endoplasmic reticulum.

16. **A.** Phospholipids are the main constituent of plasma membranes and are not found in cell walls. Chitin is found in the cell walls of fungi; polysaccharides are found in the cell walls of plants and archaea; peptidoglycans are found in the cell walls of bacteria.

17. **B.** The solute, starch, cannot pass through the dialysis tubing. The dialysis bag will gain weight because water will diffuse into it. Note that answer choice A refers to the movement of water and answer choice B refers to the movement of the solute and that both describe the gradient relative to the solute. Also note that the distilled water in the beaker is hypotonic and the solution in the dialysis bag is hypertonic when their solute concentrations are compared.

18. **B.** Since sucrose cannot pass through the membrane, no sucrose will enter the tube. However, since there is a concentration gradient, water will diffuse down the gradient. The beginning concentrations of water in the tube and in the beaker are 100% and 90%, respectively. Therefore, water will move from the tube and into the beaker. The water level in the tube will drop (and the beaker level will rise), and the concentration of sucrose in the beaker will decrease.

Free-Response Questions

1. The large surface area of the rough endoplasmic reticulum provides abundant space for the embedded ribosomes and the production and modification of proteins.

2. The flexibility of the plasma membrane is influenced by the relative numbers of saturated and unsaturated fatty acids in the phospholipids. More phospholipids with unsaturated fatty acids would be found in shade leaves because the bend caused by the double bonds in the unsaturated fatty acids increases the separation of the phospholipids, which increases the membrane's flexibility in response to cooler temperatures.

3. After the protein is produced by a ribosome on the membrane surface of the ER, the ER attaches a carbohydrate to the protein and packages the glycoprotein within a vesicle. The vesicle then transports the glycoprotein to a Golgi apparatus, where it may undergo additional modification. The modified glycoprotein is again packaged into a vesicle that transports the glycoprotein from the Golgi apparatus to the plasma membrane, where it is exported by exocytosis.

4. **a.** The plasma membrane is composed of a phospholipid bilayer. A molecule of phospholipid consists of two fatty acids and a phosphate group attached to a glycerol component. The fatty acid tails represent a hydrophobic region of the molecule, while the glycerol-phosphate head is hydrophilic. The phospholipids are arranged into a bilayer formation with the hydrophilic heads pointing to the outside and the hydrophobic tails pointing toward the inside. As a result, the plasma membrane is a barrier to most molecules. In plants, fungi, and bacteria, the membrane deposits cellulose or other polysaccharides on the outside of the membrane to create a cell wall. The cell wall provides support to the cell.

Embedded in the phospholipid bilayer are various proteins and, in animal cells, cholesterol molecules. This mixture of molecules accounts for the fluid mosaic model of the plasma membrane, that is, a highly flexible lipid boundary impregnated with various other molecules.

b. The plasma membrane is a selectively permeable membrane. Small molecules, like O_2 and CO_2, readily diffuse through the membrane. The movement of larger molecules is regulated by proteins in the plasma membrane. There are several kinds of these proteins. Channel proteins provide passage for certain dissolved substances. Transport proteins actively transport substances against a concentration gradient. The extracellular matrix in animal cells, consisting of the polysaccharides of glycolipids, recognition proteins, and other glycoproteins, provides adhesion or participates in cell-to-cell interactions. Receptor proteins recognize hormones and transmit their signals to the interior of the cell.

c. Various substances can be exported into the external environment by exocytosis. In exocytosis, substances are packaged in vesicles that merge with the plasma membrane. Once they merge with the membrane, their contents are released to the outside. In an opposite kind of procedure, food and other substances can be imported by endocytosis. In endocytosis, the plasma membrane encircles the substance and encloses it in a vesicle.

Note: When a question has two or more parts, separate your answers and identify each part with the corresponding letter.

5. *For this question, be sure to separate your answer into two parts. The first part should compare prokaryotes and eukaryotes, that is, describe characteristics they have in common. For example, they both have a plasma membrane, ribosomes, and DNA. Also, many prokaryotes have a cell wall, a structure they have in common with the eukaryotic cells of plants and fungi.*

In the second part of your answer, contrast prokaryotes and eukaryotes, that is, describe how they are different. Indicate that the DNA is packaged differently (naked DNA molecules in prokaryotes compared to DNA associated with proteins in eukaryotes), that prokaryotic ribosomes are smaller than those of eukaryotes, that the prokaryotic flagella do not contain microtubules, and that prokaryotic cells lack a nuclear membrane and the various eukaryotic organelles. Also, the cell walls of bacteria contain peptidoglycans, and those of archaea contain other polysaccharides, but not the cellulose and chitin found in the cell walls of plants and fungi.

6. *For this question, be sure to separate your answer into two parts. In the first part, describe each cell organelle and its function. In the second part, explain that partitioning metabolic functions into organelles serves primarily to separate the biochemical activities. In addition, describe how the channels among layers of endoplasmic reticulum serve to create compartments as well. Last, describe the packaging relationship between the ER and Golgi bodies.*

Chapter 4

Cellular Respiration

Review

Cellular respiration is the first major topic for which you apply your knowledge of chemistry. For the most part, however, the chemistry is descriptive—that is, you won't have to solve chemical equations or even memorize structural formulas. Instead, you need to provide names of major molecules (usually just the reactants and the products), describe their sequence in a metabolic process, and most important, describe how the process accomplishes its metabolic objectives.

Energy Basics

Energy is required to do work—to do things, to move things, to put things together. How energy does work follows two laws of thermodynamics:

- The **first law of thermodynamics** states that the total amount of energy in the universe remains constant. Energy cannot be created or destroyed, but it can be converted from one form to another. These forms of energy include **kinetic energy** (energy of motion) and **potential energy** (stored energy, such as that in a chemical bond or in something elevated in a gravitational field).
- The **second law of thermodynamics** states that when energy is converted from one form to another, some energy is "lost." The word "lost" is not intended to mean "gone" (that would be a violation of the first law of thermodynamics). Instead, it means that not all of the energy gets passed from one usable form to another, that some of the energy becomes unusable or unable to do work. The unusable energy is usually in the form of heat. Further, as additional energy conversions occur, more energy becomes unusable, things become disorganized, and disorder or randomness increases. That disorder is called **entropy.** Because energy is constantly moving from one form to another, a consequence of the second law of thermodynamics is that entropy increases in the universe.

Energy conversions are usually discussed within the context of a **system,** such as a chemical reaction, a cell, a multicellular organism, or a planet like Earth. It can be a **closed system,** where only energy transfers among specific items are considered, or it can be an **open system,** in which exchanges of energy with the surroundings are included. Living things and the earth are open systems because they receive energy from sunlight.

In an **exergonic** chemical reaction, there is a net *release* of energy. These reactions can occur spontaneously, that is, without an input of energy. When glucose is broken down to CO_2 and H_2O during cellular respiration, for example, energy that is released from the reaction is stored in ATP, a source of energy for metabolic reactions:

$$C_6H_{12}O_6 + 6\,O_2 \rightarrow 6\,CO_2 + 6\,H_2O + \text{ATP}$$

In an **endergonic** reaction, energy must be *added* to the reaction for it to occur. In photosynthesis, for example, sunlight provides the energy necessary for the production of glucose from CO_2 and H_2O:

$$6\,CO_2 + 6\,H_2O + \text{sunlight} \rightarrow C_6H_{12}O_6 + 6\,O_2$$

Note that the overall equation for photosynthesis is the reverse of that for respiration, except that the form of energy is different (ATP vs. sunlight).

Even though an exergonic reaction *can* occur spontaneously, it may not. Before most reactions can occur, **activation energy** is required to contort or destabilize the reactants. The necessary activation energy can be lowered by the presence of a catalyst (such as a metal ion or an enzyme). However, the overall energy of a reaction—the energy released or the energy required—is not changed by the presence of a catalyst (Figure 4-1).

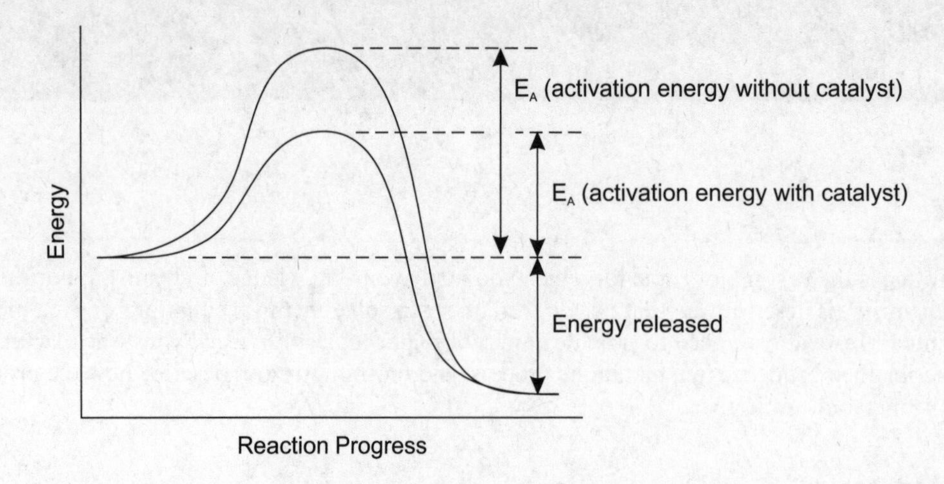

Activation Energy for an Exergonic Reaction
Figure 4-1

Most metabolic reactions are endergonic and require an input of energy. That energy usually comes from the hydrolysis of adenosine *tri*phosphate (ATP) to adenosine *di*phosphate (ADP). The hydrolysis of ATP (an *exergonic* reaction) is **coupled** with the *endergonic* metabolic reaction. The coupling usually involves the transfer of energy with one of the inorganic phosphates (P_i) from ATP to one of the reactants, as shown in Figure 4-2. Coupled reactions with ATP (and with the help of specific enzymes) are typical of most endergonic metabolic reactions. Note that added energy is expressed as a positive number, whereas released energy is expressed as a negative number.

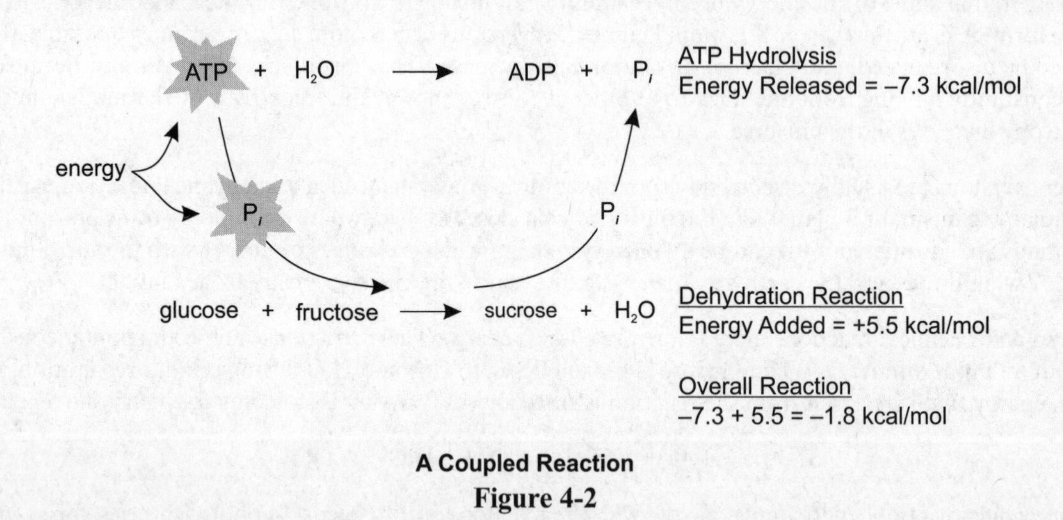

A Coupled Reaction
Figure 4-2

In order for life to persist, living things require a constant input of energy. That energy is used to maintain order in opposition to the entropy that increases as a result of chemical reactions. Without an input of energy, entropy increases, cells deteriorate, and death follows. **Photosynthesis,** the process of incorporating energy from sunlight into carbohydrates, and **respiration,** the process of extracting energy from those carbohydrates, provide the energy that allows cells to maintain order, minimize entropy, and remain alive.

Generating ATP

Phosphorylation is the process of adding energy and an inorganic phosphate to ADP to make ATP:

$$\text{energy} + P_i + \text{ADP} \rightarrow \text{ATP}$$

There are two basic mechanisms of phosphorylation in cells:

1. **Substrate level phosphorylation** occurs when a phosphate group *and* its associated energy are transferred to ADP to form ATP. The *substrate* molecule (a molecule with the phosphate group) donates the high energy phosphate group. Such phosphorylation occurs during glycolysis, the initial breakdown process of glucose (discussed later in this chapter).
2. **Oxidative phosphorylation** occurs when a phosphate group is added to ADP to form ATP, but the energy for the bond does not accompany the phosphate group. Instead, electrons give up energy for generating ATP during each step of a process, where electrons are transferred from one molecule (electron carrier) to another in a chain of reactions.

Cellular respiration is the ATP-generating process that occurs in cells. As previewed earlier, energy is extracted from energy-rich glucose to form ATP. The chemical equation describing this process is:

$$C_6H_{12}O_6 + 6\ O_2 \rightarrow 6\ CO_2 + 6\ H_2O + \text{energy}$$

$C_6H_{12}O_6$ is glucose, but sometimes you see CH_2O or $(CH_2O)_n$. These are general formulas for glucose or any carbohydrate.

When oxygen is available, most cells will generate ATP by **aerobic respiration.** There are three steps that occur in this ATP generating pathway:

1. **Glycolysis**
2. **Krebs cycle** (citric acid cycle)
3. **Oxidative phosphorylation**

When oxygen is not available, cells will generate ATP by **anaerobic respiration.** There are two slightly different pathways for this process:

1. **Alcohol fermentation**
2. **Lactic acid fermentation**

Each of these processes is discussed in detail in the following section. Refer to Figure 4-3. In addition to obtaining energy directly from glucose, other carbohydrates, such as starch and glycogen, can be hydrolyzed to glucose, and sucrose can be hydrolyzed to glucose and fructose. So they all end up being glucose or fructose and enter the glycolytic (glycolysis) pathway.

Proteins can also be a source of energy. Proteins that are eaten are digested to amino acids before they are absorbed into the bloodstream. Body proteins, if necessary, can be hydrolyzed to amino acids. Amino groups ($-NH_2$) are first stripped from the amino acids and then excreted as waste. The remainders of the amino acids are then enzymatically converted to various substances that enter intermediate steps of glycolysis or the Krebs cycle.

Fats are storage molecules for energy so they, too, can be sources of energy. Glycerol and fatty acids are obtained from fats by hydrolysis or by the digestion of fats that are eaten. After enzymatic conversions, glycerol enters glycolysis and fatty acids enter the Krebs cycle (as acetyl CoA).

Glycolysis

Glycolysis is the decomposition (lysis) of **glucose** (glyco) to **pyruvate** (or **pyruvic acid**). The steps are summarized here and in Figure 4-3.

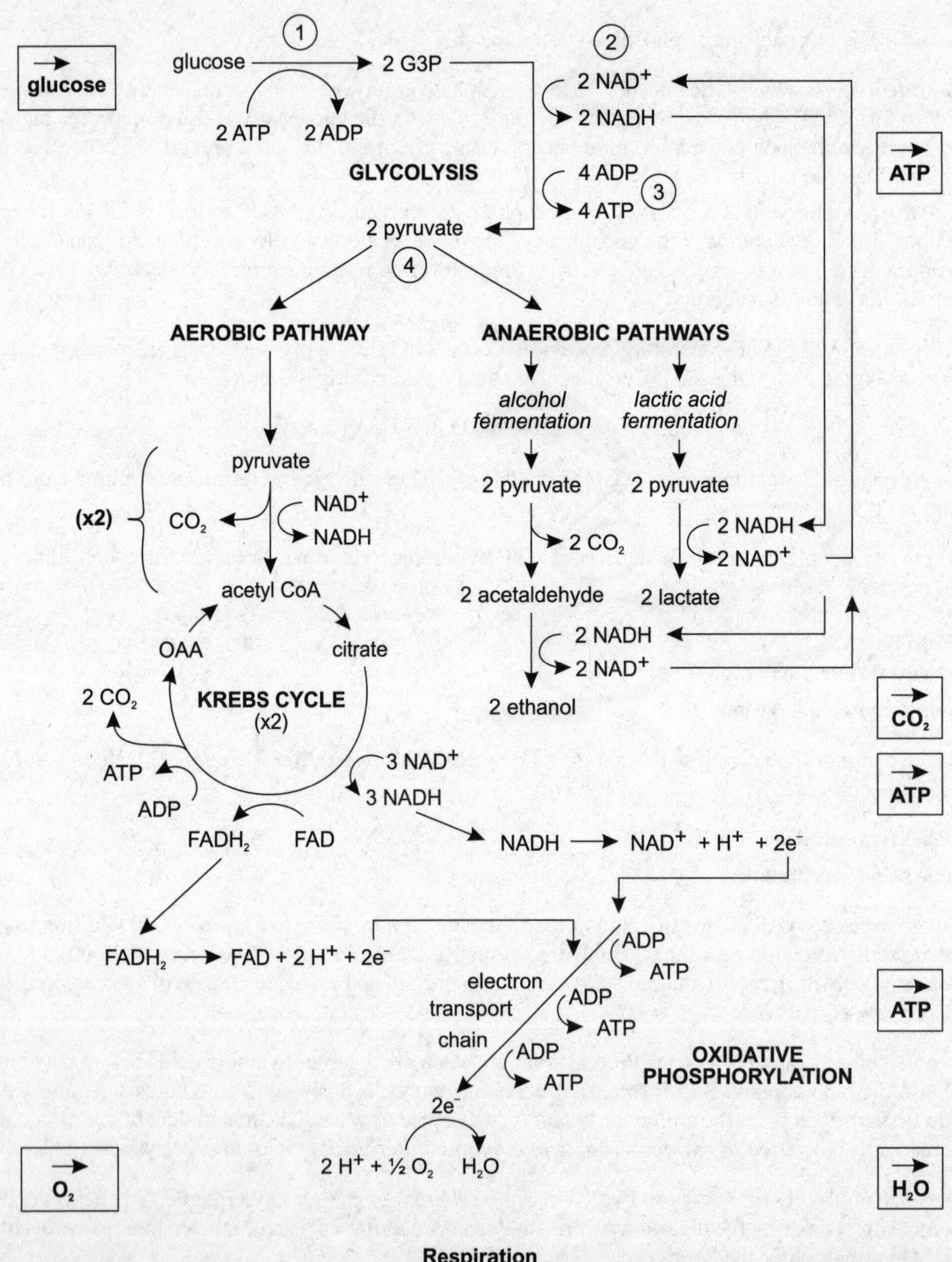

Respiration
Figure 4-3

For each molecule of glucose:

1. **2 ATP are added.** The first several steps require the input of ATP. This changes glucose in preparation for subsequent steps.
2. **2 NADH are produced.** NADH, a *coenzyme,* is an electron carrier when NAD^+ combines with two energy-rich electrons and H+ (obtained from an intermediate molecule during the breakdown of glucose). As a result, NADH is an energy-rich molecule.
3. **4 ATP are produced by substrate-level phosphorylation.**
4. **2 pyruvate are formed.**

In summary, glycolysis takes 1 glucose molecule and turns it into 2 pyruvate, 2 NADH, and a net of 2 ATP (made 4 ATP, but used 2 ATP). The process occurs in the cytosol.

The Krebs Cycle

The Krebs cycle details what happens to pyruvate, the end product of glycolysis. Although the Krebs cycle is described for 1 pyruvate, remember that glycolysis produces 2 pyruvate. In Figure 4-3, the "× 2" next to the pyruvate and the Krebs cycle is a reminder to multiply the products of this cycle by 2 to account for the products of a single glucose.

The processing of pyruvate can be summarized as follows:

1. **Pyruvate to acetyl CoA.** In a step leading up to the actual Krebs cycle, pyruvate combines with coenzyme A (CoA) to produce acetyl CoA. In that reaction, 1 NADH and 1 CO_2 are also produced.
2. **Krebs cycle: 3 NADH, 1 $FADH_2$, 1 ATP, 2 CO_2.** The Krebs cycle begins when acetyl CoA combines with OAA (oxaloacetate) to form citrate. There are seven intermediate products. Along the way, 3 NADH, 1 $FADH_2$, and 1 ATP are made, and 2 CO_2 are released. $FADH_2$, like NADH, is a coenzyme, accepting electrons during a reaction.

The CO_2 produced by the Krebs cycle is the CO_2 animals exhale when they breathe.

Oxidative Phosphorylation

Oxidative phosphorylation is the process of producing ATP from NADH and $FADH_2$. Electrons from NADH and $FADH_2$ pass along an **electron transport chain (ETC).** The chain consists of proteins that pass these electrons from one carrier protein to the next. Some carrier proteins, such as the **cytochromes,** include nonprotein parts containing iron. Along each step of the chain, the electrons give up energy used to phosphorylate ADP to ATP. NADH provides electrons that have enough energy to generate about 3 ATP, while $FADH_2$ generates about 2 ATP.

The final electron acceptor of the electron transport chain is **oxygen.** The ½ O_2 accepts the two electrons and, together with 2 H^+, forms water.

One of the carrier proteins in the electron transport chain, **cytochrome c,** is so ubiquitous among living organisms that the approximately 100-amino-acid sequence of the protein is often compared among species to assess genetic relatedness.

Mitochondria

The two major processes of aerobic respiration, the Krebs cycle and oxidative phosphorylation, occur in mitochondria. There are four distinct areas of a mitochondrion, as follows (Figure 4-4):

1. **Outer membrane.** This membrane, like the plasma membrane, consists of a double layer of phospholipids.
2. **Intermembrane space.** This is the narrow area between the inner and outer membranes. H^+ ions (protons) accumulate here.

3. **Inner membrane.** This second membrane, also a double phospholipid bilayer, has convolutions called **cristae** (singular, **crista**). Oxidative phosphorylation occurs here. Within the membrane and its cristae, the electron transport chain, consisting of a series of protein complexes, removes electrons from NADH and $FADH_2$ and transports H^+ ions from the matrix to the intermembrane space. Some of these protein complexes are indicated in Figure 4-4 (PC I, PC II, PC III, and PC IV). Another protein complex, **ATP synthase,** is responsible for the phosphorylation of ADP to form ATP.
4. **Matrix.** The matrix is the fluid material that fills the area inside the inner membrane. The Krebs cycle and the conversion of pyruvate to acetyl CoA occur here.

Chemiosmosis in Mitochondria

Chemiosmosis is the mechanism of ATP generation that occurs when energy is stored in the form of a *proton concentration gradient* across a membrane. A description of the process during oxidative phosphorylation in mitochondria follows (Figure 4-4):

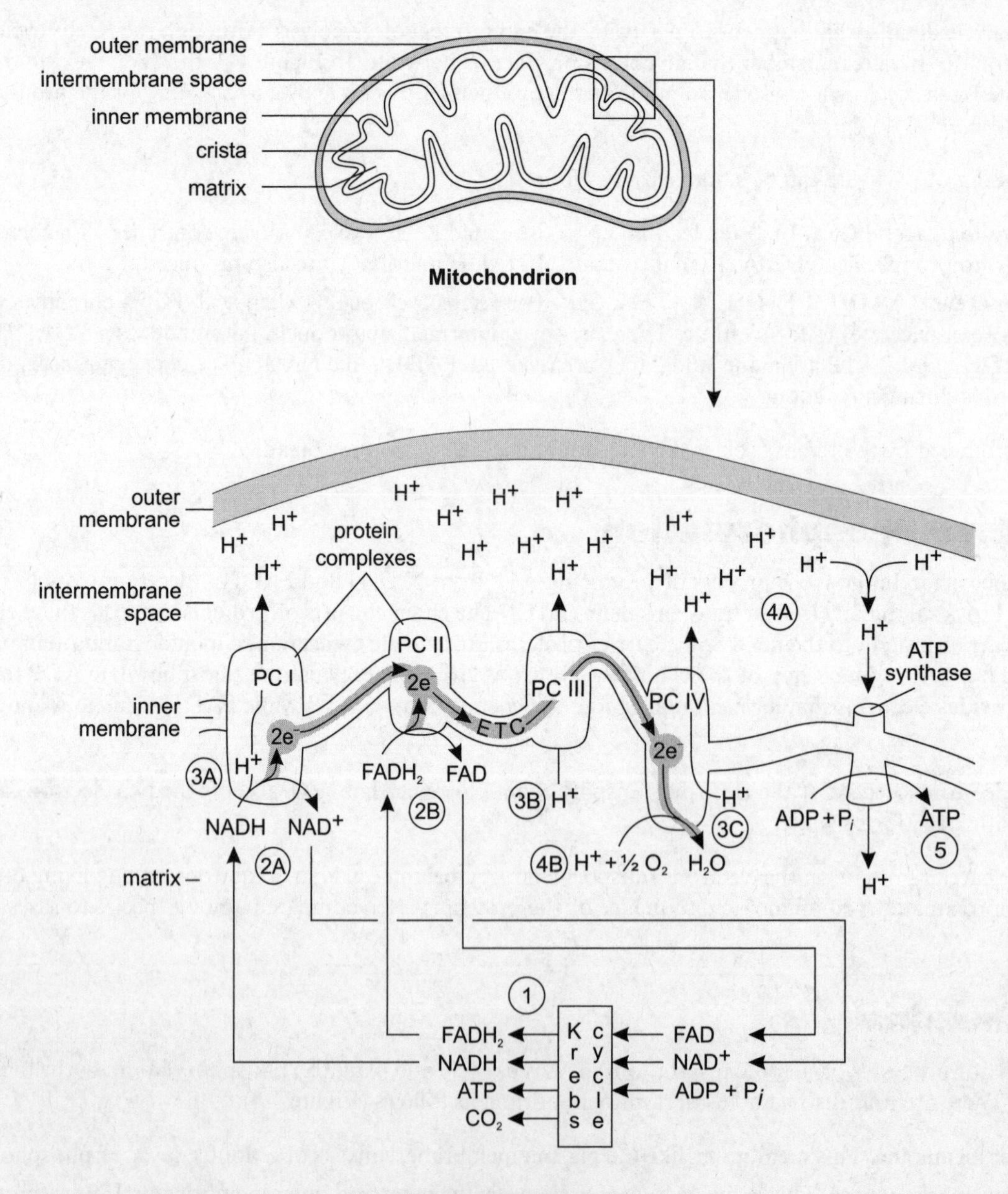

Chemiosmosis in Mitochondria

Figure 4-4

1. **The Krebs cycle produces NADH and $FADH_2$ in the matrix.** In addition, CO_2 is generated and substrate-level phosphorylation occurs to produce ATP.
2. **Electrons are removed from NADH and $FADH_2$.** Protein complexes in the inner membrane remove electrons from these two molecules (2A, 2B). The electrons move along the electron transport chain, from one protein complex to the next (shown as a shaded strip within the inner membrane).
3. **H^+ ions (protons) are transported from the matrix to the intermembrane compartment.** Protein complexes transport H^+ ions from the matrix, across the inner membrane, and to the intermembrane space (3A, 3B, 3C).
4. **A pH and electrical gradient across the inner membrane is created.** As H^+ are transferred from the matrix to the intermembrane space, the concentration of H^+ *increases* (pH decreases) in the intermembrane space (4A) and *decreases* in the matrix (pH increases). The concentration of H^+ in the matrix decreases further as electrons at the end of the electron transport chain (PC IV) combine with H^+ and oxygen to form water (4B). The result is a proton gradient (equivalent to a pH gradient) and an electric charge (or voltage) gradient. These gradients are potential energy reserves in the same manner as water behind a dam is stored energy.
5. **ATP synthase generates ATP.** ATP synthase, a channel protein in the inner membrane, provides a pathway for the protons in the intermembrane compartment to flow back into the matrix. As the protons are drawn though the channel by the voltage and pH gradients, the protons lose energy to the ATP synthase. ATP synthase uses the energy to generate ATP from ADP and P_i. It is similar to how a dam generates electricity when water passing through turbines forces them to turn.

How Many ATP?

How many ATP can *theoretically* be made from the energy released from the breakdown of 1 glucose molecule? In the cytoplasm, 1 glucose produces 2 NADH, 2 ATP, and 2 pyruvate during glycolysis. When the 2 pyruvate (from 1 glucose) are converted to 2 acetyl CoA, 2 more NADH are produced. From 2 acetyl CoA, the Krebs cycle produces 6 NADH, 2 $FADH_2$, and 2 ATP. If each NADH produces 3 ATP during oxidative phosphorylation, and $FADH_2$ produces 2 ATP, the total ATP count from 1 original glucose molecule appears to be 38. However, this number is reduced to 36 because the 2 NADH that are produced in the cytoplasm during glycolysis must be transported into the mitochondria for oxidative phosphorylation. The transport of NADH across the mitochondrial membrane reduces the net yield of each NADH to only 2 ATP (Table 4-1). However, the total yield from 1 glucose molecule *actually* hovers around 30 ATP due to variations in mitochondria efficiencies and competing biochemical processes.

Table 4-1

Process	Location of Process	Type of Phosphorylation	$FADH_2$ Produced	NADH Produced		ATP Yield
glycolysis	cytosol	substrate-level				2 ATP
glycolysis	cytosol	oxidative		2 NADH	=	4 ATP
pyruvate to acetyl CoA	mitochondrial matrix	oxidative		2 NADH	=	6 ATP
Krebs cycle	mitochondrial matrix	substrate-level				2 ATP
oxidative phosphorylation	mitochondrial inner membrane	oxidative		6 NADH	=	18 ATP
oxidative phosphorylation	mitochondrial inner membrane	oxidative	2 $FADH_2$		=	4 ATP
Total						**36 ATP**

A balance sheet accounting for ATP production from glucose by aerobic respiration. Total ATP production is theoretically 36 ATP for each glucose processed.

Anaerobic Respiration

What if oxygen is not present? If oxygen is not present, no electron acceptor exists to accept the electrons at the end of the electron transport chain. If this occurs, then NADH accumulates. After all the NAD^+ have been converted to NADH, the Krebs cycle and glycolysis both stop (both need NAD^+ to accept electrons). When this happens, no new ATP is produced, and the cell may soon die.

Anaerobic respiration is a method cells use to escape this fate. Although the two common metabolic pathways, **alcohol** and **lactic acid fermentation,** are slightly different, the objective of both processes is to replenish NAD^+ so that glycolysis can proceed once again. Anaerobic respiration occurs in the cytosol alongside glycolysis.

Alcohol Fermentation

Alcohol fermentation (or sometimes, just **fermentation**) occurs in plants, fungi (such as yeasts), and bacteria. The steps, illustrated in Figure 4-3, are as follows:

1. **Pyruvate to acetaldehyde.** For each pyruvate, 1 CO_2 and 1 acetaldehyde are produced. The CO_2 formed is the source of carbonation in fermented drinks like beer and champagne.
2. **Acetaldehyde to ethanol.** The important part of this step is that the energy in NADH is used to drive this reaction, releasing NAD^+. For each acetaldehyde, 1 ethanol is made and 1 NAD^+ is produced. The ethanol (ethyl alcohol) produced here is the source of alcohol in beer and wine.

It is important that you recognize the objective of this pathway. At first glance, you should wonder why the energy in an energy-rich molecule like NADH is removed and put into the formation of ethanol, essentially a waste product that eventually kills the yeast (and other organisms) that produce it. The goal of this pathway, however, does not really concern ethanol, but the task of freeing NAD^+ to allow glycolysis to continue. Recall that in the absence of O_2, all the NAD is bottled up in NADH. This is because oxidative phosphorylation cannot accept the electrons of NADH without oxygen. The purpose of the fermentation pathway, then, is to release some NAD^+ for use by glycolysis. The reward for this effort is 2 ATP from glycolysis for each 2 converted pyruvate. This is not much, but it's better than the alternative—0 ATP and imminent cellular death.

Lactic Acid Fermentation

Only one step occurs in lactic acid fermentation (Figure 4-3). A pyruvate is converted to lactate (or lactic acid); in the process, NADH gives up its electrons to form NAD^+. As in alcohol fermentation, the NAD^+ can now be used for glycolysis. In humans and other mammals, most lactate is transported to the liver, where it is converted back to glucose when surplus ATP is available.

Review Questions

Multiple-Choice Questions

The questions that follow provide a review of the material presented in this chapter. Use them to evaluate how well you understand the terms, concepts, and processes presented. Actual AP multiple-choice questions are often more general, covering a broad range of concepts, and often more lengthy. For multiple-choice questions typical of the exam, take the two practice exams in this book.

Directions: Each of the following questions or statements is followed by four possible answers or sentence completions. Choose the one best answer or sentence completion.

1. What is the value of the alcohol fermentation pathway?

- **A.** It produces ATP.
- **B.** It produces lactate (or lactic acid).
- **C.** It produces ADP for the electron transport chain.
- **D.** It replenishes NAD^+ so that glycolysis can produce ATP.

2. What is the purpose of oxygen in aerobic respiration?

- **A.** Oxygen accepts electrons at the end of an electron transport chain.
- **B.** Oxygen is necessary to carry away the waste CO_2.
- **C.** Oxygen is used in the formation of sugar molecules.
- **D.** The oxygen molecule becomes part of the ATP molecule.

Questions 3–7 refer to the following diagram. The three boxes represent the three major biosynthetic pathways in aerobic respiration. Arrows represent net reactants or products.

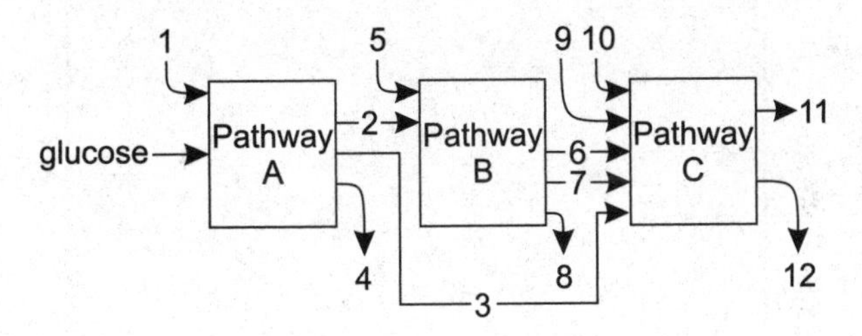

3. Arrow 2 is

- **A.** O_2
- **B.** ATP
- **C.** $FADH_2$
- **D.** pyruvate

4. Arrows 4, 8, and 12 could all be

- **A.** NADH
- **B.** ATP
- **C.** H_2O
- **D.** $FADH_2$

5. Arrows 3 and 7 could both be

A. NADH
B. ATP
C. H_2O
D. $FADH_2$

6. Arrow 9 could be

A. O_2
B. ATP
C. H_2O
D. FAD

7. Pathway B is

A. oxidative phosphorylation
B. photophosphorylation
C. the Krebs cycle
D. glycolysis

8. Which of the following sequences correctly indicates the potential ATP yield of the indicated molecules from greatest ATP yield to least ATP yield?

A. pyruvate, ethanol, glucose, acetyl CoA
B. glucose, pyruvate, acetyl CoA, NADH
C. glucose, pyruvate, NADH, acetyl CoA
D. glucose, $FADH_2$, NADH, pyruvate

Questions 9–10 refer to the following graph that shows the amount of CO_2 that is released by plant cells at various levels of atmospheric oxygen.

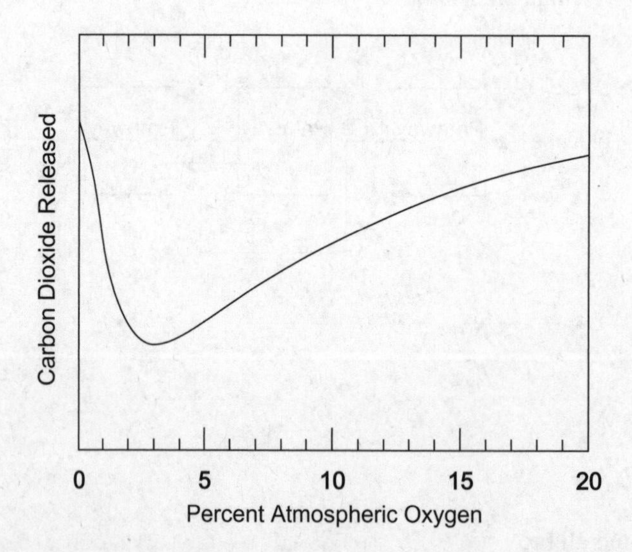

9. At levels of atmospheric O_2 below 1%, the amount of CO_2 released is relatively high. This is probably because

A. The Krebs cycle is very active.
B. O_2 is being converted to H_2O.
C. Alcohol fermentation is occurring.
D. There are insufficient amounts of coenzyme A.

10. As levels of atmospheric O_2 increase beyond 5%, the amounts of CO_2 released increase. This is probably a direct result of

A. an increase in glycolytic activity
B. a greater availability of appropriate enzymes
C. an increase in Krebs cycle activity
D. an increase in atmospheric temperature

11. All of the following processes produce ATP EXCEPT:

A. glycolysis
B. the Krebs cycle
C. lactic acid fermentation
D. oxidative phosphorylation of NADH

12. Chemiosmosis describes how ATP is generated from ADP. All of the following statements conform to the process as it occurs in mitochondria EXCEPT:

A. H^+ accumulates in the area between the membrane of the cristae and the outer membrane of the mitochondrion.
B. A voltage gradient is created across the cristae membranes.
C. A proton gradient is created across the cristae membranes.
D. Electrons flowing through the ATP synthase channel protein provide the energy to phosphorylate ADP to ATP.

13. After strenuous exercise, a muscle cell would contain increased amounts of all of the following EXCEPT:

A. ADP
B. CO_2
C. lactate (or lactic acid)
D. glucose

14. All of the following statements about cellular respiration are true EXCEPT:

A. Some of the products from the breakdown of proteins and lipids enter the Krebs cycle.
B. If oxygen is present, water is produced.
C. The purpose of oxygen in aerobic respiration is to donate the electrons that transform $NAD^+ + H^+$ to NADH.
D. Lactate or ethanol is produced when oxygen is unavailable.

15. All of the following processes release CO_2 EXCEPT:

A. the Krebs cycle
B. alcohol fermentation
C. oxidative phosphorylation
D. the conversion of pyruvate to ethanol

Free-Response Questions

The AP exam has long and short free-response questions. The long questions have considerable descriptive information that may include tables, graphs, or figures. The short questions are brief but may also include figures. Both kinds of questions have four parts and generally require that you bring together concepts from multiple areas of biology.

The questions that follow are designed to further your understanding of the concepts presented in this chapter. Unlike the free-response questions on the exam, they are narrowly focused on the material in this chapter. For free-response questions typical of the exam, take the two practice exams in this book.

Directions: The best way to prepare for the AP exam is to write out your answers as if you were taking the exam. Use complete sentences and do *not* use outline form or bullets. You may use diagrams to supplement your answers, but be sure to describe the importance or relevance of your diagrams.

---

1. In the process of alcohol fermentation, 2 NADH molecules are converted to 2 NAD^+ as energy from the NADH is used to drive the formation of ethanol. Explain why there is a need to add energy to a process whose purpose is to extract energy from glucose.

2. The mitochondrion has two phospholipid-bilayer membranes: an outer membrane and an inner membrane. Explain why two membranes are necessary.

3. Describe the Krebs cycle and oxidative phosphorylation. Specifically address the following:

- **a.** ATP and coenzyme production
- **b.** the locations where these biosynthetic pathways occur
- **c.** chemiosmotic theory

4. Describe, at the molecular level, how aerobic respiration extracts energy from each of the following:

- **a.** starches and other carbohydrates
- **b.** proteins
- **c.** lipids

5. **a.** Explain, at the molecular level, why many organisms need oxygen to maintain life.

b. Explain, at the molecular level, how some organisms can sustain life in the absence of oxygen.

Answers and Explanations

Multiple-Choice Questions

1. D. In the absence of oxygen, all of the NAD^+ gets converted to NADH. With no NAD^+ to accept electrons from the glycolytic steps, glycolysis stops. By replenishing NAD^+, alcohol fermentation allows glycolysis to continue.

2. A. At the end of the electron transport chain in oxidative phosphorylation, ½ O_2 combines with 2 electrons and 2 H^+ to form water.

3. D. You should review aerobic respiration by identifying each arrow: Pathway A is glycolysis, Pathway B is the Krebs cycle, and Pathway C is oxidative phosphorylation. Arrow 1: ADP or NAD^+; arrow 2: pyruvate; arrow 3: NADH; arrow 4: ATP; arrow 5: ADP, NAD^+, or FAD; arrows 6 and 7: $FADH_2$ and NADH (either one can be 6 or 7); arrow 8: ATP or CO_2; arrows 9 and 10: O_2 and ADP (either one can be 9 or 10); and arrows 11 and 12: H_2O and ATP (either one can be 11 or 12).

4. **B.** ATP is produced in the glycolytic pathway (glycolysis), the Krebs cycle, and by oxidative phosphorylation.

5. **A.** Arrow 3 represents the NADH produced in glycolysis (Pathway A) and is used in oxidative phosphorylation (Pathway C). In addition, NADH could also be represented by arrow 7, a product of the Krebs cycle (Pathway B). Arrow 7 could also represent $FADH_2$, but $FADH_2$ cannot be represented by arrow 3. Thus, only NADH can be represented by both arrows 3 and 7. If arrow 7 represents NADH, then arrow 6 represents $FADH_2$.

6. **A.** Arrow 9 could represent the O_2 that accepts the electrons after they pass through the electron transport chain in oxidative phosphorylation. Arrow 9 could also be ADP, but ADP is not among the answer choices.

7. **C.** Pathway B represents the Krebs cycle. The Krebs cycle uses the energy in pyruvate (arrow 2) to generate $FADH_2$ and NADH (arrows 6 and 7).

8. **B.** Each of these molecules has the potential to produce the following amounts of ATP: glucose, 36 ATP; pyruvate, 15 ATP; acetyl CoA, 12 ATP; NADH, 3 ATP (or 2 ATP if they originate in glycolysis); and $FADH_2$, 2 ATP. The metabolic pathway that breaks down ethanol to H_2O and CO_2 in the human liver is variable. However, answer choice A can be eliminated without knowing how many ATP molecules ethanol can yield because glucose produces more ATP than does pyruvate.

9. **C.** When O_2 is absent (or very low), anaerobic respiration (alcohol fermentation) is initiated. Alcohol fermentation releases CO_2. Photosynthesis, which would consume CO_2 to produce glucose, is obviously not occurring. This indicates that the plant activity illustrated by the graph is occurring at night (or during a heavily clouded day).

10. **C.** CO_2 is produced in the Krebs cycle. As in the previous question, the production of CO_2, rather than its consumption, indicates that photosynthesis is not occurring and that the plant activity is taking place at night.

11. **C.** Lactic acid fermentation, the conversion of pyruvate to lactate, removes electrons from NADH to make NAD^+. No ATP is generated by this step.

12. **D.** Protons, not electrons, pass through ATP synthase as they move down the proton gradient. An electrical gradient or voltage is produced by the greater number of positive charges (from the protons) in the intermembrane space relative to the number of positive charges inside the crista membrane.

13. **D.** During strenuous exercise, glucose is broken down to pyruvate. Aerobic respiration produces CO_2. Anaerobic respiration, which would occur during strenuous exercise, would increase lactate formation. Exercise would also consume ATP, producing ADP and P_i.

14. **C.** The purpose of O_2 is to *accept* the electrons at the end of the electron transport chain in oxidative phosphorylation. The electrons combine with O_2 and H to form water. Products from the breakdown of lipids and proteins are converted to pyruvate, acetyl CoA, or intermediate carbon compounds used in the Krebs cycle.

15. **C.** Oxidative phosphorylation describes the transfer of electrons from NADH and $FADH_2$ to electron acceptors that pump H^+ across the inner mitochondrial membrane. Oxygen is required as the final electron acceptor of these electrons. However, no CO_2 is involved. In contrast, all the remaining answer choices describe processes that release CO_2. Note that answer choices B and D describe the same process.

Free-Response Questions

1. In the absence of O_2, oxidative phosphorylation and ATP generation cannot proceed and NADH accumulates. As a result, there is no NAD^+ available for glycolysis. Fermentation regenerates NAD^+ from NADH, and this NAD^+ can be used in glycolysis to generate ATP.
2. The two membranes create an intermembrane space between them, where H^+ (protons), transported from the matrix, accumulate. This creates a proton gradient, which drives the movement of protons through ATP synthase back into the matrix. As protons pass through the ATP synthase channel, ATP is generated from ADP and P_i.

3. a. The Krebs cycle and oxidative phosphorylation are the oxygen-requiring processes involved in obtaining ATP from pyruvate. Pyruvate is derived from glucose through glycolysis, a process that does not require oxygen.

Before pyruvate enters the Krebs cycle, it combines with coenzyme A. During this initial reaction with coenzyme A, 2 electrons and 2 H^+ removed from pyruvate combine with NAD^+ to form 1 NADH + H^+. NADH is a coenzyme storing enough energy to generate 3 ATP in oxidative phosphorylation. A CO_2 molecule is also released. The end product of this reaction is acetyl CoA. To begin the Krebs cycle, acetyl CoA combines with oxaloacetate (OAA) to form citrate, releasing the coenzyme A component. A series of reactions then occurs that generates 3 molecules of the coenzyme NADH (from NAD^+), 1 molecule of the coenzyme $FADH_2$ (from FAD), and 1 ATP (from ADP + P_i) for each molecule of acetyl CoA that enters the Krebs cycle. The last product in the series of reactions, OAA, is the substance that reacts with acetyl CoA; thus, the Krebs reactions sustain a cycle.

Energy from the coenzymes NADH and $FADH_2$ is extracted to make ATP in oxidative phosphorylation. For each of these coenzymes, 2 electrons pass through an electron transport chain, passing through a series of protein carriers (some are cytochromes, such as cytochrome c). During this passage, 3 ATP are generated for each NADH originating in the Krebs cycle. $FADH_2$ generates 2 ATP. At the end of the electron transport chain, O_2 accepts the electrons (and 2 H^+) to form water. NAD^+ and FAD can be used again to receive electrons in the Krebs cycle. The total number of ATP generated from a single pyruvate is 15 ATP.

b. The Krebs cycle and oxidative phosphorylation occur in the mitochondria. The Krebs cycle occurs in the matrix of the mitochondria. The protein carriers for the electron transport chain are embedded in the inner mitochondrial membranes, called the cristae. Thus, oxidative phosphorylation occurs in these cristae membranes.

c. Chemiosmosis describes how ATP is generated from ADP + P_i. During oxidative phosphorylation, H^+ (protons) are deposited on the outside of the cristae, between the cristae and the outer membrane. The excess number of protons in this intermembrane space creates a pH and electric gradient. The gradient provides the energy to generate ATP as protons pass back into the matrix through ATP synthase, a channel protein in the cristae.

Note that the answer to each part of the question is labeled a, b, or c. Answering each part separately helps you to organize your answer and helps the grader recognize that you addressed each part of the question.

4. *This question is similar to question 1. One major difference is that you need to describe glycolysis in detail. You also need to correctly connect glycolysis with the Krebs cycle and ATP production because glycolysis produces 2 pyruvates. The explanation in part 1 is for the production of only 1 pyruvate. The second major difference is that you need to include a discussion of how starches, proteins, and lipids are tied to respiration. That part of the answer follows.*

a. Starches are polymers of glucose. Various enzymes break down starches and other carbohydrates to glucose. Disaccharides like sucrose are catalyzed to glucose and fructose. Once these carbohydrates are broken down to glucose, the glucose enters the glycolytic pathway. Fructose undergoes some intermediate steps and enters the glycolytic pathway after a couple of steps. Glycolysis then breaks down glucose, obtaining ATP and pyruvate. Pyruvate then enters the Krebs cycle, yielding more ATP, NADH, and $FADH_2$. NADH and $FADH_2$ then generate ATP during oxidative phosphorylation.

b. Lipids are hydrolyzed to glycerol and fatty acids. Both of these components undergo enzymatic reactions that eventually produce acetyl CoA. Acetyl CoA, in turn, begins the Krebs cycle, which generates NADH, $FADH_2$, and ATP.

c. Proteins are hydrolyzed to amino acids. Each of the various amino acids produces different products when broken down. Some of these products are converted to acetyl CoA; others are converted to OAA or other Krebs cycle intermediates. NH_3 is a toxic waste product from amino acid breakdown and is exported from the cell.

5. a. This question is the same as question 4, except there is an additional focus on the function of O_2. For this question, then, it is especially important that you state that the purpose of O_2 is to accept electrons at the end of the electron transport chain in oxidative phosphorylation. Then describe the consequences if oxygen is not present—no ATP, no oxidative phosphorylation, and no Krebs cycle.

b. This question requires that you describe lactic acid fermentation and alcohol fermentation, specifically indicating that the function of these two processes is to regenerate NAD^+ so that glycolysis can continue and produce 2 ATP for each glucose.

Chapter 5

Photosynthesis

Review

Like cellular respiration, photosynthesis requires that you apply your knowledge of chemistry. Again, you will need to know the names of important molecules, describe their sequence in metabolic processes, and describe how the processes accomplish their metabolic objectives.

Photosynthesis is the process of capturing free energy in sunlight and storing that energy in chemical bonds, especially glucose. The general chemical equation describing photosynthesis is

$$6\ CO_2 + 6\ H_2O + \text{light} \rightarrow C_6H_{12}O_6 + 6\ O_2$$

Note that this equation for photosynthesis is the reverse of the equation for cellular respiration, except that the energy required by photosynthesis comes from light. The processes are interconnected: Energy stored in chemical bonds by photosynthesis is extracted from those bonds by respiration to make adenosine triphosphate (ATP). The energy in ATP is then used to do cellular work through metabolic processes.

The process of photosynthesis begins with light-absorbing pigments in chloroplasts of plant cells. A pigment molecule is able to absorb the energy from light only within a narrow range of wavelengths. The dominant light-absorbing pigment is the green chlorophyll *a,* while other accessory pigments include other chlorophylls and various red, orange, and yellow carotenoids. Together, these pigments complement each other to maximize energy absorption across the sunlight spectrum. When light is absorbed by one of these pigments, the energy from the light is incorporated into electrons within the atoms that make up the molecule. These energized electrons (or **"excited" electrons**) are unstable and almost immediately re-emit the absorbed energy. The energy is then reabsorbed by electrons of a nearby pigment molecule. The process of energy absorption, followed by energy re-emission, continues, with the energy bouncing from one pigment molecule to another. The process ends when the energy is absorbed by one of two special chlorophyll *a* molecules, P_{680} or P_{700}. These two chlorophyll molecules, named with numbers that represent the wavelengths at which they absorb their maximum amounts of light (680 and 700 nanometers, respectively), are different from other chlorophyll molecules because of their association with various nearby pigments. Together with these other pigments, chlorophyll P_{700} forms a pigment cluster called photosystem I (PS I). Chlorophyll P_{680} forms photosystem II (PS II).

Photosynthesis includes the following processes:

1. **Noncyclic photophosphorylation** and **cyclic photophosphorylation** use H_2O and the energy in sunlight to generate ATP, NADPH, and O_2.
2. The **Calvin cycle** uses CO_2 and the energy in ATP and NADPH to make glucose.

The chemical reactions for photosynthesis are illustrated in Figure 5-1. Refer to the figure as you read the following descriptions of the metabolic processes.

Noncyclic Photophosphorylation

Photophosphorylation is the process of using energy derived from light (photo) to generate ATP from adenosine diphosphate (ADP) and P_i (phosphorylation). Noncyclic photophosphorylation begins with PS II and follows these seven steps:

1. **Photosystem II.** Electrons trapped by P_{680} in photosystem II are energized by light. In Figure 5-1, two electrons are shown moving "up," signifying an increase in their energy.
2. **Primary electron acceptor.** Two energized electrons are passed to a molecule called the primary electron acceptor. This electron acceptor is called "primary" because it is the first in a chain of electron acceptors.
3. **Electron transport chain.** Electrons pass through an electron transport chain **(ETC).** This chain consists of proteins in the thylakoid membrane of the chloroplast that pass electrons from one carrier protein to the next. Some carrier proteins, such as the **cytochromes,** include nonprotein parts containing iron. The electron transport chains in photosynthesis are analogous to those in the inner mitochondrial membrane in oxidative phosphorylation.
4. **Phosphorylation.** As the two electrons move "down" the electron transport chain, they lose energy. The energy lost by the electrons as they pass along the electron transport chain is used to phosphorylate, on average, about 1.5 ATP molecules.
5. **Photosystem I.** The electron transport chain terminates with PS I (with P_{700}). Here the electrons are again energized by sunlight and passed to a primary electron acceptor (different from the one associated with PS II).
6. **NADPH.** The two electrons pass through a short electron transport chain. At the end of the chain, the two electrons combine with the coenzyme $NADP^+$ and H^+ to form NADPH. Like NADH in respiration, $NADP^+$ is an electron acceptor that becomes energy-rich when it accepts these electrons. (You can keep the two coenzymes NADH and NADPH associated with the correct processes by using the P in NADPH as a reminder of the P in photosynthesis. The P in NADPH, however, actually represents phosphorus.)
7. **Splitting of water.** The two electrons that originated in PS II are now incorporated into NADPH. The loss of these two electrons from PS II is replaced when H_2O is split into two electrons, 2 H^+ and ½ O_2. The two electrons from H_2O replace the lost electrons from PS II, one of the H^+ provides the H in NADPH and the H in NADPH, and the ½ O_2 contributes to the oxygen gas that is released.

In summary, photophosphorylation takes the energy in light and the electrons in H_2O to make the energy-rich molecules ATP and NADPH. Because the reactions require light, they are often called **light-dependent reactions** or, simply, **light reactions.** The following equation informally summarizes the process:

$$H_2O + ADP + P_i + NADP^+ + \text{light} \rightarrow ATP + NADPH + O_2 + H^+$$

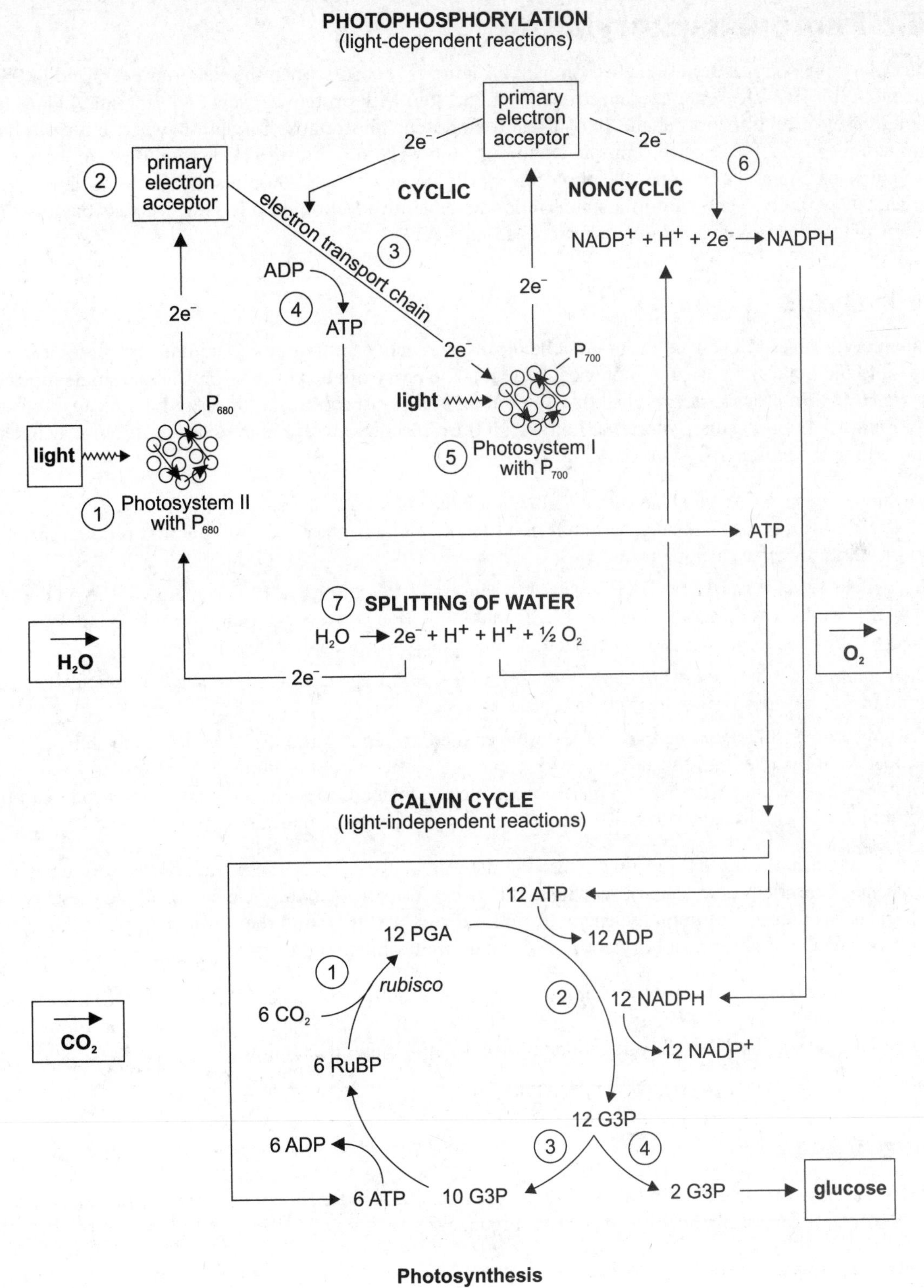

Photosynthesis

Figure 5-1

Cyclic Photophosphorylation

A second photophosphorylation sequence (shown in Figure 5-1) occurs when the electrons energized in PS I are "recycled." In this sequence, energized electrons from PS I join with protein carriers and generate ATP as they pass along the electron transport chain. In contrast to noncyclic photophosphorylation, where electrons become incorporated into NADPH, electrons in cyclic photophosphorylation return to PS I. Here they can be energized again to participate in cyclic or noncyclic photophosphorylation. Cyclic photophosphorylation occurs simultaneously with noncyclic photophosphorylation to generate additional ATP. Two electrons passing through cyclic photophosphorylation generate, on average, about 1 ATP.

Calvin Cycle

The Calvin cycle "fixes" CO_2. That is, it takes chemically unreactive, inorganic CO_2 and incorporates it into an organic molecule that can be used in biological systems. The biosynthetic pathway produces a single molecule of glucose ($C_6H_{12}O_6$). In order to accomplish this, the Calvin cycle must repeat six times, and use 6 CO_2 molecules. Thus, in Figure 5-1 and the discussion that follows, all the molecules involved have been multiplied by 6. Only the most important molecules are discussed:

1. **Carbon fixation: 6 CO_2 combine with 6 RuBP to produce 12 PGA.** The enzyme **rubisco** catalyzes the merging of CO_2 and RuBP. The Calvin cycle is referred to as **C_3 photosynthesis** because the first product formed, PGA, contains three carbon atoms.
2. **Reduction: 12 ATP and 12 NADPH are used to convert 12 PGA to 12 G3P.** The energy in the ATP and NADPH molecules is incorporated into G3P, making G3P a very energy-rich molecule. ADP, P_i, and $NADP^+$ are released and then re-energized in noncyclic photophosphorylation.
3. **Regeneration: 6 ATP are used to convert 10 G3P to 6 RuBP.** Regenerating the 6 RuBP originally used to combine with 6 CO_2 allows the cycle to repeat.
4. **Carbohydrate synthesis.** Note that 12 G3P were created in step 2, but only 10 were used in step 3. What happened to the remaining two? These two remaining G3P are used to build glucose. Other monosaccharides, like fructose and maltose, can also be formed. In addition, glucose molecules can be combined to form disaccharides, like sucrose, and polysaccharides, like starch and cellulose.

You should recognize that no light is directly used in the Calvin cycle. Thus, these reactions are often called **light-independent reactions** or even **dark reactions.** But be careful—the Calvin cycle occurs in the presence of light. This is because it is dependent upon the energy from ATP and NADPH, and these two energy-rich molecules can be created only during photophosphorylation, which can occur only in light.

In summary, the Calvin cycle takes CO_2 from the atmosphere and the energy in ATP and NADPH to create a glucose molecule. Of course, the energy in ATP and NADPH represents energy from the sun captured during photophosphorylation. The Calvin cycle can be informally summarized as follows:

$$6\ CO_2 + 18\ ATP + 12\ NADPH \rightarrow 18\ ADP + 12\ NADP^+ + 1\ \text{glucose}$$

Chloroplasts

Chloroplasts are the sites where both the light-dependent and light-independent reactions of photosynthesis occur. Chloroplasts consist of the following areas (Figure 5-2):

1. **Outer membrane.** This membrane, like the plasma membrane, consists of a double layer of phospholipids.
2. **Intermembrane space.** This is the narrow area between the inner and outer membranes.
3. **Inner membrane.** This second membrane is also a double phospholipid bilayer.

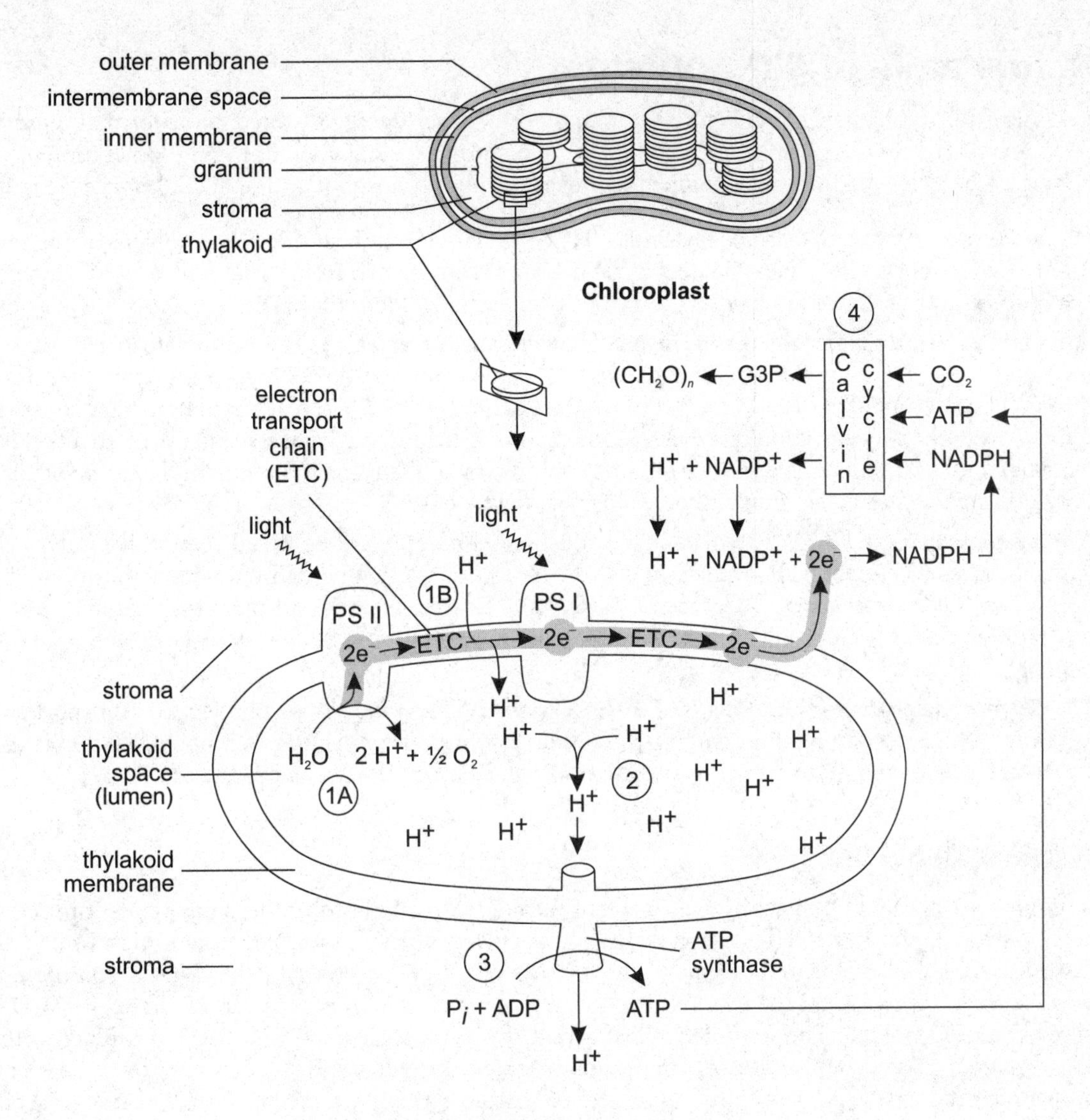

Chemiosmosis in Chloroplasts

Figure 5-2

4. **Stroma.** The stroma is the fluid material that fills the area inside the inner membrane. The Calvin cycle occurs here, fixing carbon from CO_2 to generate G3P, the precursor to glucose.
5. **Thylakoids.** Suspended within the stroma are stacks of pancake-like membranes. Individual membrane layers (the "pancakes") are **thylakoids;** an entire stack of thylakoids is a **granum** (plural, **grana**). The membranes of the thylakoids contain the protein complexes (including the photosystems PS I and PS II), cytochromes, and other electron carriers of the light-dependent reactions.
6. **Thylakoid lumen.** This is the inside, or lumen, of the thylakoid; H^+ ions (protons) accumulate here.

Note how the spatial arrangement of the photosynthetic processes in chloroplasts is similar to that for the respiratory processes in mitochondria. In both cases, carrier proteins for electron transport chains are *embedded in membranes.* Also, enzymes for CO_2 processing occur in *fluids* adjacent to the membranes. Specifically, the enzymes for CO_2 utilization in the Calvin cycle occur in the stroma (*outside* the thylakoid membranes), while the enzymes for CO_2 generation in the Krebs cycle occur in the matrix (*inside* the cristae membranes).

Chemiosmosis in Chloroplasts

Chemiosmosis is the mechanism of ATP generation that occurs when energy is stored in the form of a *proton concentration gradient* across a membrane. The process in chloroplasts is analogous to ATP generation in mitochondria. The following four steps outline the process of photophosphorylation in chloroplasts (Figure 5-2):

1. **H^+ ions (protons) accumulate inside thylakoids.** H^+ are released into the inside space (lumen) of the thylakoid when water is split by PS II (see Figure 5-2, 1A). Also, H^+ are carried from the stroma into the lumen by a cytochrome in the electron transport chain (ETC) between PS II and PS I (1B).
2. **A pH and electrical gradient across the thylakoid membrane is created.** As H^+ accumulate inside the thylakoid, the pH *decreases.* Since some of these H^+ come from outside the thylakoids (from the stroma), the H^+ concentration decreases in the stroma and its pH *increases.* This creates a pH gradient consisting of differences in the concentration of H^+ across the thylakoid membrane from a stroma pH 8 to a thylakoid pH 5 (a factor of 1,000). Since H^+ ions are positively charged, their accumulation on the inside of the thylakoid creates an electric gradient (or voltage) as well.
3. **ATP synthase generates ATP.** The pH and electrical gradient represent potential energy, like water behind a dam. The **channel protein ATP synthase** allows the H^+ to flow through the thylakoid membrane and out to the stroma. The energy generated by the passage of the H^+ (like the water through turbines in a dam) provides the energy for the ATP synthase to phosphorylate ADP to ATP. The passage of about 3 H^+ is required to generate 1 ATP.
4. **The Calvin cycle produces G3P using NADPH, CO_2, and ATP.** At the end of the electron transport chain following PS I, electrons combine with $NADP^+$ and H^+ to produce NADPH. With NADPH, ATP, and CO_2, 2 G3P are generated and subsequently used to make glucose or other carbohydrates, $(CH_2O)_n$.

Photorespiration

Because of its critical function in catalyzing the fixation of CO_2 in all photosynthesizing plants, rubisco is the most common protein on Earth. However, it is not a particularly efficient molecule. In addition to its CO_2-fixing capabilities, it is also able to fix oxygen. The fixation of oxygen, a process called **photorespiration,** leads to two problems. The first is that the CO_2-fixing efficiency is reduced because, instead of fixing only CO_2, rubisco fixes some O_2 as well. The second problem is that the products formed when O_2 is combined with RuBP do not lead to the production of useful, energy-rich molecules like glucose. Instead, energy must be spent to break down the products of photorespiration inside specialized cellular organelles. Thus, considerable effort is made by plants to rid the cell of the products of photorespiration. Since the early atmosphere in which primitive plants originated contained very little oxygen, it is hypothesized that the early evolution of rubisco was not influenced by its O_2-fixing handicap.

Capturing Free Energy Without Light

Organisms that use *sunlight* as a source of free energy to drive photosynthesis and to produce carbohydrates are called **photoautotrophs.** Some prokaryotes, called **chemoautotrophs,** are able to use *inorganic substances* as a source of energy to generate organic molecules. This process, generally referred to as **chemosynthesis,** uses H_2S (hydrogen sulfide), NH_3 (ammonia), or NO_2^- (nitrite) as a source of energy.

One example of chemosynthesis occurs among symbiotic bacteria found growing in giant tube worms at deep ocean depths near hydrothermal vents. The bacteria use energy from H_2S to produce carbohydrates (shown here as CH_2O):

$$CO_2 + 4\,H_2S + O_2 \rightarrow CH_2O + 4\,S + 3\,H_2O$$

Review Questions

Multiple-Choice Questions

The questions that follow provide a review of the material presented in this chapter. Use them to evaluate how well you understand the terms, concepts, and processes presented. Actual AP multiple-choice questions are often more general, covering a broad range of concepts, and often more lengthy. For these types of questions, you should take the two practice exams in this book.

Directions: Each of the following questions or statements is followed by four possible answers or sentence completions. Choose the one best answer or sentence completion.

1. Which of the following statements about photosynthetic pigments is true?

A. There is only one kind of chlorophyll.
B. Chlorophyll absorbs mostly green light.
C. Chlorophyll is found in the membranes of thylakoids.
D. P_{700} is a carotenoid pigment.

2. When deciduous trees drop their leaves in the fall, the leaves turn to various shades of red, orange, and yellow. The source of these colors is

A. chlorophyll
B. carotenoids
C. fungal growth
D. natural decay of cell walls

3. A product of noncyclic photophosphorylation is

A. NADPH
B. H_2O
C. CO_2
D. ADP

4. Which of the following molecules contains the most stored energy?

A. ATP
B. NADPH
C. glucose
D. starch

5. All of the following occur in cyclic photophosphorylation EXCEPT:

A. Electrons move along an electron transport chain.
B. Electrons in chlorophyll become excited.
C. ATP is produced.
D. NADPH is produced.

Questions 6–10 refer to the following lettered answer choices. Each answer may be used once, more than once, or not at all.

A. cyclic photophosphorylation
B. noncyclic photophosphorylation
C. photorespiration
D. Calvin cycle

6. Combines O_2 with RuBP

7. Stores energy obtained from light into NADPH

8. Produces ATP without the need for H_2O and CO_2

9. Occurs in the stroma of a chloroplast

10. Requires electrons that are obtained by splitting water

Questions 11–14 refer to the following diagram. The two boxes represent the two major biosynthetic pathways in C_3 photosynthesis. Arrows represent reactants or products.

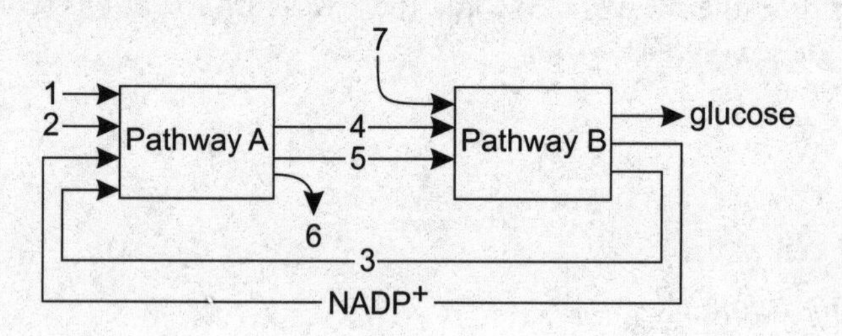

11. Arrow 1 could represent

A. ATP
B. H_2O
C. O_2
D. CO_2

12. Arrow 3 could represent

A. NADPH
B. ADP
C. O_2
D. electrons

13. Arrow 4 could represent

A. NADPH
B. ADP
C. glucose
D. electrons

14. Arrow 7 represents

A. ATP
B. NADPH
C. light
D. CO_2

15. All of the following are true about photosynthesis EXCEPT:

A. The Calvin cycle usually occurs in the dark.
B. The majority of the light reactions occurs on the thylakoid membranes in the chloroplast.
C. Light energy is stored in ATP.
D. A proton gradient drives the formation of ATP from ADP + P_i.

Questions 16–17 refer to the following graph that shows the relationship between CO_2 uptake by leaves and the concentrations of O_2 (percent of atmosphere in growth chambers) and CO_2 (ppm in growth chambers).

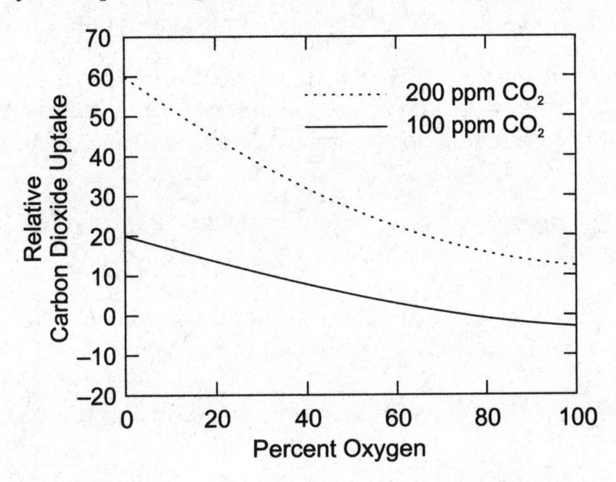

16. The relative CO_2 uptake is a measure of

A. photosynthetic rate
B. light intensity
C. leaf size
D. leaf temperature

17. According to the graph, the relative CO_2 uptake is best under which of the following conditions?

A. 100 ppm CO_2, 20% O_2
B. 100 ppm CO_2, 80% O_2
C. 200 ppm CO_2, 20% O_2
D. 200 ppm CO_2, 80% O_2

Free-Response Questions

The AP exam has long and short free-response questions. The long questions have considerable descriptive information that may include tables, graphs, or figures. The short questions are brief but may also include figures. Both kinds of questions have four parts and generally require that you bring together concepts from multiple areas of biology.

The questions that follow are designed to further your understanding of the concepts presented in this chapter. Unlike the free-response questions on the exam, they are narrowly focused on the material in this chapter. For free-response questions typical of the exam, take the two practice exams in this book.

Directions: The best way to prepare for the AP exam is to write out your answers as if you were taking the exam. Use complete sentences for all your answers and do *not* use outline form or bullets. You may use diagrams to supplement your answers, but be sure to describe the importance or relevance of your diagrams.

1. There are a variety of light-absorbing pigments in chloroplasts. Explain the purpose of having a variety of light-absorbing pigments.

2. During noncyclic photophosphorylation, electrons from the splitting of water populate photosystem II. Explain where these electrons from H_2O eventually end up.

3. Protons concentrate inside the thylakoids. Explain the purpose of this proton buildup.

4. Describe the biochemical pathways of the light-independent and light-dependent reactions in C_3 photosynthesis. Begin with a molecule of H_2O and CO_2 and end with a molecule of glucose.

Answers and Explanations

Multiple-Choice Questions

1. **C.** All of the light-absorbing pigments and most of the enzymes for the light reactions are found in the thylakoid membranes. There are several kinds of chlorophyll, including chlorophyll *a* and *b* (and *c* and *d* in certain algae). P_{700} and P_{680} are special chlorophyll *a* molecules, differing from other chlorophyll *a* molecules because of their special arrangement among nearby proteins and thylakoid membrane constituents. Chlorophyll is green because it *reflects* green light, not absorbs it. It looks green because the green light it reflects is the light we see. Similarly, carotenoids look orange or yellow because they reflect those colors.

2. **B.** The leaves turn to these colors because, as the leaves age, the tree begins to break down the chlorophyll (to extract its valuable components, like magnesium). In the absence of chlorophyll, the carotenoids are visible. As the various carotenoids break down, other colors become visible from carotenoids that still remain intact.

3. **A.** NADPH, ATP, and O_2 are the products of noncyclic photophosphorylation.

4. **D.** The molecules referenced in the question, in order of decreasing potential energy, are starch, glucose, NADPH, and ATP. Starch is a polymer of glucose. A single glucose molecule can provide about 30 ATP molecules, and a single NADPH can provide about 3 ATP molecules.

5. **D.** NADPH is produced only by noncyclic photophosphorylation.

6. **C.** The first step of carbon fixation occurs during the Calvin cycle when CO_2 combines with RuBP. Photorespiration, in contrast, occurs when O_2, instead of CO_2, combines with RuBP. The product of this reaction is transported to a peroxisome, where H_2O_2 is generated and subsequently broken down.
7. **B.** The energy-rich products of noncyclic photophosphorylation are ATP and NADPH.
8. **A.** During cyclic photophosphorylation, electrons energized by sunlight pass along an electron transport chain. About 1 ATP is generated from ADP and P_i with the energy from 2 electrons. This process can repeat as long as sunlight is available. H_2O (necessary for noncyclic photophosphorylation) and CO_2 (necessary for the Calvin cycle) are not needed for cyclic photophosphorylation.
9. **D.** Rubisco and the other enzymes that catalyze reactions in the Calvin cycle occur in the stroma. Most of the pigments and enzymes for photophosphorylation are embedded in the thylakoid membranes. The manganese-containing protein complex that catalyzes the splitting of water is embedded on the inner side of the thylakoid membrane, so the splitting of water occurs there.
10. **B.** The splitting of H_2O provides the electrons for noncyclic photophosphorylation. These electrons are incorporated in NADPH and ultimately find their way into glucose during the Calvin cycle.
11. **B.** It could also be light, but that is not an answer choice. You should test yourself on all aspects of this diagram: Pathway A is noncyclic photophosphorylation, and Pathway B is the Calvin cycle. Arrows 1 and 2: H_2O and light; arrow 3: ADP; arrows 4 and 5: ATP and NADPH; arrow 6: O_2; and arrow 7: CO_2.
12. **B.** ADP is recycled. ADP is used in photophosphorylation to produce ATP. In the Calvin cycle (Pathway B), the energy in ATP is used, releasing ADP and P_i. The answer could also be $NADP^+$, but that choice is not given (and, besides, $NADP^+$ is already shown in the figure).
13. **A.** It could also be ATP, but that is not an answer choice.
14. **D.** Arrow 7 represents CO_2, which is necessary for Pathway B, the Calvin cycle. The Calvin cycle "fixes" CO_2 and produces G3P, the precursor to glucose.
15. **A.** Although the Calvin cycle is light-independent, it requires ATP and NADPH from photophosphorylation, which occurs only in light.
16. **A.** CO_2 uptake is a measure of Calvin cycle activity. The Calvin cycle generates glucose, the end product of photosynthesis. Many factors, including those provided in the other answer choices, influence photosynthetic rate. However, CO_2 uptake alone only indicates the amount of photosynthetic activity, not why that photosynthetic activity is occurring at some measured rate.
17. **C.** According to this graph, CO_2 uptake is greater when there is more CO_2 in the growth chamber (the 200 ppm curve), and when O_2 is minimum (the left side of the graph). Lower concentrations of O_2 minimize photorespiration.

Free-Response Questions

1. Each kind of light-absorbing pigment absorbs light energy over a different range of wavelengths. This allows the cell to absorb a greater amount of the incident light energy than if there were only one light-absorbing pigment.
2. After becoming energized in photosystem I, the electrons contribute to the formation of NADPH. Then, NADPH passes those electrons to the Calvin cycle, where they contribute partly to the regeneration of RuBP (so the Calvin cycle can repeat) and partly to the formation of carbohydrates.
3. The accumulation of protons in the thylakoid lumen creates a pH and electrical gradient across the thylakoid membrane. The gradient drives the movement of protons through ATP synthase, which, in turn, drives the phosphorylation of ADP to ATP.

4. Photosynthesis is the process by which water and light energy are used to fix inorganic carbon dioxide into glucose. The complete equation is

$$6\ H_2O + 6\ CO_2 + \text{light} \rightarrow C_6H_{12}O_6 + 6\ O_2$$

Various pigments exist in the photosynthetic membranes of plants. These include chlorophyll *a* and *b* and various carotenoids. The purpose of a variety of pigments is to absorb light energy (photons) of different wavelengths. The absorption spectrums of the pigments overlap so that they maximize the energy absorbed. When light energy is absorbed, the pigment becomes energized. The energy is then bounced around among pigments until it is absorbed by either of two special kinds of chlorophyll *a*, P_{700} or P_{680}.

P_{700} and P_{680} are organized into separate photosystems (pigment systems), I and II. Cyclic photophosphorylation involves photosystem I. In cyclic photophosphorylation, the energy absorbed by the pigment system is captured by the P_{700}. As a result, two electrons are excited to a higher energy level, where they are absorbed by a primary electron acceptor. From here, the electrons are passed through an electron transport chain from electron acceptor to electron acceptor (some of which are cytochromes). During this transit, the energy loss of the electrons is used to bond a phosphate group to ADP, making it ATP (adenosine triphosphate). The process is called phosphorylation, and the result is that energy (originally from light) from the two electrons is trapped in an ATP molecule. The cycle is completed when the two electrons return to the photosystem pigments.

Noncyclic photophosphorylation is a more advanced system involving both photosystems I and II. It begins when chlorophyll P_{680} in photosystem II traps photon energy and energizes two electrons. These two electrons are passed to a primary electron acceptor, then through an electron transport chain, producing, on average, 1.5 ATP and are then finally returned to photosystem I. Here the electrons are energized again (by light) and are received by another primary electron acceptor. These two electrons then combine with 2 H^+ to form NADPH.

Meanwhile, a water molecule is split producing 2 electrons, 2 H^+, and ½ O_2. The 2 electrons replace the electrons energized from photosystem II. The oxygen is released. One of the H^+ combines with $NADP^+$ and the 2 electrons from noncyclic photophosphorylation to produce NADPH. NADPH is then used to supply energy to the Calvin cycle.

The light-independent reactions (Calvin cycle) combine CO_2, NADPH, and ATP to form G3P and RuBP. It takes 6 CO_2 to create 2 G3P, so the cycle repeats six times to produce 2 G3P. The 2 G3P form glucose. Glucose can then be used to make various other carbohydrates, such as sucrose and starch, or it can be broken down to release its store of energy in the form of ATP to drive metabolic activities.

This is a thorough answer. Another way to answer this question would be to describe these processes as they occur in the chloroplast. In other words, focus on the activities of chemiosmosis. Illustrations such as those in this chapter would also get you points.

Chapter 6

Cell Communication

Review

All cells in a multicellular organism have the same genetic makeup. How then do cells become different, and how are the various cell types able to carry out different cellular processes? Chapter 9, "Gene Expression and Regulation," will review how genes, the units of heredity, are able to produce traits. In short, the information stored in DNA is used to make another nucleic acid, messenger RNA (mRNA). The information in mRNA is then used to make proteins. As enzymes, proteins regulate chemical reactions that generate traits. Other proteins may serve to regulate gene expression. In summary:

genes (DNA) → mRNA → proteins (enzymes) → structure, physiology, behavior

Still, this does not explain how the same set of genes can produce different kinds of structures or promote different cellular processes. Cells become different because gene expression varies among cells; that is, in some cells genes are turned on, while in others, those same genes are turned off. The major factor that influences which genes will be expressed is the environment. Each cell receives chemical and physical signals from its surroundings that trigger metabolic activities that direct or influence the expression of its genes.

Signals from the external environment come from biological sources, such as chemicals from a pathogen or a bee sting, or from physical sources, such as light or heat. Within a multicellular organism, cells also receive signals from other cells. This can happen in several ways, depending upon how far the signal needs to travel:

1. **Direct contact** between animal cells allows proteins, carbohydrates, and lipids of the plasma membranes to transmit information. This kind of communication is common among cells during early development. Information is also transmitted through two kinds of communication junctions:
   - **Gap junctions** in *animal* cells allow for chemical and electrical signaling between cells. Ions and small molecules can pass through gap junctions, but larger molecules, like proteins and nucleic acids, cannot.
   - **Plasmodesmata** in *plant* cells are tunnels of cytoplasm between cells. They provide passageways across plasma membranes and cell walls for the movement of ions, amino acids, sugars, small proteins (including transcription factors), and micoRNA (miRNA). Transcription factors and microRNA regulate gene expression. (Details are in Chapter 9, "Gene Expression and Regulation.")
2. **Synaptic signaling** occurs between junctions of nerve cells or between nerve and muscle cells. **Neurotransmitters,** short-lived chemical signals, cross a very small space between these cells to stimulate or inhibit a nerve impulse or muscle contraction.
3. **Paracrine signaling** is a mechanism for local communication. Cells secrete substances that affect only nearby cells because the substances are either readily absorbed by adjacent cells or rapidly broken down in the extracellular fluid. Growth factors, for example, are paracrine signals secreted during early animal development.
4. **Endocrine signaling** provides a mechanism for distributing signals throughout a multicellular organism. For example, hormones produced in one part of the body target cells in another part of the body.

The remainder of this chapter focuses on how a signal, once recognized by a cell, actually manifests a change in the cell. The mechanism for this process is a signal transduction pathway.

Signal Transduction Pathways

A **signal transduction pathway** is a sequence of molecular interactions that transforms an extracellular signal into a specific cellular response. The process can be summarized like this:

Signal (1st messenger) → Receptor → Proteins or other 2nd messengers → Cellular responses

Signaling molecules, or **ligands,** are first messengers. They are small molecules that bind to larger receptor proteins of specific target cells. When the specific ligand binds to a receptor protein, it induces a change in the three-dimensional shape of the receptor protein. That change initiates some kind of activity in the receptor protein. There are two types of signaling molecules:

- **Hydrophilic ligands** are signaling molecules that cannot cross the phospholipid bilayer of a membrane. They bind to **membrane receptors** at the membrane surface.
- **Hydrophobic ligands** and certain small molecules are signaling molecules that are able to cross the membrane unaided. They bind to **intracellular receptors** in the cytoplasm or nucleus.

Receptor proteins are molecules that have binding sites for signaling molecules. When activated by a specific signaling molecule, they initiate a series of reactions that activates a cellular process. There are two kinds of receptor proteins:

- **Membrane receptors** are transmembrane proteins, extending from the outside to the inside of a membrane. The part of the receptor protein that faces away from the cell presents a binding site for a specific signaling molecule. The other end of the protein, facing the cytoplasm, initiates a chemical reaction.
- **Intracellular receptors** are proteins that occur in the cytoplasm or nucleus.

Second messengers are molecules that relay a signal from the inside face of a receptor protein to other molecules that may initiate a cellular response or may act as additional second messengers. They have these characteristics:

- They are small, *nonprotein* molecules.
- They are hydrophilic, hydrophobic, or gaseous molecules.
- Examples include Ca^{2+}, IP_3 (inositol trisphosphate), cAMP (cyclic adenosine monophosphate), and DAG (diacylglycerol).

Various other membrane-bound or cytoplasmic proteins are often additional components of a transduction pathway, carrying the signal response from the receptor protein to other target proteins that may initiate a cellular response. Although these proteins have specific names, they are generally referred to as **relay proteins, response proteins, substrate proteins,** or **effector proteins.**

A **signaling cascade** is a series of enzymatic reactions. The first enzyme in the series activates a second enzyme, the second enzyme activates a third, and so forth. Because each enzyme can be used repeatedly, the products of each reaction magnify as the sequence progresses, like a chain reaction. Ultimately, a signal that may have begun with a single signaling molecule may be amplified to produce a huge number of molecules that elicit a strong cellular response.

- A **kinase cascade,** or **phosphorylation cascade,** is a signaling cascade consisting of a number of different kinase enzymes (Figure 6-1). A kinase is an enzyme that phosphorylates its substrate, that is, adds a phosphate group to it. In a kinase cascade, each kinase phosphorylates and, thus, activates the next kinase in the sequence, ultimately phosphorylating and activating a protein that initiates a cellular response. The kinase cascade amplifies the signaling response.
- **Scaffold proteins** improve the efficiency of a signaling cascade by holding all the participating enzymes in close proximity. The scaffold also serves to keep the members of one signaling cascade isolated from members of another cascade.
- A **protein phosphatase** is an enzyme that dephosphorylates its substrate, that is, removes a phosphate group from it. When these enzymes dephosphorylate the kinases in a kinase cascade, they serve to terminate the signaling response.

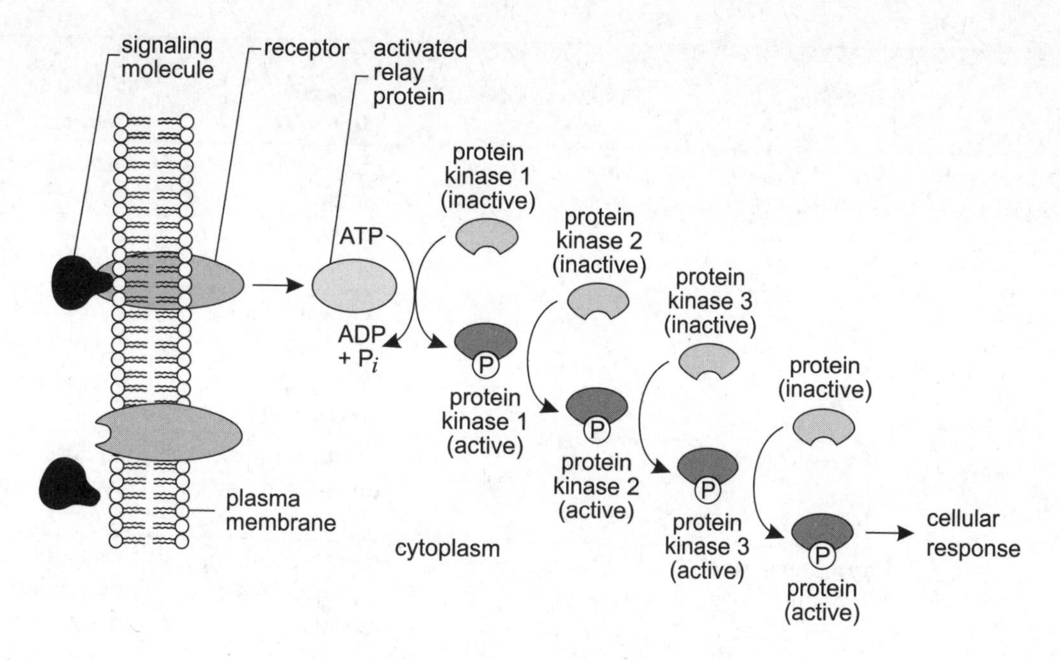

A Kinase (or Phosphorylation) Cascade

Figure 6-1

There is both great specificity and great flexibility in transduction pathways. Signaling molecules are specific for the binding sites of receptor proteins. However, the response that a specific signaling molecule initiates varies with cell type and the influence of cytoplasmic substances.

The complexity of many signal transduction pathways has several advantages:

- **Amplification.** It provides a mechanism for amplifying the effect of a signaling molecule.
- **Control.** It gives the cell more control over the accuracy of the signaling pathway. Because all components of the pathway must be functioning properly, there is a smaller chance that the transduction might occur in error.
- **Multiplicity.** A single signaling molecule can activate multiple cytoplasmic proteins, each generating a different biochemical response. As a result, multiple processes can be coordinated to produce a single cellular response.

Table 6-1 summarizes four kinds of receptor proteins—three membrane receptors and one intracellular receptor—and the signal transduction pathways they initiate. You should be prepared to describe at least one of these receptor proteins in detail in a free-response question on the AP exam. In other cases, questions may provide you with a full description of a pathway and ask you to use the descriptions to make conclusions. Familiarity with all four kinds of receptor proteins will work in your favor.

Also note that some of the examples of these pathways are pursued in more detail in subsequent chapters.

Table 6-1: Summary of Receptors and Their Functions

Receptor Type	Receptor Description	Ligand (1st Messengers) Examples	Supporting Mechanisms	Cellular Response Examples
Gated ion receptor	Ligand-gated ion channel	Acetylcholine		Na^+ gate opens; nerve impulse or muscle contraction
	Voltage-gated ion channel	Change in membrane voltage		Na^+, K^+ gates open; nerve transmission

continued

Receptor Type	Receptor Description	Ligand (1st Messengers) Examples	Supporting Mechanisms	Cellular Response Examples
G protein-coupled receptor (GPCR)	GPCR + G protein + effector protein	Various	Enzymatic effector protein	Enzyme activity
			cAMP (2nd messenger)	Glycogen → glucose
			IP_3/DAG (2nd messengers)	IP_3 releases Ca^{2+} as 2nd messenger
			Ca^{2+} (2nd messenger)	Muscle contraction
Protein kinase receptor	Receptor tyrosine kinase (RTK)	Insulin	Multiple kinase cascades	Glucose → glycogen; glucose export
	Receptor serine/threonine kinase	Mitogens, growth factors	Ras (G protein); mitogen-kinase cascade	Activation of transcription factors that promote growth and cell differentiation
Intracellular receptor	Cytoplasmic or nuclear receptors	Steroid hormones (testosterone, estrogen)		Development of primary and secondary sex characteristics

Gated Ion Receptors

A **gated ion receptor** is a transmembrane protein that contains a gated channel, that is, a channel that opens and closes in response to a specific signal. When open, the channel allows a specific ion to pass through. A **ligand-gated ion receptor** responds to a specific ligand, that is, a *molecular* signal. There are also **voltage-gated-ion receptors** that open or close in response to *voltage* differences across the membrane. A typical ligand-gated ion sequence follows (Figure 6-2a):

1. **Ligand-gated ion receptor receives signal.** A specific messenger ligand binds to the outward-facing surface of the receptor.
2. **Receptor channel opens and ions pass through.** In response to the binding of the ligand, the three-dimensional shape of the receptor changes, opening (or closing) a channel that allows a specific ion to pass through and enter the cytoplasm.
3. **Ions initiate chemical response.** Once in the cytoplasm, the ions initiate a chemical response.
4. **Ligand-gated ion receptor deactivated when ligand detaches from receptor.** The ligand-ion receptor is deactivated when the messenger ligand is broken down by another enzyme, the binding site for the ligand is blocked by an allosteric ligand, or the ion passage is obstructed by a channel blocker.

An important example of **ligand-gated** ion channel involves **acetylcholine,** a neurotransmitter that is the signaling molecule that transmits nerve impulses *between* nerve cells (neurons). A neuron that is transmitting a signal releases acetylcholine into the extracellular space between neurons (the **synaptic cleft,** or **synapse**). When acetylcholine binds to the ligand-gated receptor molecules of the receiving neuron, the receptor molecules open a gated channel that allows sodium ions (Na^+) to enter the cell. As Na^+ enters the cell, the inside of the cell becomes more positive. This change in membrane voltage (called an action potential) initiates a nerve impulse. Neurons stimulate muscle contraction in a similar process.

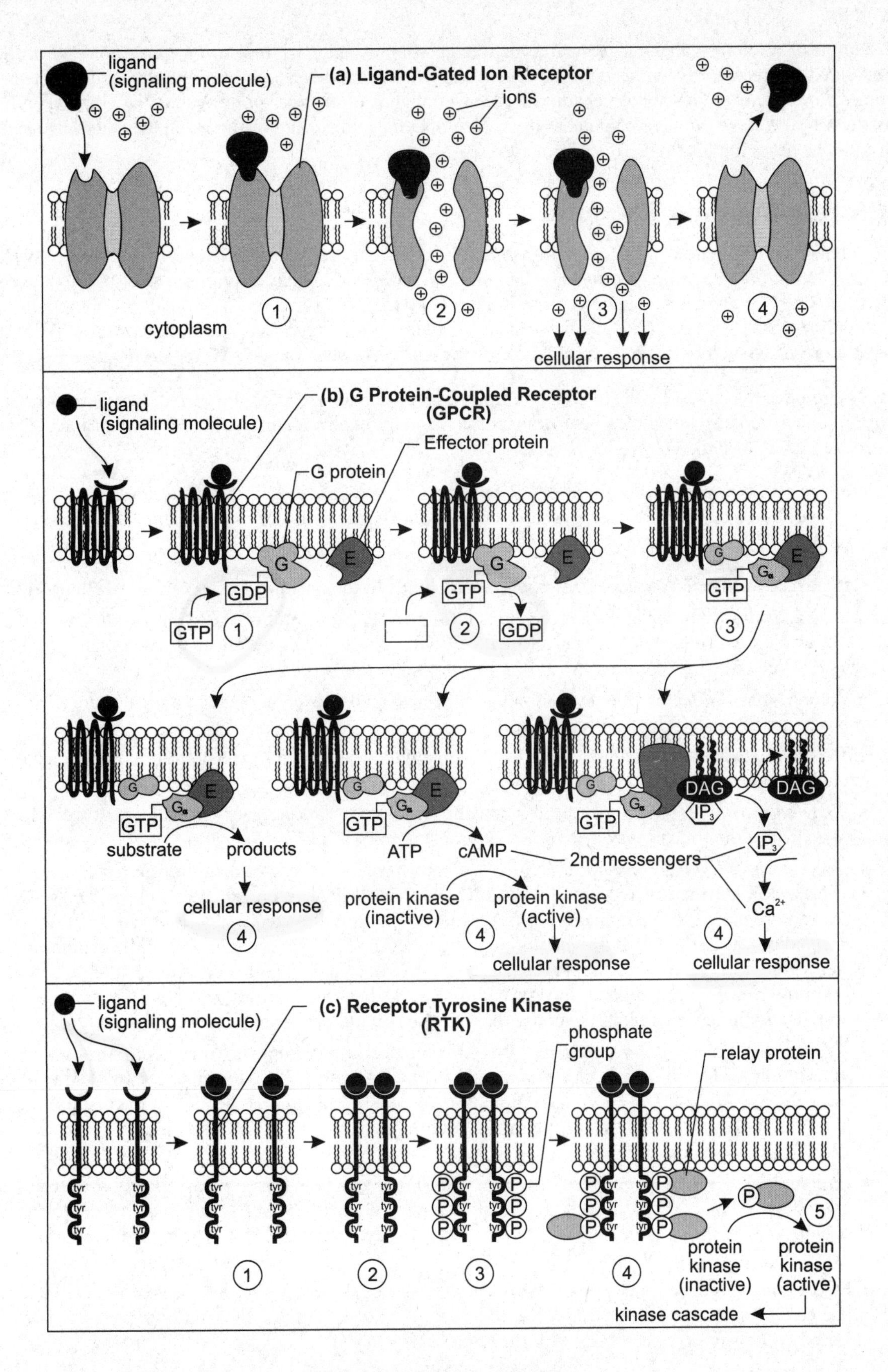

Membrane Receptor Proteins

Figure 6-2

An example of a **voltage-gated** ion channel is the transmission of a nerve impulse *along* a neuron. When the ligand-gated-ion receptor responds to acetylcholine and Na^+ enters the cytoplasm, the voltage inside the neuron becomes more positive. This voltage change, if strong enough, stimulates a *voltage*-gated Na^+ channel and, subsequently, a *voltage*-gated K^+ channel to open. The opening and closing of these voltage-gated ion channels transmits the nerve impulse along the neuron.

G Protein-Coupled Receptors

A **G protein-coupled receptor (GPCR)** is a transmembrane protein that activates a **G protein.** The G protein, in turn, activates another membrane protein, which, in turn, triggers a cellular response or activates a second messenger. The G protein is so named because it has a GTP (or GDP) attached to it. GTP is functionally and structurally like ATP, except with a guanine instead of an adenine nitrogen base. In its inactive (or "off") state, a GDP is attached to the G protein. It is activated (or turned "on") when the GDP is replaced with a GTP.

GPCRs comprise the largest family of signal receptors. They include receptors for vision, taste, airborne signals (odors and pheromones), hormones, neurotransmitters, and immune system activity. Many pharmaceuticals and opiates are GPCR ligands.

A description of a typical sequence for a GPCR follows (Figure 6-2b). Keep in mind, however, that the GPCR pathway varies based on cell type, the particular GPCR that is activated, and the biochemical makeup of individual cells.

1. **GPCR receives signal.** A specific messenger ligand binds to the outward-facing surface of the receptor.
2. **GPCR activates G protein by exchanging a GTP for a GDP.** As a result of the ligand binding to the GPCR, a conformational change occurs, activating the GPCR. The activated GPCR, in turn, exchanges a GTP for the GDP on a nearby G protein. A GTP is now bound to the G protein.
3. **G protein binds to effector protein.** A subunit of the activated G protein binds to a membrane effector protein (usually an enzyme), activating it.
4. **Effector protein initiates response.** The effector protein, now activated by the bound G protein, elicits a cellular response. Some of the possible responses include the following:
   - *Enzymatic activity.* The effector protein may be an enzyme that catalyzes a specific substrate. More specifically, the enzyme may be a protein kinase and initiate a kinase cascade.
   - *Produce the second messenger* ***cAMP.*** If the effector protein is the transmembrane enzyme adenylyl cyclase, the enzyme converts an ATP to a cAMP by removing two phosphate groups (ATP $\rightarrow$ AMP + $2P_i$) and binding the remaining phosphate group so that it is attached to the ribose sugar in two places instead of one (wrapping it around and making it "cyclic"). Many cAMPs are rapidly produced, generating a very strong and rapid response. The cAMP signaling pathway then activates a cytoplasmic response protein, such as an enzyme protein kinase. Depending upon the cell type, the response protein initiates some specific cellular response that may be stimulatory or inhibitory.
   - *Produce the second messengers* ***IP_3*** *and* ***DAG.*** If the effector protein is the transmembrane enzyme phospholipase C, the enzyme cleaves a membrane phospholipid, PIP_2, to generate two second messengers—**IP_3** and **DAG.** DAG, the lipid portion of PIP_2, remains embedded in the membrane, while the soluble IP_3 moves into the cytoplasm. As second messengers, IP_3 and DAG initiate a variety of cellular responses, depending upon cell type.
   - *Produce the second messenger* ***Ca^{2+}***. In many cells, IP_3 triggers the transport of calcium ions (Ca^{2+}) across membranes. For example, in secretory cells of salivary glands, IP_3 binds and activates receptor proteins in membranes of the smooth ER to release Ca^{2+} into the cytoplasm. Here, Ca^{2+} is a second messenger, triggering the release of saliva.
5. **GPCR signaling is deactivated when GTP is hydrolyzed** (not shown in Figure 6-2). When the GTP that is attached to the effector protein is hydrolyzed (GTP $\rightarrow$ GDP + P_i), the GPCR pathway is deactivated. The released GDP is free to reassociate with the G protein.

Glycogen breakdown in muscle and liver cells is an example of a cAMP signaling pathway. When a G protein is activated by a signaling molecule (the hormone epinephrine), the G protein exchanges GTP for GDP on the effector protein (adenylyl cyclase), which, in turn, converts ATP to cAMP, a second messenger. The cAMP phosphorylates (thus, activating) a protein kinase (called PKA). Protein kinase activity leads to the activation of an enzyme that removes single units of glucose from glycogen.

Protein Kinase Receptors

A **protein kinase receptor** is a transmembrane-protein enzyme. These enzymes are **kinases,** enzymes that add a phosphate group to a protein. Phosphorylation replaces the hydroxyl (OH^-) group that occurs in an R group of an amino acid with a phosphate (PO_4^{3-}) group. Such hydroxyl groups only occur in three amino acids: tyrosine, serine, and threonine. The best understood of these receptors is the **receptor tyrosine kinase (RTK).**

A typical RTK sequence follows (Figure 6-2c):

1. **RTK receives signal.** At the outer surface of the membrane, the RTK binds to a signaling molecule.
2. **RTK forms dimer.** Two RTKs associate, forming a pair (dimer).
3. **RTK is activated by autophosphorylation.** On the inner surface of the membrane, each of the two RTKs in the dimer phosphorylates the other RTK using phosphate groups from ATPs. This process, called autophosphorylation, activates the protein complex. Multiple phosphate groups can attach, each to a tyrosine amino acid.
4. **Relay protein is phosphorylated.** Relay proteins bind to the tyrosine-phosphate domains of the RTK. The phosphates are then transferred from the tyrosines to the relay proteins. *More than one kind of protein may be phosphorylated by a single RTK dimer and each kind of phosphorylated protein can serve as a relay protein that initiates a different transduction pathway.*
5. **Relay protein initiates transduction pathway.** The relay proteins, now activated by the addition of a phosphate group, are released. A relay protein can activate a cellular response or initiate a protein kinase transduction pathway that leads to a cellular response. Each kind of relay protein participates in a different cellular response.
6. **RTK pathway is deactivated by dephosphorylation or receptor protein isolation** (not shown in Figure 6-2). The RTK pathway is deactivated when dephosphorylating enzymes remove phosphate groups from the kinases or when the membrane folds to encircle and internalize the receptor protein in a vesicle (endocytosis).

There are two fundamental differences between an RTK receptor pathway and the GPCR pathway discussed earlier:

1. The RTK receptor is usually *directly* responsible for initiating a transduction pathway. In contrast, the GPCR *indirectly* activates a transduction pathway via a G protein and an effector molecule.
2. The RTK receptor may trigger multiple transduction pathways, directing a host of coordinated cellular responses. In contrast, a typical GPCR triggers a single transduction pathway, ultimately activating a single final product that leads to a specific cellular response.

Insulin signal transduction is an example of an RTK pathway. Insulin, a protein hormone produced in the pancreas, is secreted into the blood in response to excess glucose in the blood. The hormone regulates the cellular intake and utilization of glucose. Insulin, the signaling molecule, binds to the insulin receptor, an RTK of target cells. Binding stimulates conformational changes that activate the receptor, which then trigger the formation of an RTK dimer and autophosphorylation. Then, the complex binds to and phosphorylates an insulin response protein. This response protein, now activated, initiates several signaling cascades. In muscle cells, one cascade leads to glycogen synthesis (the formation of glycogen from glucose monomers) for short-term energy storage, another to the transport of glucose into the cell. In liver cells, glycogen synthesis is also stimulated, and in addition, glucose synthesis from smaller molecules is inhibited. In fat cells, the pathway leads to triglyceride formation (for long-term energy storage) instead of glycogen formation.

The **mitogen-activated protein (MAP)** kinases are an example of a kinase cascade activated by a G protein. **Mitogens** (mitosis + generating) are substances that stimulate cell division. After a protein kinase receptor is activated by a signaling molecule, the receptor dimerizes, autophosphorylates, and activates **Ras,** a G protein, by exchanging a GTP for a GDP. The Ras protein then activates a mitogen-kinase (MK) cascade, where one mitogen kinase catalyzes the phosphorylation of the next mitogen kinase in the sequence. In this fashion, mitogen kinase kinase kinase (MKKK) activates MKK, which activates MK. This final MK activates nuclear regulatory proteins, called **transcription factors,** which turn genes on or off. These genes manage cell division and cell differentiation. Note that the Ras protein associated with the RTK receptor and the G protein associated with a GPCR are both G proteins because they are activated when a GTP replaces a GDP, but the series of steps that define the pathways of these two receptors is very different. Ras is the first *cytoplasmic* enzyme in a kinase cascade, whereas the G protein associated with a GPCR is *membrane bound* and activates another *membrane-bound* effector protein.

Intracellular Receptors

In contrast to the membrane-bound receptor proteins discussed earlier, **intracellular receptors** are positioned in the cytoplasm or nucleus. The ligands for these intracellular receptor proteins are small molecules or lipid-soluble, nonpolar molecules that can passively diffuse across the plasma membrane. Ligands also include second messengers, like IP_3, that are products of a signal transduction pathway generated by a membrane-bound receptor protein. Like other kinds of transduction pathways, the cellular response elicited by a particular receptor protein varies among cell types. Also, various molecules specific to individual cells may act as coactivators to direct the target of the transduction pathway. The target of receptor protein activity may be in the cytoplasm or the nucleus. When the target is in the nucleus (typically, the DNA), the receptor is often called a **nuclear receptor.** A description of a typical intracellular receptor pathway follows (Figure 6-3):

1. **A signaling molecule enters the cytoplasm.** The signaling molecule can be a first messenger lipid-soluble molecule or small molecule that diffuses across the plasma membrane or a second messenger molecule that is introduced into the cytoplasm as a product of an intracellular transduction pathway.
2. **The signaling molecule binds to the intracellular receptor, activating it.** The receptor may be in the cytoplasm or in the nucleus. In some cases, activation of the receptor triggers the release of an inhibitor that prevented the receptor from functioning.
3. **The receptor-signaling molecule complex acts as a transcription factor.** The receptor-signal complex binds to the DNA, promoting (or suppressing) the transcription of genes.
4. **Deactivation of the pathway can occur when signaling molecules or receptor proteins are enzymatically degraded.** In some cases, phosphorylation of the receptor protein results in deactivation. In addition, the release of hormones into the blood is typically shut down by negative feedback mechanisms.

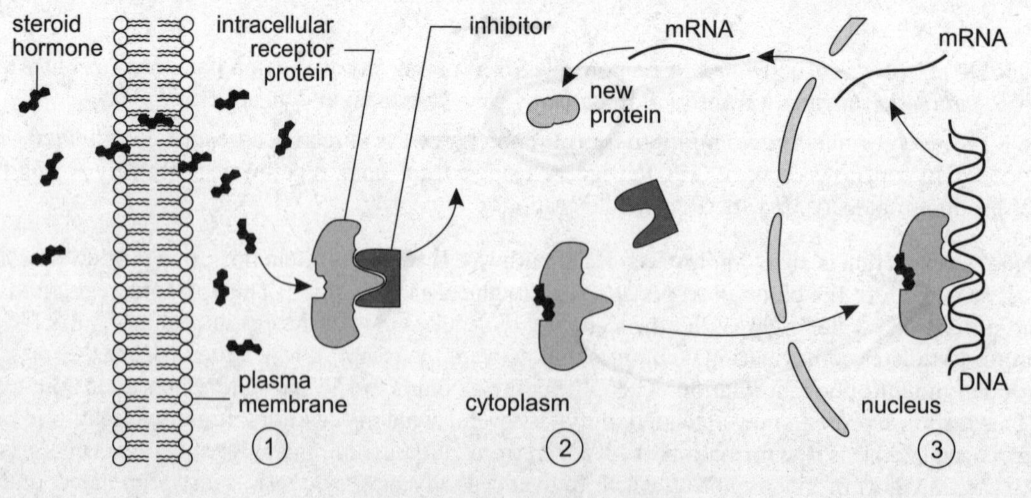

Intracellular Receptor Protein

Figure 6-3

Steroid hormones, such as **testosterone** or **estrogen,** are examples of ligands that bind to intracellular receptors. These signaling molecules diffuse across the plasma membrane and bind to a specific receptor protein in the cytoplasm. The now activated complex (hormone + receptor protein) moves to the nucleus, where it binds to DNA and promotes transcription of genes that direct cellular activities. Gene expression varies depending on cell type and gender. In males, for example, testosterone activates genes in the testes that direct the development of sperm cells, but in muscle cells, it stimulates the production of muscle fibers. In females, estrogen activates genes that direct cells in the uterus to prepare for pregnancy, but in mammary cells, estrogen inactivates those same genes.

Disease and Cancer

Although genes contain the information for what a cell will become and how that cell will function, external signals strongly influence how those genes actually express that information. For various reasons, signals are sometimes inaccurately acted upon because the signal transduction pathway is distorted and does not operate properly. Here are two examples.

Cholera is a waterborne disease caused by bacteria. When contaminated water is ingested, the bacteria secrete a toxin that disrupts the normal activity of GPCRs of intestinal cells. In particular, the GTP attached to the G protein cannot be converted back to a GDP, so the protein cannot be deactivated. In these cells, this G protein regulates the concentration of Cl^- and, when locked in its active state, continuously generates cAMP. In response, Cl^- is continuously transported out of the cell. Water follows the Cl^- by osmosis into the lumen of the intestines. If not treated, the resulting diarrhea can lead to dehydration and death, and the severe diarrhea assists the bacteria in returning to the water supply.

Cancer is the result of uncontrolled cell division. Normally, cell division is highly regulated, with multiple checkpoints during a cell cycle to ensure that the process is progressing correctly. Some cells initiate cell division in response to **growth factors,** proteins released by cells to stimulate other cells to divide. Growth factors activate a protein kinase receptor, which, in turn, activates the Ras protein. The Ras protein then initiates a MAK cascade, which ultimately activates a transcription factor. One such transcription factor binds to DNA at the gene called *p53.* The product of *p53*, protein 53, checks for DNA damage. If the DNA is damaged, p53 directs enzymes to repair the DNA. Once repaired, p53 permits cell division to proceed. If repair is unsuccessful, it directs other enzymes to kill the cell, thus preventing the proliferation of damaged cells. Such programed cell death is called **apoptosis.** If a member protein in the MAP signal transduction pathway is damaged or if the DNA contains a mutated version of *p53,* the p53 protein product may be critically altered or nonexistent. As a result, DNA surveillance does not occur and cell division progresses even if the DNA is damaged. Continued uncontrolled cell division leads to a proliferation of cancer cells. Although mutations in *p53* and other genes can be inherited, they are also caused by the ultraviolet radiation in sunlight and by chemicals in tobacco smoke.

Review Questions

Multiple-Choice Questions

The questions that follow provide a review of the material presented in this chapter. Use them to evaluate how well you understand the terms, concepts, and processes presented. Actual AP multiple-choice questions are often more general, covering a broad range of concepts, and often more lengthy. For multiple-choice questions typical of the exam, take the two practice exams in this book.

Directions: Each of the following questions or statements is followed by four possible answers or sentence completions. Choose the one best answer or sentence completion.

1. Insulin is a signaling molecule that

A. is a ligand for a membrane receptor protein
B. is a ligand for an intracellular receptor protein
C. enters the nucleus and acts as a transcription factor
D. is a second messenger that activates cAMP

2. Cortisol is a steroid signaling molecule that communicates its signal by

- **A.** binding to a membrane receptor protein
- **B.** binding to an intracellular receptor protein
- **C.** binding to DNA
- **D.** binding to mRNA

3. All of the following are second messengers EXCEPT:

- **A.** cAMP
- **B.** Ca^{2+}
- **C.** IP_3
- **D.** FOXP2, a transcription factor

4. Gap junctions and plasmodesmata allow signaling

- **A.** by direct contact
- **B.** across synapses that span the synaptic cleft between nerve cells
- **C.** among nearby cells during early animal development
- **D.** between different organs of a multicellular organism

5. Receptor protein activation occurs when

- **A.** ADP is phosphorylated to ATP
- **B.** ATP is converted to cAMP
- **C.** there is a conformation change in the receptor protein
- **D.** the receptor protein binds to a second messenger

6. A consequence of a signaling cascade is that it

- **A.** supplies energy to the cell
- **B.** accelerates mRNA activity
- **C.** is less susceptible to the impact of mutations
- **D.** amplifies the signaling response

Questions 7–12 refer to the following. Each answer in the key may be used once, more than once, or not at all.

- **A.** a protein kinase receptor
- **B.** a G protein-coupled receptor (GPCR)
- **C.** an intracellular receptor
- **D.** a ligand-gated ion receptor

7. Nonpolar ligands bind to

8. An exchange of a GTP for a GDP in a membrane-bound protein is characteristic of

9. Dimerization and autophosphorylation are characteristic of

10. A second membrane-bound protein is activated by

11. Passageways allowing movement of substances across the membrane are characteristic of

12. Second messengers are generated by the action of

Free-Response Questions

The AP exam has long and short free-response questions. The long questions have considerable descriptive information that may include tables, graphs, or figures. The short questions are brief but may also include figures. Both kinds of questions have four parts and generally require that you bring together concepts from multiple areas of biology.

The questions that follow are designed to further your understanding of the concepts presented in this chapter. Unlike the free-response questions on the exam, they are narrowly focused on the material in this chapter. For free-response questions typical of the exam, take the two practice exams in this book.

Directions: The best way to prepare for the AP exam is to write out your answers as if you were taking the exam. Use complete sentences for all your answers and do *not* use outline form or bullets. You may use diagrams to supplement your answers, but be sure to describe the importance or relevance of your diagrams.

1. Although there are significant differences among the four receptor protein mechanisms, two aspects of their activity are the same. Describe the two aspects of their mechanisms that are the same.
2. Although both a protein kinase receptor and a G protein-coupled receptor (GPCR) can phosphorylate a cytoplasmic protein kinase, they do it in very different ways. Contrast how these two signaling mechanisms phosphorylate a protein kinase.
3. There are four major kinds of signal transduction pathways, each employing a different kind of receptor protein:
   - ligand-gated ion receptor
   - G protein-coupled receptor (GPCR)
   - receptor tyrosine kinase (RTK) receptor
   - intracellular receptor

   For each of these pathways, answer each of the following questions:

   a. Describe how the receptor protein is activated.
   b. Describe the signal transduction pathway.
   c. Describe how the signal transduction pathway is deactivated.
   d. Provide an example of the pathway.

Answers and Explanations

Multiple-Choice Questions

1. **A.** Because insulin is a protein, it is both a large and a charged molecule. As a result, it is unable to cross the plasma membrane. Therefore, it must bind to a membrane receptor protein, never actually entering the cell.
2. **B.** Steroids are nonpolar molecules able to traverse the plasma membrane unaided. After crossing the membrane, cortisol binds to an intracellular receptor protein. Note that signaling molecules do not themselves bind to DNA or mRNA, but rather activate receptor proteins that initiate the appropriate chemical responses.
3. **D.** Second messengers are small nonprotein molecules. Transcription factors are proteins.
4. **A.** Gap junctions in animals and plasmodesmata in plants provide a passageway between adjacent cells for signaling molecules to pass. Synaptic signaling occurs across synapses, the small gaps between nerve cells. Paracrine signaling occurs among nearby cells, and endocrine signaling occurs for cells separated by relatively large distances.

5. **C.** A receptor protein becomes activated when it undergoes a three-dimensional conformational change. The new arrangement of atoms in the protein opens passageways or exposes active sites for binding.

6. **D.** At each step of the signaling cascade, a kinase enzyme can catalyze multiple reactions. Each of those reactions is then the beginning of the next step of the cascade, now multiplied many times over. Each step, then, has a multiplier effect, amplifying the signal. Also, because there are multiple participants, the signaling cascade is more susceptible to the influence of mutations, as a mutation in any member of the cascade can have a deleterious effect on the ultimate product of the signal.

7. **C.** Nonpolar ligands can cross the plasma membrane without assistance. They enter the cytoplasm and bind to intracellular receptor proteins.

8. **B.** An activated GPCR exchanges a GTP for a GDP on a nearby membrane-bound G protein. The exchange activates the G protein. A GTP exchange for GDP can also occur for an RTK receptor pathway (like the one involving Ras), but that exchange occurs on a cytoplasmic protein, not a membrane-bound protein.

9. **A.** When a ligand binds to a protein kinase receptor (such as receptor tyrosine kinase, RTK), it causes it to form a dimer with a second protein kinase receptor. Once the dimer forms, it attaches phosphate groups to itself (autophosphorylates).

10. **B.** An activated GPCR exchanges a GTP for a GDP on the G protein. The G protein is a nearby membrane-bound protein.

11. **D.** A ligand-gated ion receptor opens a gate, providing a passageway for ions to enter or exit the cell.

12. **B.** Second messengers are activated by the protein that is activated by the G protein of a GPCR.

Free-Response Questions

1. Each of the receptor proteins requires activation by the binding of a ligand. Also, the binding of the ligand to the receptor protein causes a conformational change in its three-dimensional structure.

2. Once activated, a protein kinase receptor, like RTK, *directly* phosphorylates and, thus, activates a cytoplasmic protein kinase. In contrast, a GPCR is *indirectly* responsible for the phosphorylation event. For a GPCR, phosphorylation of a protein kinase begins when the GPCR activates a membrane-bound G protein, which, in turn, activates a second membrane-bound enzyme, which then can phosphorylate a protein kinase.

3. **a.** The binding of a specific ligand (signaling molecule) to the receptor protein triggers a conformational change in the receptor of each of these signal transduction pathways, putting it in its activated state.
   b. *Step-by-step descriptions for each of these four signal transduction pathways are provided in the text.*
   c. *Deactivation is the last step in the step-by-step descriptions of the pathways in the text.*
   d. *Examples are provided in the text.*

Chapter 7

Cell Cycle

Review

Cell division consists of two phases, **nuclear division** followed by **cytokinesis.** Nuclear division divides the genetic material in the nucleus, while cytokinesis divides the cytoplasm. There are two kinds of nuclear division—**mitosis** and **meiosis.** Mitosis divides the nucleus so that both daughter cells are genetically identical. In contrast, meiosis is a reduction division, producing genetically variable daughter cells that contain half the genetic information of the parent cell.

The first step in either mitosis or meiosis begins with the condensation of the genetic material, **chromatin,** into tightly coiled bodies, the **chromosomes.** Each chromosome is made of two identical halves called **sister chromatids** joined at the **centromere** (Figure 7-1). Each chromatid consists of a single, tightly coiled molecule of DNA, the genetic material of the cell. In diploid cells, there are two copies of every chromosome, forming a pair, called homologous chromosomes. In a homologous pair of chromosomes, one homologue originated from the maternal parent, the other from the paternal parent. Humans have 46 chromosomes, 23 homologous pairs, consisting of a total of 92 chromatids.

The **cell cycle** describes the sequence of events that occurs during the life of most eukaryotic cells (Figure 7-2). During **interphase,** the period during which the cell is *not* dividing, the chromatin is enclosed within a clearly defined nuclear envelope. Within the nucleus, one or more **nucleoli** are visible. Outside the nucleus, two **microtubule organizing centers (MTOCs)** lie adjacent to one another. In animals, the MTOCs are the centrosomes, and each contains a pair of **centrioles.** When cell division begins, these features change, as described in the "Mitosis and Cytokinesis" section that follows.

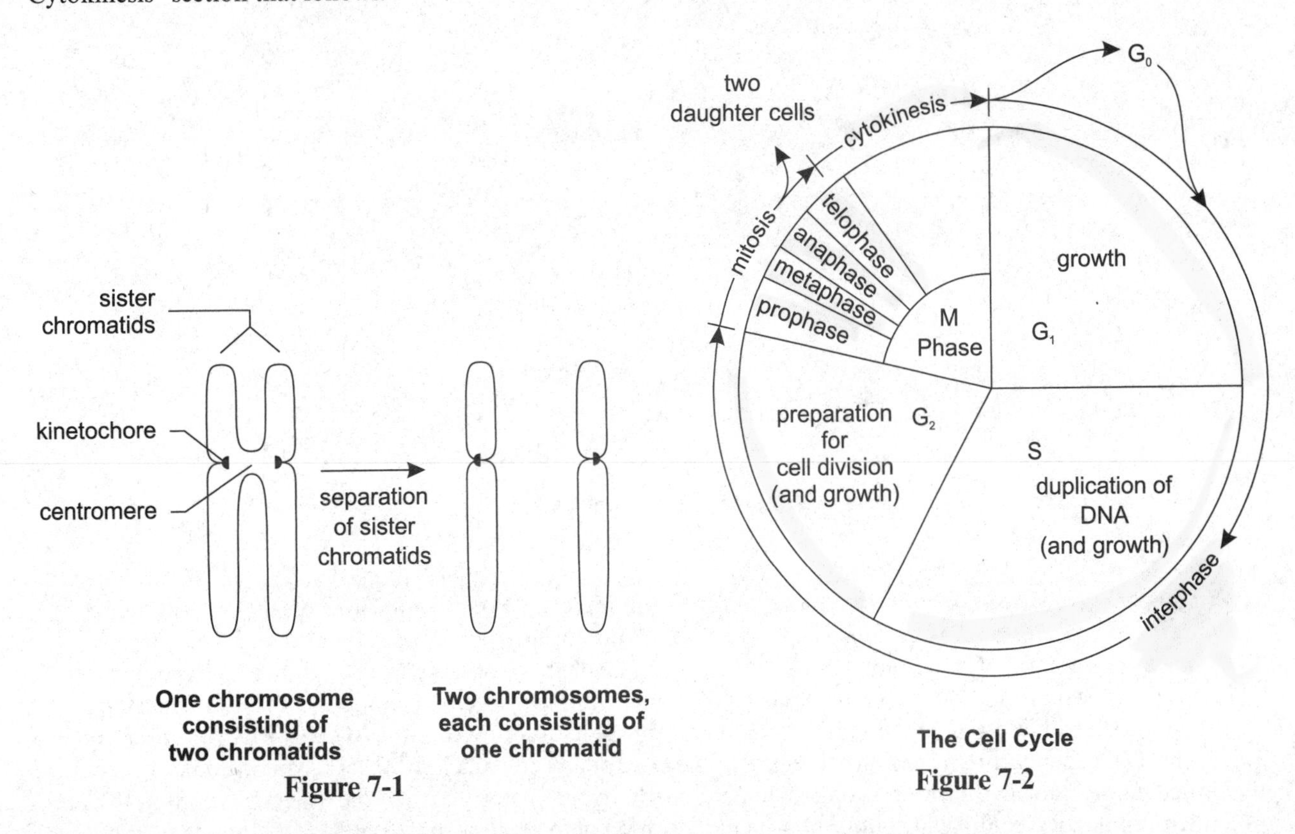

Figure 7-1

The Cell Cycle

Figure 7-2

Mitosis and Cytokinesis

Mitosis is the division of the nucleus to form two nuclei, and **cytokinesis** is the division of the cytoplasm to form two new cells, each with one nucleus. Together, they occur during the **M phase** of the cell cycle.

Mitosis

There are four phases in mitosis (adjective, **mitotic**): **prophase, metaphase, anaphase,** and **telophase** (Figure 7-3). Cytokinesis begins during telophase. A description of each mitotic stage follows. A single word in parentheses highlights the main feature of each phase.

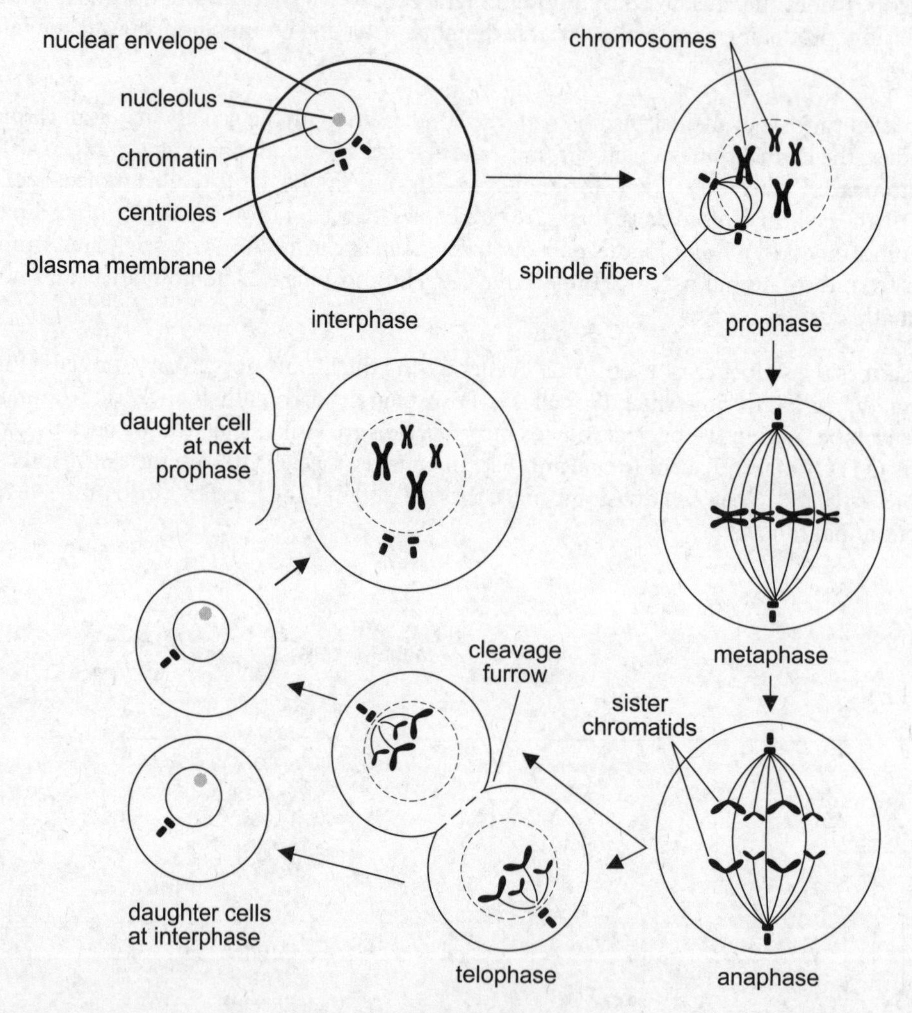

Mitosis in an Animal Cell
Figure 7-3

1. In **prophase (condensation)**, three activities occur simultaneously. First, the nucleoli disappear and the chromatin condenses into chromosomes. Second, the nuclear envelope breaks down. Third, the **mitotic spindle** is assembled. The development of the mitotic spindle begins as the MTOCs move apart to opposite ends (or poles) of the nucleus. As they move apart, microtubules develop from each MTOC, increasing in length by the addition of tubulin units to the microtubule ends away from the MTOC. Microtubules from each MTOC connect to a specialized region in the centromere called a **kinetochore.** Microtubules tug on the kinetochore, moving the chromosomes back and forth, toward one pole, then the other. In addition to these microtubules, the completed spindle also includes other microtubules from each MTOC that overlap at the center of the spindle and do not attach to the chromosomes.

2. **Metaphase (alignment)** begins when the chromosomes are aligned across the **metaphase plate,** a plane lying between the two poles of the spindle. Metaphase ends when the microtubules, still attached to the kinetochores, pull each chromosome apart into two chromatids. Each chromatid is complete with a centromere and a kinetochore. Once separated from its sister chromatid, each chromatid is called a chromosome. (To count the number of chromosomes at any one time, count the number of centromeres.)
3. **Anaphase (separation)** begins after the chromosomes are separated into chromatids. During anaphase, the microtubules connected to the chromatids (now chromosomes) shorten, effectively pulling the chromosomes to opposite poles. The microtubules shorten as tubulin units are uncoupled at their chromosome ends. Overlapping microtubules originating from opposite MTOCs, but not attached to chromosomes, interact to push the poles farther apart. At the end of anaphase, each pole has a complete set of chromosomes, the same number of chromosomes as the original cell. (Since they consist of only one chromatid, each chromosome contains only a single copy of the DNA molecule.)
4. **Telophase (restoration)** concludes the nuclear division. During this phase, a nuclear envelope is restored around each pole, forming two nuclei. The chromosomes within each of these nuclei disperse into chromatin, and the nucleoli reappear.

Cytokinesis

Whereas mitosis divides the nucleus into two daughter nuclei, cytokinesis divides the cytoplasm to form two cells. Cytokinesis differs in plants and animals by the formation of two kinds of structures, as follows:

- **Cell plate.** In plants, vesicles originating from Golgi bodies migrate to the plane between the two newly forming nuclei. The membranes of the vesicles fuse to form two new plasma membranes, and the contents of the vesicles form a **cell plate,** which develops into the new cell wall.
- **Cleavage furrow.** In animals, actin filaments (microfilaments) form a ring inside the plasma membrane between the two newly forming nuclei. As the actin filaments shorten, they act like purse strings to pull the plasma membrane into the center, dividing the cell into two daughter cells. The groove that forms as the purse strings are tightened is called a **cleavage furrow.**

When mitosis and cytokinesis are completed and interphase begins, the cell begins a period of growth. This growth period is divided into three phases, designated **G_1, S,** and **G_2,** to distinguish special activities that occur. Although you can associate the labels G_1 and G_2 with growth and S with synthesis, it is important to recognize that growth takes place during all three phases. The S phase marks the time during which the second DNA molecule for each chromosome is synthesized. As a result of this DNA replication, each chromosome that appears at the beginning of the next mitotic division will appear as two sister chromatids. During the G_2 period of growth, materials for the next mitotic division are prepared. The time span through mitosis and cytokinesis (M phase), through G_1, S, and G_2 **(interphase),** is the cell cycle (Figure 7-2).

A diploid cell is a cell with two copies of every chromosome (designated by $2n$). A cell that begins mitosis in the diploid state will end mitosis with daughter cells still in the diploid state, each with two identical copies of every chromosome. However, each of these chromosomes will consist of only one chromatid (one DNA molecule). During the S phase of interphase, the second DNA molecule is replicated from the first, so when the next mitotic division begins, each chromosome will, again, consist of two chromatids.

Meiosis

Meiosis (adjective, **meiotic**) is very similar to mitosis. Because of the similarity, however, the two processes are easily confused. The major distinction is that meiosis consists of two groups of divisions: meiosis I and meiosis II (Figure 7-4). In meiosis I, homologous chromosomes pair at the metaphase plate, and then the homologues migrate to opposite poles. In meiosis II, chromosomes spread across the metaphase plate and sister chromatids separate and migrate to opposite poles. Thus, meiosis II is analogous to mitosis. A description of each meiotic stage follows. A single word in parentheses highlights the main feature of each phase.

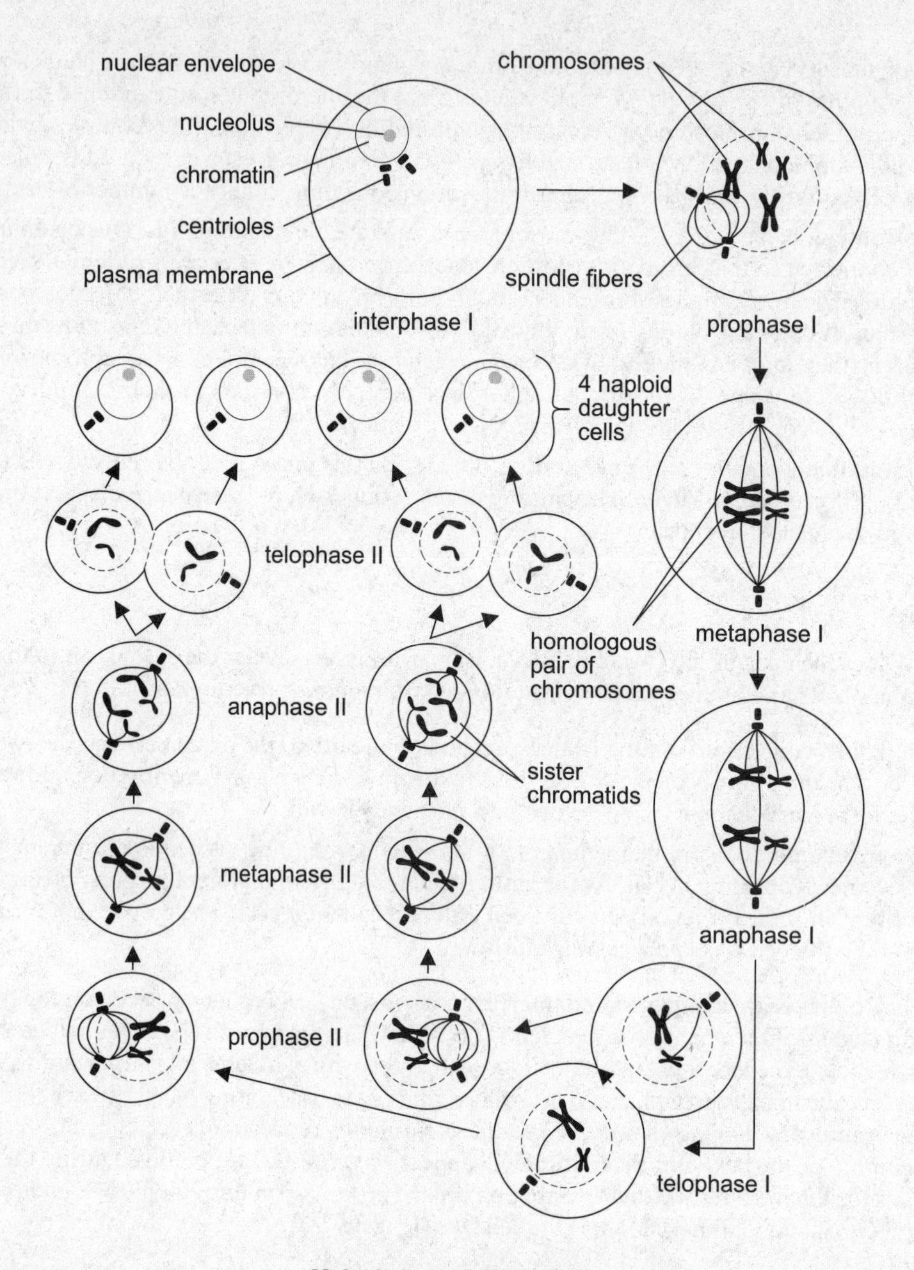

Meiosis in an Animal Cell

Figure 7-4

1. **Prophase I (condensation)** begins like prophase of mitosis. The nucleolus disappears, chromatin condenses into chromosomes, the nuclear envelope breaks down, and the spindle apparatus develops. Unlike mitosis, however, once the chromosomes are condensed, *homologous chromosomes pair with each other,* a process called **synapsis.** During synapsis, corresponding regions along *non*sister chromatids form close associations called **chiasmata** (singular, **chiasma**). Chiasmata are sites where genetic material is exchanged between nonsister homologous chromatids, a process called **crossing over.**
2. At **metaphase I (alignment),** homologous pairs of chromosomes are spread across the metaphase plate. Microtubules extending from one pole are attached to the kinetochore of one member of each homologous pair. Microtubules from the other pole are connected to the second member of each homologous pair.
3. **Anaphase I (separation)** begins when homologues uncouple as they are pulled to opposite poles.

4. In **telophase I (restoration),** the chromosomes have reached their respective poles, and a nuclear membrane develops around them. Note that each pole will form a new nucleus that will have half the number of chromosomes, but each chromosome will contain two chromatids. Since daughter nuclei will have half the number of chromosomes, cells that they eventually form will be **haploid.**

   Beginning in telophase I, the cells of many species begin cytokinesis and form cleavage furrows or cell plates. In other species, cytokinesis is delayed until after meiosis II. Also, a short interphase II may begin. In any case, no replication of chromosomes occurs during this period. Instead, part II of meiosis begins in both daughter nuclei.
5. In **prophase II (condensation),** the nuclear envelope disappears and the spindle develops. There are no chiasmata and no crossing over of genetic material as in prophase I.
6. In **metaphase II (alignment),** the chromosomes align singly on the metaphase plate (not in pairs, as in metaphase I). Single alignment of chromosomes is exactly what happens in *mitosis* except here, in meiosis, there is only half the number of chromosomes.
7. **Anaphase II (separation)** begins as each chromosome is pulled apart into two chromatids by the microtubules of the spindle apparatus. The chromatids (now chromosomes) migrate to their respective poles. Again, this is exactly what happens in mitosis, except that now there is only half the number of chromosomes.
8. In **telophase II (restoration),** the nuclear envelope reappears at each pole and cytokinesis occurs. The end result of meiosis is four haploid cells (chromosome makeup of each daughter cell designated by n). Each cell contains half the number of chromosomes, and each chromosome consists of only one chromatid.

Mitosis versus Meiosis

Comparing mitosis and meiosis (Table 7-1), you will find that mitosis ends with two *diploid* daughter cells, each with a complete set of chromosomes. True, each chromosome is composed of only one chromatid, but the second chromatid is regenerated during the S phase of interphase. Mitosis, then, merely duplicates cells; the two daughter cells are essentially clones of the original cell. As such, mitosis occurs during growth and development of multicellular organisms and for repair (replacement) of existing cells. Mitosis is also responsible for asexual reproduction, common among plants and single-celled eukaryotes. Mitosis occurs in **somatic** cells, all body cells except those that produce eggs and sperm (or pollen).

In contrast, meiosis ends with four *haploid* daughter cells, each with half the number of chromosomes (one chromosome from every homologous pair). In order for one of these haploid cells to produce a "normal" cell with the full set of chromosomes, it must first combine with a second haploid cell to create a diploid cell. In other words, meiosis produces **gametes**, that is, eggs and sperm (or pollen), for sexual reproduction. The fusing of an egg and a sperm, **fertilization** (or **syngamy**), gives rise to a diploid cell, the zygote. The single-celled **zygote** then divides by mitosis to produce a multicellular organism. Note that one copy of each chromosome in the zygote originates from one parent, and the second copy originates from the other parent. Thus, a pair of homologous chromosomes in the diploid zygote represents both maternal and paternal heritage.

Table 7-1

<table>
<tr><th>Characteristics in a Human Cell</th><th>Mitosis</th><th>Meiosis I</th><th>Meiosis II</th></tr>
<tr><td>Chromosome number in a parent cell before division begins</td><td>46</td><td colspan="2">46</td></tr>
<tr><td>Chromatid number in a parent cell before division begins</td><td>92</td><td colspan="2">92</td></tr>
<tr><td>Crossing over at prophase</td><td>No</td><td>Yes</td><td>No</td></tr>
</table>

continued

Characteristics in a Human Cell	Mitosis	Meiosis I	Meiosis II
Chromosome arrangement on metaphase plate	Chromosomes line up	Homologues pair	Chromosomes line up
Number of daughter cells at end of division	2	2	4
Chromosome notation for daughter cells	$2n$	n	n
Number of chromosomes in each daughter nucleus	46	23	23
Number of chromatids in each daughter nucleus before replication	46	46	23
Genome notation for daughter cells	Diploid	Haploid	Haploid
Purpose of division	Cell replacement, organism growth, asexual reproduction	Sexual reproduction	
Genetics of daughter cells	Genetically identical (clones)	Genetically variable	
Type of cells where division occurs	Somatic cells	Reproductive cells (ovaries, testes, anthers)	
Type of cells produced	Somatic cells	Gametes: eggs, sperm, pollen	

The **life cycle** of a human illustrates the production of gametes by meiosis and subsequent growth by mitosis (Figure 7-5). Note that the number of chromosomes in diploid and haploid cells is indicated by $2n$ and n, respectively. Human cells (except gametes) contain 46 chromosomes (23 homologous pairs). Thus, $2n = 46$. For human gametes, $n = 23$. In humans, gametes are produced in the reproductive organs—the ovaries and the testes.

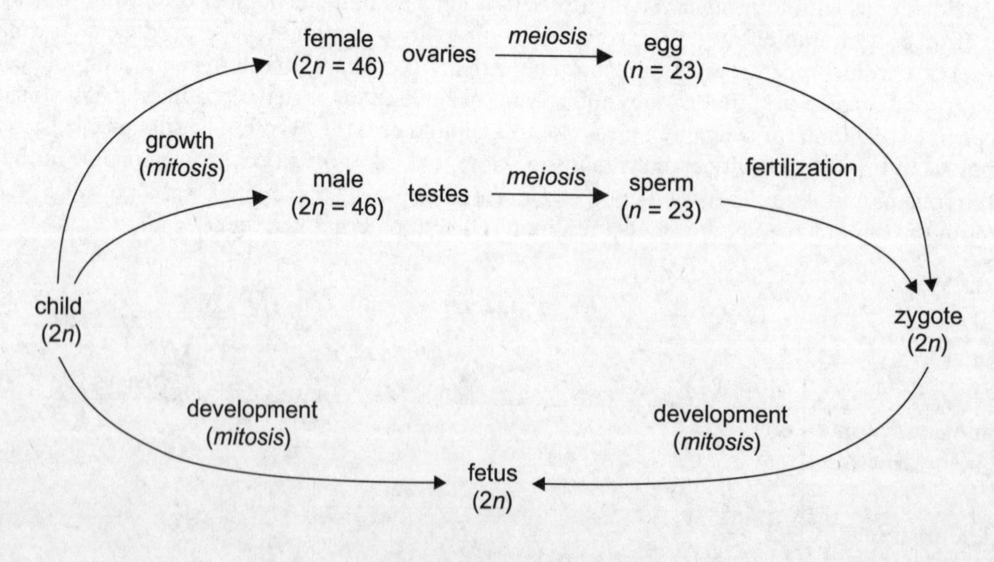

The Human Life Cycle
Figure 7-5

Genetic Variation

In mitosis, barring an error in DNA replication (mutation), every daughter cell is exactly like the parent cell. Meiosis and sexual reproduction, however, result in a reassortment of the genetic material. This reassortment, called **genetic recombination,** originates from three events during the reproductive cycle:

1. **Crossing over.** During prophase I, nonsister chromatids of homologous chromosomes exchange pieces of genetic material. As a result, each homologue no longer entirely represents a single parent.
2. **Independent assortment of homologues.** During metaphase I, the homologues of each pair of homologous chromosomes separate and go to opposite poles. Which chromosome goes to which pole depends upon the orientation of a chromosome pair at the metaphase plate. This orientation and subsequent separation is random for each homologous pair. For some chromosome pairs, the chromosome that is mostly maternal may go to one pole, but for another pair, the maternal chromosome may go to the other pole.
3. **Joining of gametes.** Because sexual reproduction requires the gametes of two individuals, new and variable combinations are created. Further variation is introduced because which sperm fertilizes which egg is to a large degree a random event. In many cases, however, this event may be affected by the genetic composition of a gamete. For example, some sperm may be faster swimmers and have a better chance of fertilizing the egg.

Regulation of the Cell Cycle

Various factors determine whether and when a cell divides. Two *functional* limitations for cell size limit growth or influence the start of a new cell division, as follows:

1. **Surface-to-volume ratio (S/V).** When a cell grows, the volume of a cell increases faster than the surface area of the plasma membrane enclosing it. This is because volume increases by the cube of the radius (volume of a sphere $= \frac{4}{3}\pi r^3$, where r is the radius), whereas the surface area increases by only the square of the radius (surface area $= 4\pi r^2$). When S/V is large, the surface area is large relative to the volume. Under these conditions, the cell can efficiently react with the outside environment. For example, adequate amounts of oxygen (for respiration) can diffuse into the cell, and waste products can be rapidly eliminated. When S/V is small, the surface area is small compared to the volume. When this occurs, the surface area might be unable to exchange enough substances with the outside environment to service the large volume of the cell. At this point, cell growth stops or cell division begins.
2. **Genome-to-volume ratio (G/V).** The genetic material (chromosomes) in the nucleus, collectively called its **genome,** controls the cell by producing substances that make enzymes and other biosynthetic substances. These substances, in turn, regulate cellular activities. The capacity of the genome to do this is limited by its finite amount of genetic material. As the cell grows, its volume increases, but its genome size remains constant. As the G/V decreases, the cell's size exceeds the ability of its genome to produce sufficient amounts of materials for regulating cellular activities.

At the molecular level, the cell cycle is strictly controlled by various signal molecules within the cell. These signals respond to *internal* factors, ensuring that the necessary steps in the cell cycle have been accurately completed before going on to the next step in the cell cycle.

1. **Checkpoints.** At specific points during the cell cycle, the cell evaluates internal and external conditions to determine whether to continue through the cell cycle. The three checkpoints are as follows (Figure 7-6):
   - The **G_1 checkpoint** occurs near the end of the G_1 phase. Here, the quality of the DNA is evaluated; if DNA damage is detected, DNA repair is attempted. If that fails, **apoptosis,** a program for self-destruction, ensues. If nutrients or growth factors are absent, the cell proceeds no further through the cell cycle, remaining in an extended G_1 phase until conditions are appropriate. Some cells, like nerve or muscle cells, are genetically programmed not to divide, and they remain in a **G_0 phase,** rarely dividing after they have matured. Liver cells, on the other hand, can leave the G_0 phase and return to dividing if they need to replace injured liver tissue.

- The **G_2 checkpoint,** occurring at the end of the G_2 phase of the cell cycle, evaluates the accuracy of DNA replication and signals whether to begin mitosis. If DNA damage is detected, DNA repair is attempted. If repair is unsuccessful, apoptosis ensues.
- The **M checkpoint,** occurring during metaphase, ensures that microtubules are properly attached to all kinetochores at the metaphase plate before division continues with anaphase.

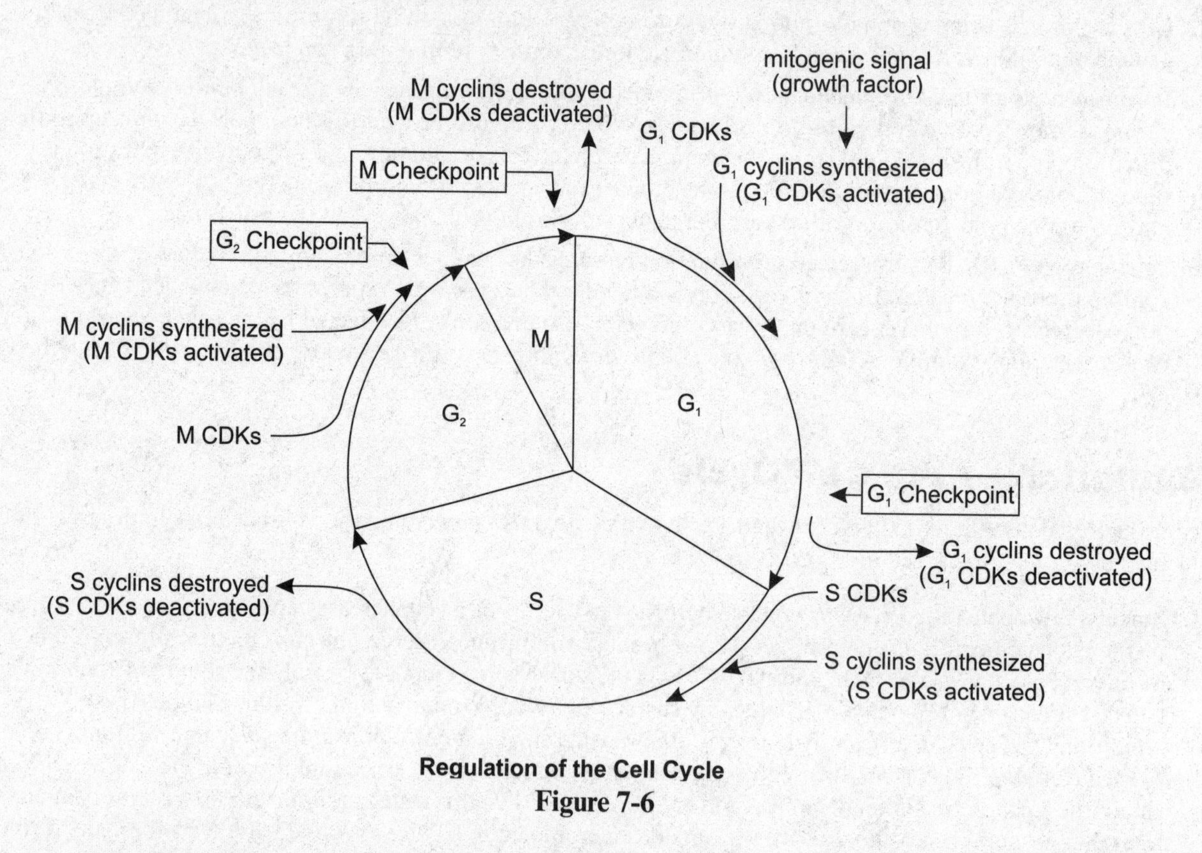

Regulation of the Cell Cycle
Figure 7-6

2. **Cyclin-dependent kinases (CDKs).** CDKs are proteins responsible for advancing the cell past the checkpoints and through the cell cycle. CDKs have the following attributes:
   - **CDKs are kinases.** Kinases are enzymes that phosphorylate other proteins. Once phosphorylated, the protein is energized and ready to act. Proteins can be *inactivated* by dephosphorylation or by destruction by other enzymes.
   - **CDKs are activated by cyclins.** Cyclins are proteins that attach to CDKs, altering their conformation and readying them for activation. Complete activation requires phosphorylation. Without a cyclin attached, a CDK is inactive (that's why it's described as "cyclin dependent").
   - **Mitosis-promoting factor (or maturation-promoting factor) (MPF)** is a cyclin-CDK complex that advances the cell cycle through the G_2 checkpoint. Each checkpoint has its own combination of a specific CDK and a specific cyclin that advances the cell cycle through the checkpoint.

Concentrations of different cyclins vary ("cycle") during the cell cycle with a regular pattern. As the cell cycle approaches each checkpoint, a specific cyclin combines with a specific CDK (Figure 7-6). The conformation change that results unblocks an active site on the CDK, readying it for activation. If checkpoint conditions are met, the CDK is activated, often by phosphorylation, and the activated cyclin-CDK complex initiates activity that advances the cell cycle through the checkpoint. Once through the cycle phase, the cyclins are destroyed, and new cyclins, specific for the next checkpoint, begin to accumulate.

Various *external* factors also influence the cell cycle:

1. **Growth factors.** The plasma membranes of cells have receptors for growth factors that stimulate a cell to divide. For example, platelets are cell fragments that circulate in the blood and contribute to the clotting mechanism. When platelets encounter damaged tissue, they release **platelet-derived growth factor (PDGF),** which binds to the plasma membrane of fibroblasts (a connective tissue) and stimulates its cell division. The new fibroblasts contribute to the healing of damaged tissue. More than 50 different growth factors are known.
2. **Density-dependent inhibition.** Many cells stop dividing when the surrounding cell density reaches a certain maximum.
3. **Anchorage dependence.** Most cells only divide when they are attached to an external surface, such as the flat surface of a neighboring cell (or the side of a culture dish).

Cancer is characterized by uncontrolled cell growth and division. **Transformed** cells, cells that have become cancerous, proliferate without regard to cell cycle checkpoints, density-dependent inhibition, anchorage dependence, and other regulatory mechanisms. Thus, cancer is a disease of the cell cycle.

Review Questions

Multiple-Choice Questions

The questions that follow provide a review of the material presented in this chapter. Use them to evaluate how well you understand the terms, concepts, and processes presented. Actual AP multiple-choice questions are often more general, covering a broad range of concepts, and often more lengthy. For multiple-choice questions typical of the exam, take the two practice exams in this book.

Directions: Each of the following questions or statements is followed by four possible answers or sentence completions. Choose the one best answer or sentence completion.

1. If a cell has 46 chromosomes at the beginning of mitosis, then at the separation phase (anaphase) there would be a total of

A. 23 chromatids
B. 46 chromosomes
C. 46 chromatids
D. 92 chromosomes

2. If a cell has 46 chromosomes at the beginning of the first meiotic division, then at the separation phase of the first meiotic division (anaphase I) there would be a total of

A. 23 chromosomes
B. 46 chromosomes
C. 46 chromatids
D. 92 chromosomes

3. All of the following statements are true EXCEPT:

A. Spindle fibers are composed largely of microtubules.
B. Centrioles consist of nine triplets of microtubules arranged in a circle.
C. All eukaryotic cells have centrioles.
D. All eukaryotic cells have a spindle apparatus.

Questions 4–7 refer to a mitotically dividing cell and to the lettered answer choices below. Each answer may be used once, more than once, or not at all.

A. anaphase (separation)
B. telophase (restoration)
C. prophase (condensation)
D. interphase (nondividing)

4. Cytokinesis begins.

5. Chromosomes begin migrating to opposite poles.

6. MTOCs migrate to opposite poles.

7. Chromosomes replicate.

Questions 8–9 refer to the following figures. Figure A represents a normal diploid cell with 2n = 8. Each of the four symbols inside the cell represents a unique chromosome.

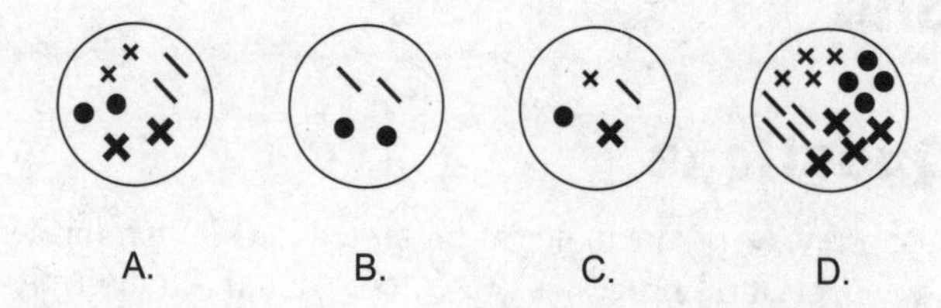

8. a zygote

9. a gamete

10. Crossing over occurs during which of the following prophase (condensation) events?

A. prophase of mitosis
B. prophase I of meiosis
C. prophase II of meiosis
D. prophase I and II of meiosis

11. In typical cell divisions, all of the following contribute to genetic variation EXCEPT:

A. anaphase (separation) of mitosis
B. anaphase (separation) of meiosis I
C. fertilization
D. crossing over

Questions 12–16 refer to the cell illustrations below. The normal diploid number for the cells illustrated is four chromosomes. Each answer may be used once, more than once, or not at all.

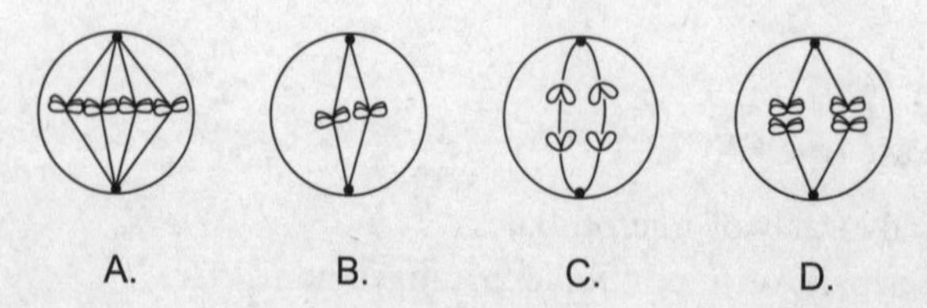

12. anaphase (separation) II

13. metaphase (alignment) II

14. metaphase (alignment) of mitosis

15. metaphase (alignment) I

Free-Response Questions

The AP exam has long and short free-response questions. The long questions have considerable descriptive information that may include tables, graphs, or figures. The short questions are brief but may also include figures. Both kinds of questions have four parts and generally require that you bring together concepts from multiple areas of biology.

The questions that follow are designed to further your understanding of the concepts presented in this chapter. Unlike the free-response questions on the exam, they are narrowly focused on the material in this chapter. For free-response questions typical of the exam, take the two practice exams in this book.

Directions: The best way to prepare for the AP exam is to write out your answers as if you were taking the exam. Use complete sentences and do *not* use outline form or bullets. You may use diagrams to supplement your answers, but be sure to describe the importance or relevance of your diagrams.

1. Many of the activities that occur during mitosis and meiosis are similar. Describe one stage of meiosis that is distinctively different from mitosis.

2. Genetic variation is a product of meiosis. Describe how genetic variation is created by meiosis.

3. The cell cycle is strictly regulated by various processes. Describe what goes wrong when a cell is overtaken by cancer.

4. Describe the process of cell division in plants and animals, giving specific attention to the following:

- **a.** the stages of mitosis, cytokinesis, and other phases of the cell cycle (do not include meiosis)
- **b.** factors that induce cells to divide
- **c.** factors that might contribute to abnormal cell divisions, such as cancer

5. Describe meiosis in animal and plant cells, giving specific attention to the following:

- **a.** the stages of meiosis
- **b.** the function of meiotic daughter cells and the organs where meiosis takes place
- **c.** contributions to genetic variation

Answers and Explanations

Multiple-Choice Questions

1. D. At metaphase, there are 46 chromosomes with 2 chromatids each, for a total of 92 chromatids. Metaphase ends and anaphase begins when the 2 chromatids of each chromosome separate. Once separated, these 92 chromatids become 92 chromosomes because each chromatid is now considered a complete chromosome, since it consists of a complete DNA molecule. (To count the number of chromosomes at any point during the cell cycle, count the centromeres.) Thus, there are now 92 chromosomes destined for 2 daughter cells (46 chromosomes per cell). After cytokinesis, each daughter cell contains 46 chromosomes, each consisting of 1 chromatid.

2. **B.** During metaphase I, homologous chromosomes pair at the metaphase plate. One member of each pair migrates to opposite poles during anaphase I. If the cell started with 46 chromosomes, 23 chromosomes move to each pole during anaphase I, so the total is still 46 chromosomes.

3. **C.** Most plants do not have centrioles.

4. **B.** During telophase, cytokinesis begins, chromosomes uncoil into chromatin, and the nuclear membrane and nucleolus reappear.

5. **A.** At metaphase, the chromosomes are arranged on the metaphase plate. The end of metaphase and the beginning of anaphase are defined by the separation of the chromosomes into chromatids (which are now considered chromosomes by themselves). Once anaphase begins, the chromosomes move to one pole or the other.

6. **C.** During prophase, the two MTOCs (and centrioles, if present) migrate to opposite poles as the spindle apparatus develops between them. Also during prophase, the nucleolus and nuclear membrane disappear, and the chromatin condenses into chromosomes.

7. **D.** Chromosomes replicate during the S phase of interphase.

8. **A.** Since the zygote consists of the union of two haploid gametes, the zygote would have the same number of chromosomes as the parent. As in the parent cell, the eight chromosomes would consist of four homologous pairs.

9. **C.** A gamete would possess four chromosomes, half the number of chromosomes as the parent cell. Also, these four chromosomes would consist of one member of each pair of homologous chromosomes. Figure B also has four chromosomes, but it does not present one homologue of each homologous pair.

10. **B.** Homologous chromosomes pair (synapsis) during prophase I and form chiasmata. Exchanges of genetic material occur within chiasmata.

11. **A.** There are generally no events during normal mitosis that would produce genetic differences between the two daughter cells. The daughter cells are clones, genetically identical. In contrast, the independent assortment of homologues during anaphase I, the random union of egg and sperm during fertilization, and crossing over during prophase I all contribute to genetic variation.

12. **C.** At the end of meiosis I, each daughter cell would have two chromosomes, each composed of two chromatids. At metaphase II, these two chromosomes would line up on the metaphase plate. When anaphase II begins, each of the chromosomes would split into two chromatids. One of each of these chromatids (now called chromosomes) would migrate to one or the other pole.

13. **B.** If the cell began with four chromosomes, then after meiosis I, each daughter cell would have two complete chromosomes. At metaphase II, the two chromosomes would align, unpaired, on the metaphase plate.

14. **A.** Only in mitosis would you see four chromosomes spread out, unpaired, on the metaphase plate. If this were metaphase I, the chromosomes would appear in pairs; if it were metaphase II, there would be only two chromosomes.

15. **D.** The pairing of homologous chromosomes occurs only in meiosis and at metaphase I.

Free-Response Questions

1. During chromosome alignment in meiosis I (metaphase I), homologous chromosomes pair on the equatorial plate and subsequently *homologues* separate to opposite poles. In contrast, during the same phase of mitosis, the chromosomes all line up and *chromatids* separate to opposite poles.

2. Crossing over during the condensation phase of meiosis I (prophase I) produces chromosomes with genetic material from both parents. Also, the joining of gametes during sexual reproduction creates genomes in offspring that are unique.

3. When the DNA of a cell is damaged by radiation or chemicals, genes that produce proteins that regulate cell division may have mutated. If, for example, the gene product is a damaged protein that normally monitors the integrity of the DNA, it may not trigger DNA repair and damaged DNA will be passed on to daughter cells. Accumulation of damaged DNA may result in a further breakdown of cell cycle regulation, and the cell line may begin dividing without control, thus becoming cancerous.

4. a. Mitosis consists of four phases—prophase, metaphase, anaphase, and telophase. In prophase, the chromatin condenses into chromosomes and the nuclear envelope and nucleolus disappear. As centrioles (or MTOCs, in plants) migrate to opposite poles, microtubules develop between them to form the spindle apparatus. The microtubules attach to the kinetochores in the centromeres of the chromosomes and pull on the chromosomes. At metaphase, the microtubules have pulled the chromosomes so that they are all lined up on the metaphase plate. In anaphase, the sister chromatids of each chromosome are separated and pulled to opposite poles by the microtubules of the spindle apparatus. In telophase, the chromatids are well segregated to opposite poles. Nuclear membranes appear around each pole and chromosomes diffuse into chromatin. Cytokinesis, the dividing of the cytoplasm, begins during telophase. The cell is divided by a cleavage furrow in animals or a cell plate in plants. If the mother cell began with a diploid number of chromosomes, then the two nuclei that form at each pole would also both be diploid, though at this point each chromosome would consist of only a single chromatid.

The entire cell cycle includes both mitosis and interphase. Interphase is a period of growth and is divided into three stages, identified as G_1, S, and G_2. The G_1 phase describes the first period of growth following mitosis. During the S phase, a second DNA molecule (chromatid) is replicated from each chromosome. During the G_2 phase, the cell prepares for mitosis. The M phase describes mitosis and cytokinesis.

b. Various factors induce a cell to divide. Cell size is functionally limited by surface-to-volume and genome-to-volume ratios. As the surface-to-volume ratio becomes progressively smaller as the cell grows, the ability of the plasma membrane to provide a surface large enough to meet the import and export requirements of the cell diminishes. Also, when the cell increases in size, the amount of genetic material remains the same. As a result, the ability of the nucleus to control the cell decreases.

Various checkpoints during the cell cycle evaluate conditions to determine whether cell activities in preparation for cell division should continue. The G_1 checkpoint during the G_1 phase determines if preparations should continue or if the cell should enter the G_0 state with no subsequent cell division. The G_2 checkpoint during the G_2 phase checks for accurate DNA replication. The M checkpoint during metaphase ensures that the chromosomes are properly attached to the spindle fiber microtubules.

Cell division also depends on the presence of cyclin-dependent kinases, which activate, by phosphorylation, proteins that regulate the cell cycle. Cell division is influenced by the detection of external molecules, or growth factors, that are produced by other cells. Cell division is also promoted when neighboring cells are available for attachment (anchorage dependence) or prevented by the presence of too many neighboring cells (density-dependent inhibition).

c. If and when a cell divides is determined much like other metabolic activities, by enzymes. Enzymes are produced at specific points of the cell cycle that induce specific activities that prepare the cell for division. The production of these enzymes is affected by environmental factors (such as carcinogens or cell density), by internal conditions, and by genetic factors. In a transformed cell, a cell that has become cancerous, the normal cell-cycle checkpoints and other regulatory mechanisms fail. For example, most cancerous cells lack anchorage dependence, growing and dividing without the need to attach to nearby cells.

Be sure to write your answer in sections, as was done in this answer, responding separately to each part of the question and labeling each response with the appropriate letter. Also, since your time is very limited, you should not spend too much of it defining words, unless specifically requested to do so. The free-response section of the exam is not a vocabulary test. Rather, these questions are designed to evaluate your understanding of biological processes. Thus, you should focus on describing the process, using (but not defining) as much of the appropriate vocabulary as you can. By doing so, you demonstrate both an understanding of terminology and the biological process.

5. a. Meiosis consists of two groups of divisions: meiosis I and II. In prophase I, the nuclear membrane breaks down, the nucleolus disappears, and chromatin condenses into chromosomes. The MTOCs (which contain centrioles in animals) migrate to opposite poles, developing microtubules and the spindle apparatus between them. Synapsis occurs when homologous chromosomes pair. During synapsis, crossing over between nonsister chromatids of homologous pairs results in an exchange of genetic material. The microtubules connect to the kinetochores in the centromeres of the chromosomes and pull on the chromosomes to the metaphase plate. Metaphase I occurs when the chromosomes are aligned on the metaphase plate as homologous pairs. Anaphase I begins as each member of a homologous pair of chromosomes is pulled by the microtubules to opposite poles. Telophase I follows when nuclear membranes appear. Cytokinesis and a short interphase II may occur at this point. Prophase II begins in each daughter cell in the same manner as prophase I. However, synapsis does not occur, and at metaphase II, the chromosomes are spread over the metaphase plate without any kind of pairing. Anaphase II begins as each chromosome is separated into two chromatids (now called chromosomes) and pulled by the microtubules of the spindle apparatus to opposite poles. During telophase II, meiosis is concluded as cytokinesis separates the nuclei into four haploid cells, each containing half the number of chromosomes of the original parent cell.

b. Meiosis is a reduction division that occurs in sexual reproduction. It halves the number of chromosomes so that daughter cells are haploid. In humans, the daughter cells are the gametes (sperm and eggs) formed in the testes and ovaries. Gametes fuse to form a diploid zygote, which then grows into a multicellular organism by mitotic divisions. In other organisms, meiosis may produce haploid spores, which divide by mitosis to grow into multicellular haploid organisms.

c. There are three points during meiosis and sexual reproduction where genetic material is rearranged to create genetic variation. First, crossing over during metaphase I results in an exchange of genetic material between nonsister chromatids of homologous chromosomes. Chromosomes, previously of either paternal or maternal origin, now contain genetic material from both parents. Second, homologous chromosome pairs randomly align across the metaphase plate in metaphase I. As a result, chromosomes migrating to one pole are a random mixture of paternal and maternal chromosomes. Third, the zygote is a combination of a randomly selected egg and a sperm. As a result of these random arrangements of chromosomes, daughter cells are genetically variable.

Chapter 8

Heredity

Review

Much of the material in this chapter prepares you for solving genetics problems. A genetics problem is an analysis of the characteristics (traits) of parents and offspring (progeny). Given the traits of one of these generations, you are required to determine the traits of the other generation.

Genetics problems require the application of probability rules. If a coin is tossed, there is a $\frac{1}{2}$ (50%) chance, or probability, that it will be heads. If a coin is tossed again, there is, again, a $\frac{1}{2}$ chance that it will be heads. The first toss does not affect the second toss; that is, the two tosses are independent. To determine the probability of two or more independent events occurring together, you merely multiply the probabilities of each event happening separately. This is the **multiplication rule** of probability. For two consecutive tosses of a coin, the probability of getting two heads is $\frac{1}{2} \times \frac{1}{2} = \frac{1}{4}$. For three tosses, the probability of three heads would be $\frac{1}{2} \times \frac{1}{2} \times \frac{1}{2} = \frac{1}{8}$.

The following terms are used in genetics:

1. A **gene** represents the genetic material on a chromosome that contains the instructions for creating a particular trait. Since the formula for carrying out these instructions is described by a genetic code, a gene is often said to code for a trait. In pea plants, for example, there is a gene that codes for stem length.
2. An **allele** is one of several varieties of a gene. In pea plants, there are two alleles of the gene for stem length—the tall allele, which codes for tall plants, and the dwarf allele, which codes for dwarf plants.
3. A **locus** refers to the location on a chromosome where a gene is located. Every gene has a unique locus on a particular chromosome.
4. **Homologous chromosomes** refer to a pair of chromosomes (a **homologous pair**) that contains the same genetic information, gene for gene. Each parent contributed one of the chromosomes in the pair (Figure 8-1). At any one particular locus, the two genes on a pair of homologous chromosomes (a **gene pair**) might represent two different alleles for that gene because they originated from different parents. For example, the allele for stem length on one pea plant chromosome (inherited from one parent) might code for tall plants, whereas the allele on the homologue of that chromosome (inherited from the second parent) might code for dwarf plants.

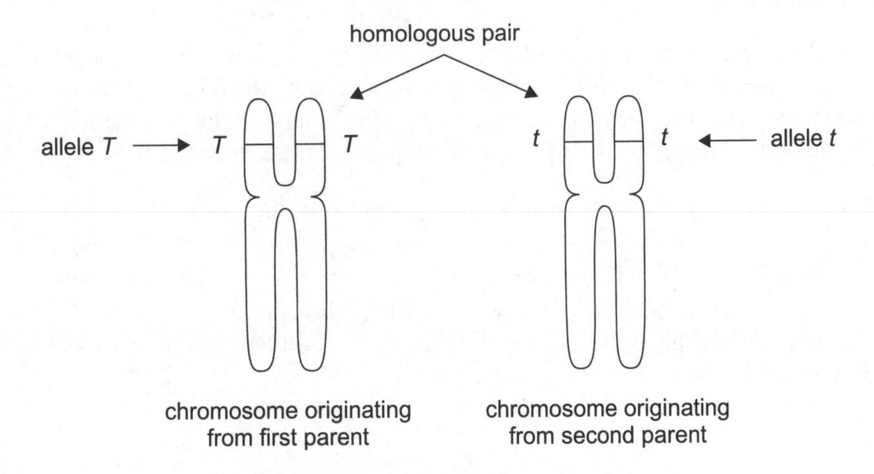

Homologous Pair of Chromosomes

Figure 8-1

5. If the two alleles inherited for a gene (each on one of the two homologous chromosomes) are different, one allele may be **dominant,** while the other is **recessive.** The trait encoded by the dominant allele is the actual trait expressed. In pea plants, the tall stem allele is dominant and the dwarf allele is recessive. Therefore, if a pea plant inherits one of each of these alleles, only the dominant allele is expressed, producing tall plants. In genetics problems, a dominant allele is often represented by a capital letter, while the recessive allele is represented by the lowercase form of the same letter. In addition, the first letter of the dominant allele is often used to represent the gene. Thus, *T* and *t* represent the dominant (tall) and recessive (dwarf) alleles, respectively, of the same gene. (It is easiest to read these two alleles as "big" *T* and "small" *t*.)
6. **Homozygous dominant** refers to the inheritance of two dominant alleles (*TT*). In this condition, the dominant trait is expressed. In the **homozygous recessive** condition, two recessive alleles are inherited (*tt*) and the recessive trait is expressed. **Heterozygous** refers to the condition where the two inherited alleles are different (*Tt*—it is normal convention to write a pair of alleles with the dominant allele first). In this condition, only the dominant allele is expressed.
7. The **phenotype** is the expression of an allele. Tall stems, blue eyes, and brown hair each represent the phenotype of their respective alleles. On the other hand, the **genotype** represents the actual alleles. For example, *TT* describes the genotype for the homozygous dominant condition. If *T* represents the allele for tall stems and *t* is the allele for dwarf stems, the *genotype Tt* would express the *phenotype* of tall stems. To help you remember their meanings, think *gene*s for genotype and *ph*ysical trait for phenotype (although phenotypes might describe physiological or behavioral traits, as well). The term "genotype" is also used to describe just those alleles under discussion (or even all of an individual's alleles), and its phenotype is used to describe the expression of those alleles.

Gregor Mendel, a nineteenth-century monk, is credited with the discovery of the laws of segregation and independent assortment. The following laws describe the separation of chromosomes during meiosis (Figure 8-2). Because the distribution of alleles among gametes is random, the rules of probability can be used to describe how the different chromosomes (and their alleles) in parents assemble in offspring.

1. The **law of segregation** refers to the segregation (separation) of alleles (and their chromosomes) to individual gametes. In other words, one member of each chromosome pair migrates to an opposite pole so that each gamete contains only one copy of each chromosome (and each allele).
2. The **law of independent assortment** refers to the independent assortment of alleles (and their chromosomes). The process is independent because the migration of homologues within one pair of homologous chromosomes to opposite poles does not influence the migration of homologues of other homologous pairs.

Mendel mated, or **crossed,** two varieties of pea plants to form offspring, or **hybrids.** In these kinds of experiments, the **P generation** represents the parents, the **F_1 generation** represents the offspring from the crossing of the parents, and the **F_2 generation** represents the offspring produced from crosses among the F_1. (The letter F stands for *filial,* which refers to sons and daughters.) Two kinds of genetic experiments are common:

1. A **monohybrid cross** is an experiment in which only one trait is being investigated. For example, a cross between a tall pea plant and a dwarf pea plant is a monohybrid cross because it is investigating a gene for only one trait, the gene for stem length.
2. A **dihybrid cross** occurs when two traits are involved. A cross investigating the traits of stem length (tall or dwarf) and flower color (purple or white) is a dihybrid cross. Another example is a cross investigating seed color (yellow, green) and seed texture (round, wrinkled).

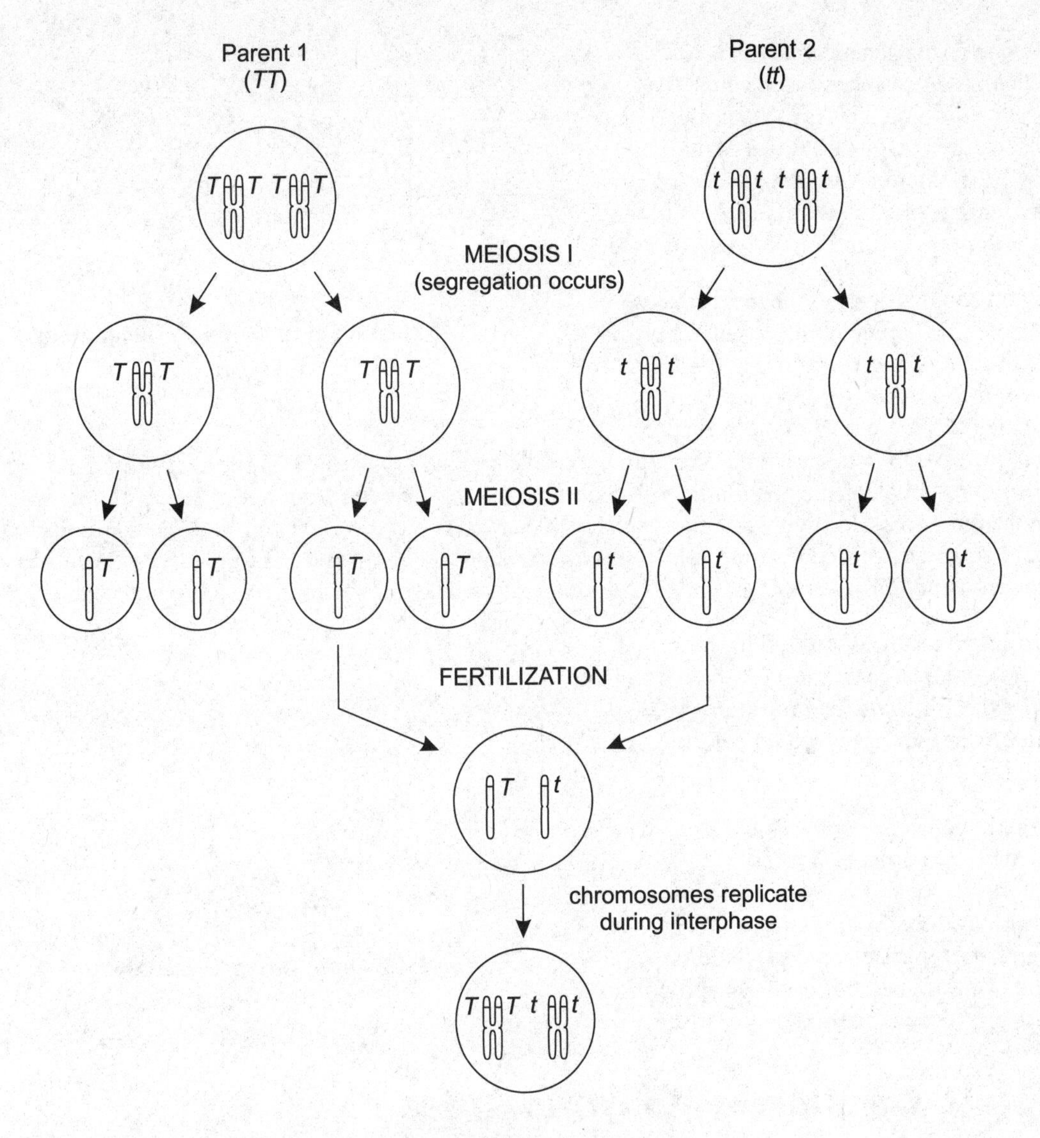

Meiosis and Fertilization

Figure 8-2

Complete Dominance, Monohybrid Cross

When traits are expressed as if one allele is dominant to a second allele, the inheritance pattern is called **complete** (or **full) dominance.** Take, for example, a **monohybrid cross,** in which a single trait (originating from a single gene) is examined. For pea plants, let *T,* a dominant allele, represent the allele for tall stems and *t,* the recessive allele, represent the allele for dwarf stems. For inheritance by complete dominance, *TT* and *Tt* produce tall plants, and *tt* produces dwarf plants. Suppose that a plant heterozygous for tall stems (*Tt*) is crossed with a dwarf-stemmed plant (*tt*). The cross can be represented as *Tt* × *tt.* Now, what predictions can be made for the genotypes of the offspring?

The first step in analyzing this genetics problem is to determine the genotypes for all possible gametes produced by both parents. Thus, the tall-stemmed parent (genotype *Tt*) produces a gamete with the *T* allele or a gamete with the *t* allele. This is because the genotype *Tt* is represented by a pair of homologous chromosomes, one with the *T* allele and the other with the *t* allele. Each chromosome (with its respective allele) migrates to an opposite pole and ends up in a separate gamete (the law of segregation). Similarly, the dwarf-stemmed parent produces a gamete with the *t* allele and another gamete with the *t* allele.

The next step in the genetic analysis is to determine all the possible ways in which the gametes can combine. This is most easily accomplished by creating a Punnett square (Figure 8-3). (Sometimes, you might see a Punnett square rotated so that it appears in the shape of a diamond; either method can be used.)

	t	*t*
T	*Tt*	*Tt*
t	*tt*	*tt*

Tt × *tt* P generation

½ *Tt* + ½ *tt* F_1 generation

Monohybrid Cross (F_1 Generation)

Figure 8-3

In a Punnett square for a monohybrid cross, the gametes from one parent are represented in two spaces at the top of the diagram (*t* and *t* in Figure 8-3). The gametes of the second parent are represented at the left side (*T* and *t* in Figure 8-3). In the middle are four boxes, each box combining the allele found at the top with the allele found to the left. The four boxes represent all of the possibilities of combining the two gametes from one parent with the two gametes from the second parent. In Figure 8-3, the results of the cross *Tt* × *tt* are ½ *Tt* and ½ *tt*. These results represent the genotypic frequencies of the offspring. The phenotypic results are ½ tall and ½ dwarf plants. Results can be given as frequencies (fractions), as percents (½ = 50%), or as ratios. The ratio of dwarf to tall plants is 1:1.

Suppose that you were asked to find the frequencies of the F_2 generation for the cross *TT* × *tt*. Following the procedures above, all of the F_1 progeny would have the genotype *Tt*.

To find the F_2 generation, cross the F_1 offspring with themselves, *Tt* × *Tt* (Figure 8-4). The genotypic results of this cross would produce ¼ *TT*, ½ *Tt*, and ¼ *tt*, while the phenotypic frequencies would be ¾ tall and ¼ dwarf because both *TT* and *Tt* genotypes produce tall plants.

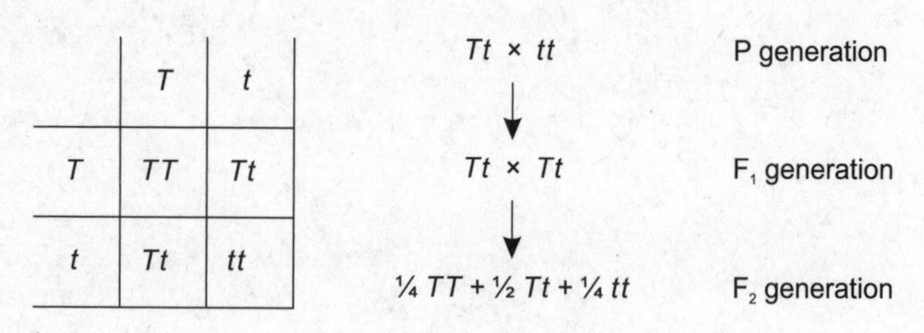

Monohybrid Cross (F_2 Generation)

Figure 8-4

Complete Dominance, Dihybrid Cross

For a **dihybrid cross,** genes for two different traits are observed at the same time. In peas, seed color can be yellow (*Y*) or green (*y*), and seed texture can be round (*R*) or wrinkled (*r*). Thus, *Y* and *y* are used to represent the two alleles (yellow and green) for the gene for seed color, and *R* and *r* (round and wrinkled) are used for the alleles for seed texture. A cross between one pea plant homozygous dominant for both traits and a second plant homozygous recessive for both traits would be given as *YYRR* × *yyrr*.

The first step in analyzing this cross is to determine the alleles of all possible gametes. The *YYRR* plant can produce only one kind of gamete, *YR*. This is determined by the law of segregation; that is, one allele of each allele pair migrates to opposite poles and ends up in separate gametes. Thus, one *Y* and one *R* end up in a single gamete. All segregation possibilities produce gametes that are *YR*. For the second plant, there is also only one kind of gamete produced, *yr*. The next step is to determine the different ways that the gametes from one parent can combine with the gametes from the second parent. Usually you would make a Punnett square for this, but since each parent has only one kind of gamete, you can quickly conclude, without a Punnett square, that all the F_1 offspring will result from the union of a *YR* gamete and a *yr* gamete. The result of that union produces individuals that have the genotype *YyRr*, which bears the phenotype yellow and round.

Now, analyze the F_2 generation produced by *YyRr* × *YyRr*. Each parent can produce four kinds of gametes: *YR*, *Yr*, *yR*, and *yr*. The Punnett square, illustrated in Figure 8-5, shows the gametes of one parent in spaces on the top line and the gametes of the second parent in spaces along the left margin. In the 16 boxes of the

P $YYRR \times yyrr$

F_1 $YyRr \times YyRr$

	YR	Yr	yR	yr
YR	YYRR	YYRr	YyRR	YyRr
Yr	YYRr	YYrr	YyRr	Yyrr
yR	YyRR	YyRr	yyRR	yyRr
yr	YyRr	Yyrr	yyRr	yyrr

F_2

genotypic frequencies	phenotypic frequencies
YYRR = 1, YYRr = 2, YyRR = 2, YyRr = 4	9 yellow round
yyRR = 1, yyRr = 2	3 green round
YYrr = 1, Yyrr = 2	3 yellow wrinkled
yyrr = 1	1 green wrinkled

Dihybrid Cross

Figure 8-5

square, alleles from gametes at the top and left are combined. By convention, both alleles from the color gene are arranged together (for example, *Y* and *y*), and both alleles of the texture gene are arranged together (for example, *R* and *r*). The next step is to list each kind of genotype and count the number of times each genotype appears.

This information is shown at the right side of Figure 8-5. The final step is to identify the phenotype of each genotype and to count how many times each phenotype appears. You will see that some phenotypes have more than one genotype. For example, *YYRR, YYRr, YyRR,* and *YyRr* all code for a yellow and round seed. The conclusion is that the F_2 progeny consist of nine plants with yellow and round seeds, three plants with green and round seeds, three plants with yellow and wrinkled seeds, and one plant with green and wrinkled seeds. This ratio, 9:3:3:1, is the same ratio Mendel observed in his experiments for this dihybrid cross.

Test Crosses

Suppose that you wanted to know the genotype of a dwarf pea plant. That would easily be identified as *tt* because dwarf-stemmed plants must have two copies of the recessive allele. Suppose, however, you wanted to know the genotype of a plant with the dominant trait, tall stems. Would it be *TT* or *Tt*? To determine which genotype is correct, you would perform a test cross. A **test cross** is a mating of an individual whose genotype you are trying to determine with an individual whose genotype is known. You will always know the genotype of the individual that expresses the recessive trait (dwarf stems). So, the cross is *T*____ × *tt*. Since you do not know the second allele for the first individual, represent it with an underscore, leaving a blank space for the unknown allele. The next step is to perform both possible crosses, *TT* × *tt* and *Tt* × *tt*. For the first cross, all F_1 will be tall (*Tt*). For the second cross, ½ will be tall (*Tt*) and ½ will be dwarf (*tt*). A farmer would perform a test cross if he or she wanted to know if the tall plant in question was *TT* or *Tt*. If there were any dwarf-stemmed offspring, the farmer could be sure that the tall parent was *Tt*, since you cannot obtain dwarf-stemmed offspring unless both parents contribute the dwarf allele (*t*). In contrast, if all offspring were tall, the farmer could reasonably conclude that the tall parent was *TT*. It is possible, though not likely, that the tall-stemmed parent could be *Tt*, and that, due to chance, no short-stemmed offspring were produced.

A coin toss presents an analogous situation. If you toss a coin six times, the prediction is that half the time it will be tails. However, there is a small chance, $\frac{1}{64}$ ($\frac{1}{2} \times \frac{1}{2} \times \frac{1}{2} \times \frac{1}{2} \times \frac{1}{2} \times \frac{1}{2}$), that all six tosses will be tails. As the number of coin tosses increases, though, the probability of obtaining all tails gets smaller and smaller. For the farmer who finds all tall offspring in the test cross, there is a small possibility that the tall parent will be *Tt*. However, the identification of the genotype as *TT* will become more conclusive as the number of offspring observed increases.

Incomplete Dominance

Sometimes, the alleles for a gene do not exhibit the dominant and recessive behaviors discussed above. Instead, the combined expression of two different alleles in the heterozygous condition produces a blending of the individual expressions of the two alleles called **incomplete dominance.** In snapdragons, for example, the heterozygous condition consisting of one allele for red flowers (*R*) and one allele for white flowers (*r*) results in a pink phenotype (*Rr*). Sometimes, both alleles are written with the same uppercase or lowercase letter but with a prime or a superscript or subscript number or letter to differentiate the two. As an example, *R* and *R'* might represent red and white alleles for snapdragons. As another example, H_1 and H_2 might represent straight-hair and curly-hair alleles in humans. The $H_1 H_2$ phenotype is expressed as an intermediate trait, wavy hair. In still other cases, there may be no apparent rationale, except perhaps historical, for the notation used to indicate different alleles.

Codominance

Another kind of inheritance pattern is termed **codominance.** In this pattern, both inherited alleles are completely expressed. For example, the M and N blood types produce two molecules that appear on the surface of human red blood cells. The *M* (sometimes written as L^M) allele produces a certain blood-cell molecule. The *N* (sometimes written as L^N) allele produces another molecule. Individuals who are *MM* ($L^M L^M$) produce one kind of molecule; those who are *NN* ($L^N L^N$) produce a second kind of molecule; and those who are *MN* ($L^M L^N$) produce both kinds of molecules.

To help you distinguish the three kinds of inheritance, imagine a continuum. At one extreme, there is complete dominance by a dominant allele over a recessive allele. At the other extreme, both alleles are expressed (codominance). Between the two extremes, a blending of two different alleles produces an intermediate phenotype (incomplete dominance).

Multiple Alleles

In the blood group that produces A, B, and O blood types, there are three possible alleles, represented by I^A, I^B, or *i*. Superscripts are used because the two alleles, *A* and *B*, are codominant. A lowercase *i* is used for the third allele because it is recessive when expressed with I^A or I^B. There are six possible genotypes representing all possible combinations of two alleles: $I^A I^A$ and $I^A i$ (A blood type), $I^B I^B$ and $I^B i$ (B blood type), $I^A I^B$ (AB blood type), and *ii* (O blood type). The four phenotypes (A, B, AB, and O types) correspond to the presence or absence of an A or B sugar component attached to proteins of the plasma membrane of red blood cells. Thus, the $I^A I^A$ and $I^A i$ genotypes have proteins with the A sugar attached, the $I^B I^B$ and $I^B i$ genotypes have proteins with the B sugar attached, and the $I^A I^B$ genotype has both kinds of proteins (half with the A sugar attached and half with the B sugar attached). For the *ii* genotype, neither sugar is attached.

You should be aware of why blood transfusions must be made between individuals of like phenotypes. If an individual with $I^B I^B$, $I^B i$, or *ii* blood is given type A blood, then the immune system of the recipient will identify the A sugar on the introduced red blood cells as a foreign substance. The immune system responds to foreign substances **(antigens)** by producing **antibodies** that attack the antigens. The result is clumping, or **agglutination,** of the blood and possibly death. Individuals with AB type blood can accept any blood type because both A and B sugars are recognized as "self." Also, anyone can accept O type blood because it contains neither A nor B sugars. Thus, a person with O type blood is a universal donor for the ABO blood group (other blood-group types, such as Rh, also need to match).

Polygenic Inheritance

Many traits are not expressed in just two or three varieties, such as yellow and green pea seeds or A, B, and O blood types, but as a range of varieties. The heights of humans, for example, are not just short or tall but are displayed as a **continuous variation** from very short to very tall. Continuous variation usually results from **polygenic inheritance,** the interaction of many genes to shape a single phenotype.

Linked Genes and Crossing Over

If two genes are on different chromosomes, such as the seed color and seed texture genes of pea plants, the genes segregate independently of one another (law of independent assortment). **Linked genes** are genes that reside on the *same* chromosome and, thus, cannot segregate independently because they are physically connected. Genes that are linked do not obey the law of independent assortment and are usually inherited together.

In the fruit fly *Drosophila melanogaster,* flies reared in the laboratory occasionally exhibit mutations in their genes. Two such mutations, affecting body color and wing structure, are linked. The normal, or wild, body color is gray (*B*), while the mutant allele is expressed as black (*b*). The second mutation, for wing structure, transforms normal wings (*V*) into vestigial wings (*v*) (small, underdeveloped, and nonfunctional). (Note that for *Drosophila* mutations, the gene notation uses letters that denote the name for the mutation.) Since these two genes are linked, a fly heterozygous for a gray body and normal wings (called gray-normal), indicated by *BbVv,* would have the *BV* on one chromosome and the *bv* on the homologous chromosome. If the linkage between these genes were not known, then the expected results from a cross between this gray-normal fly (*BbVv*) and a black fly with vestigial wings (called black-vestigial, *bbvv*) would be ¼ *BbVv*, ¼ *bbvv,* ¼ *Bbvv,* and ¼ *bbVv* (Figure 8-6). However, since

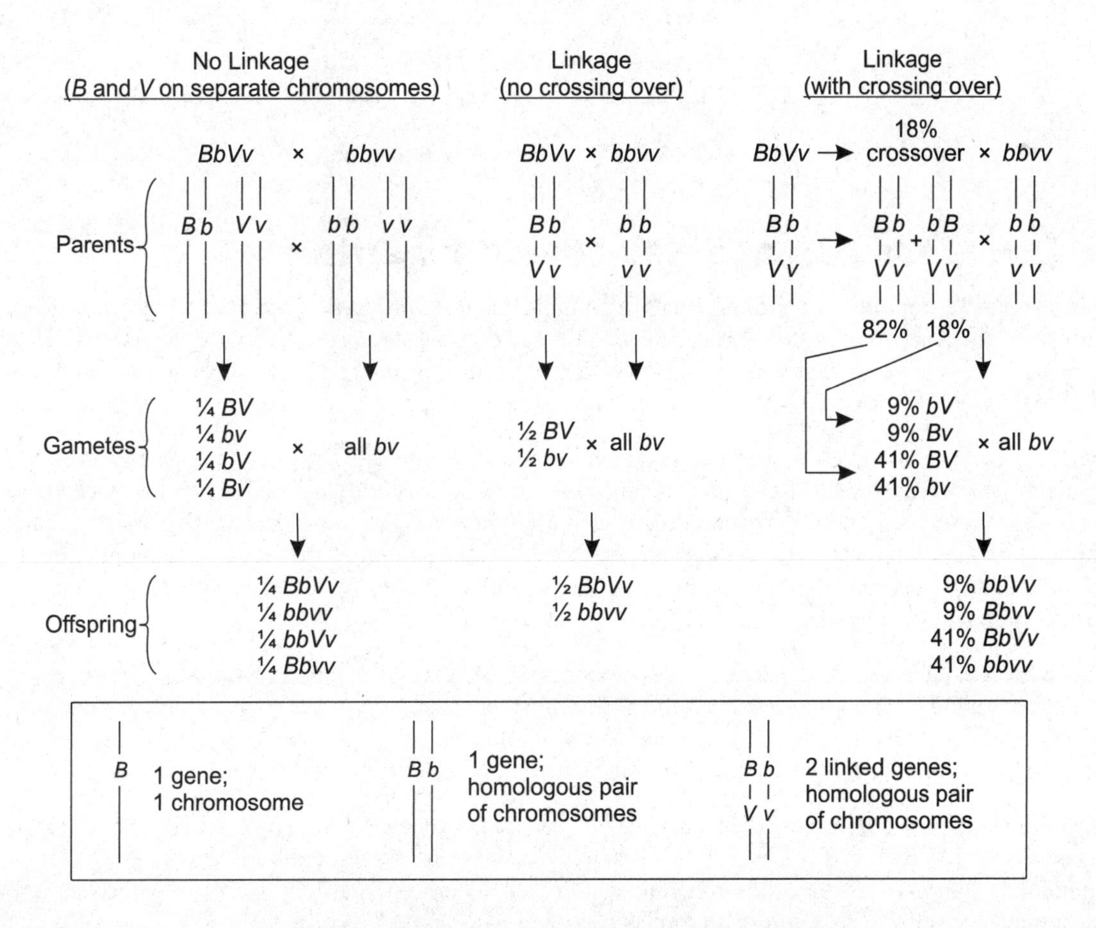

Dihybrid Cross with Linked Genes and Crossing Over

Figure 8-6

the two genes are on the same chromosome and cannot assort independently, the gray-normal fly produces only two kinds of gametes, *BV* and *bv*. *Bv* and *bV* gametes are not produced. Taking linkage into consideration, the expected offspring would be ½ *BbVv* and ½ *bbvv* (Figure 8-6). If this cross were actually carried out, however, the results would produce a ratio among the four offspring *BbVv*:*bbvv*:*Bbvv*:*bbVv* closer to 41:41:9:9. This is because linked genes cross over (recombine) during prophase I—in this case, about 18% of the time. Instead of 50% of the gametes being *BV* and 50% *bv*, an 18% crossover rate would produce 41% *BV* and 41% *bv* (which sum to 82%) and 9% *Bv* and 9% *bV* (which sum to 18%) (Figure 8-6).

The greater the distance between two genes on a chromosome, the more places between the genes that the chromosome can break and, thus, the more likely the two genes are to cross over during synapsis. As a result, recombination frequencies are used to give a picture of the arrangement of genes on a chromosome. Suppose you knew that for a fly with genotype *BBVVAA* (where *A* is the apterous, or wingless, mutant) the crossover frequency between *B* and *V* was 18%, between *A* and *V* was 12%, and between *B* and *A* was 6%. Since greater recombination frequencies indicate greater distances between genes, *B* and *V* are separated by the greatest distance. Using the frequencies as a direct measure of distance (so that a 1% crossover frequency represents 1 map unit), *B* and *V* are 18 map units apart, *A* and *V* are 12 map units apart, and *B* and *A* are 6 map units apart (Figure 8-7). This suggests that the three genes are arranged in the order *B-A-V* (or alternatively, *V-A-B*), with *B* and *A* separated by 6 units and *A* and *V* separated by 12 units. The sum of these two distances, 18 (or 6 + 12), is the map distance between *B* and *V* (other positions for *A* do not preserve the frequencies of *AV* and *AB*). A chromosome map created in this fashion is a **linkage map** and is a portrayal of the sequence of genes on a chromosome. A map portraying the true relative positions of the genes, a **cytological map,** requires additional experimental analyses.

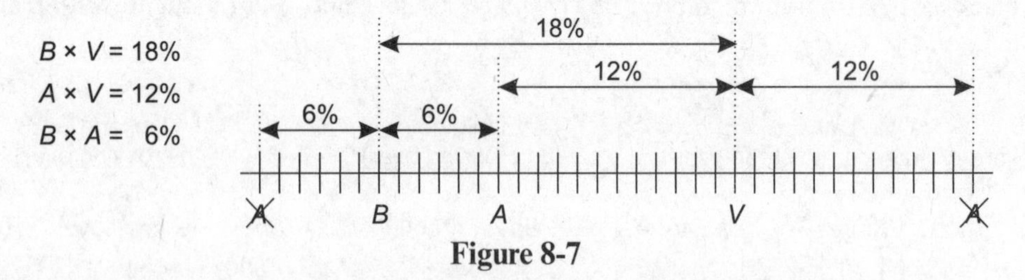

Figure 8-7

Sex Chromosomes and Sex-Linked Inheritance

There is one pair of homologous chromosomes in animals that does not have exactly the same genes. These two chromosomes, the X and Y chromosomes, are called the **sex chromosomes.** All other chromosomes are called **autosomes.** In mammals and fruit flies, inheritance of two X chromosomes (XX) produces a female and inheritance of one X and one Y (XY) produces a male.

Although inheritance of a Y chromosome makes a male, the Y chromosome is small (compared to the X) and has relatively few genes. Some of its genes code for testes and sperm development. One gene in particular, the *SRY* (sex-determining region of Y), regulates gene activities on other chromosomes, which, in turn, stimulate development of male characteristics. As a result, many traits that distinguish the sexes (milk production, male pattern baldness) are **sex limited,** that is, they can only develop in one or the other sex, even though the genes for those traits are on autosomes.

Sex-linked (or **X-linked**) genes are genes that reside on the X, or sex, chromosome. **Y-linked** genes are also possible, but because so few genes reside on the Y chromosome, Y-linkage is rarely encountered. Note the different uses of the word "linkage": Used alone, the word "linkage" refers to two or more genes that reside on the same chromosome; "sex linkage" refers to a single gene residing specifically on a sex chromosome.

You have additional considerations when working with sex-linked genes. When females (XX) inherit a sex-linked gene, they receive two copies of the gene, one on each X chromosome. This situation is similar to that for autosomal inheritance. In contrast, however, a male (XY) will inherit only one copy of the gene because only the X chromosome delivers the gene. There is no similar gene delivered by the Y chromosome. As a result, the allele on the X chromosome of a male is the allele whose trait is expressed, regardless of whether it is dominant or recessive.

Color blindness is caused by a sex-linked, recessive gene (n) in humans. Females and males who inherit the normal allele (N) are $X^N X^N$ and $X^N Y$, respectively, and both are normal. In order for a female to be color blind, she must have two copies of the defective allele ($X^n X^n$). A male, however, has to inherit only one copy of the defective allele ($X^n Y$) to be color blind. (Note how the Y chromosome is written with no allele superscript because it has none of the genes that are on the X chromosome.) As a result, color blindness, as well as other sex-linked genetic defects, are much more common in males. Heterozygous individuals, who possess a recessive allele for a genetic disorder, do not express the disorder, but they are said to be **carriers** since they can pass the defective allele to their offspring. Thus, heterozygous females, females who are $X^N X^n$, have normal vision but are carriers.

X-Inactivation

During embryonic development in female mammals, one of the two X chromosomes in each cell does not uncoil into chromatin. Instead, **X-inactivation** occurs, and one chromosome remains coiled as a dark, compact body, called a **Barr body.** Barr bodies are mostly inactive X chromosomes—most of the genes are not expressed, nor do they interact (in a dominant/recessive or codominant manner) with their respective alleles on the X chromosome that is expressed. Thus, only the alleles of the genes on the one active X chromosome are expressed by that cell. When X-inactivation begins, one of the two chromosomes in each embryonic cell randomly and independently becomes inactive. Subsequent daughter cells will have the same X chromosome inactivated as did the embryonic parent cell from which they originated. In the fully developed fetus, then, some groups of cells will have one X chromosome inactivated, while other groups will have the other X chromosome inactivated. Thus, all of the cells in a female mammal are not functionally identical.

A very visible example of X-inactivation can be seen in the different groups of cells producing different patches of color in an individual calico cat. Calico cats have orange, black, and white hair, randomly arranged in patches over their bodies. The orange and black colors are determined by a gene on the X chromosome (the white color is controlled by a different gene). When the X chromosome with the orange allele is inactivated, the black color allele on the active chromosome is expressed, and the hair is black. In other patches, the chromosome with the black allele may be inactivated, and those patches will be orange.

What does this mean for sex-linked genetic disorders in humans, such as color blindness? A carrier female ($X^N X^n$) should usually be normal with respect to this trait, because all or at least some cells will have X^N activated and produce normal functioning color vision. However, in the very unlikely case that all of the relevant cells have X^N inactivated, the carrier female should express the same symptoms of color blindness as a male.

Nondisjunction

Nondisjunction is the failure of one or more chromosome pairs or chromatids of a single chromosome to properly move to opposite poles during meiosis or mitosis. Some examples of nondisjunction follow:

1. During meiosis, the failure of two homologous chromosomes (during anaphase I) or two chromatids of a single chromosome (during anaphase II) to separate produces gametes with extra or missing chromosomes.
2. During mitosis, the failure of two chromatids of a single chromosome (during anaphase) to separate produces daughter cells with extra or missing chromosomes. This happens most often during embryonic development and results in **mosaicism,** in which a fraction of the body cells, those descendent of a cell where nondisjunction occurs, have an extra or missing chromosome.
3. **Polyploidy** occurs if all of the chromosomes undergo meiotic nondisjunction and produce gametes with twice the number of chromosomes. If a polyploid gamete is fertilized with a similar gamete, then a polyploidy zygote and individual can form. Polyploidy is common in plants.

Human Genetic Disorders

Some of the causes of genetic disorders follow:

1. **Point mutations** occur when a single nucleotide in the DNA of a gene is incorrect. This can occur if a different nucleotide is substituted for the correct one **(substitution),** if a nucleotide base-pair is omitted **(deletion),** or if an extra base-pair is inserted **(insertion).** Most point mutations have deleterious effects on gene function. Two examples follow:
   - **Sickle-cell disease,** caused by a nucleotide substitution, results in the production of defective hemoglobin, the oxygen-carrying protein in red blood cells. The defective hemoglobin molecule causes the red blood cell, usually circular, to become sickle shaped when low-oxygen conditions occur (high altitudes, strenuous exercise). In response, red blood cells do not flow through capillaries freely and oxygen is not adequately delivered throughout the body (anemia). For individuals who are homozygous for the defective allele (sickle-cell *disease*), inadequate oxygen supplies can lead to organ damage, bone abnormalities, and impaired mental functioning. Heterozygous individuals (sickle-cell *trait*) are generally without symptoms, as the normal allele is sufficient to produce adequate amounts of normal hemoglobin.
   - **Tay-Sachs disease,** usually caused by a nucleotide insertion, results when lysosomes lack the functional enzyme to break down certain fats (glycolipids). When these fats accumulate in the nerve cells of the brain, brain cells die and death usually follows early in childhood.
2. **Aneuploidy** is a genome with extra or missing chromosomes. It is most often caused by nondisjunction. Although most aneuploid gametes do not produce viable offspring, some zygotes, with certain chromosome imbalances, survive. These almost always lead to genetic disorders. Three examples follow:
   - **Down syndrome** occurs when an egg or sperm with an extra number 21 chromosome fuses with a normal gamete. The result is a zygote with three copies of chromosome 21 **(trisomy 21).** Down syndrome individuals bear various abnormalities, including mental retardation, heart defects, respiratory problems, and deformities in external features.
   - **Turner syndrome** results when there is nondisjunction of the sex chromosomes. Sperm will have either both sex chromosomes (XY) or no sex chromosomes (O, used to indicate the absence of a chromosome). Similarly, eggs will be either XX or O. When a normal X egg or sperm combines with an O sperm or egg, a Turner syndrome zygote, **XO,** results. XO individuals are sterile females with physical abnormalities. Although the absence of a single chromosome is usually fatal, missing the Y chromosome, with its few, male-necessary genes, is not nearly so deleterious (Figure 8-8).
   - **Klinefelter syndrome** occurs when an XY or XX gamete, produced as a result of nondisjunction (as in Turner syndrome), combines with a normal X gamete to produce an **XXY** individual. Because of the presence of a Y chromosome, these individuals are male but may be sterile. In addition, individuals may mildly express a variety of female secondary sex characteristics (for example, reduced facial and chest hair and breast development). Although the extra X chromosome is mostly inactive (forming a Barr body), some activity apparently remains. An XXX individual can result when an XX egg combines with a normal X sperm, but these trisomic X females are usually without serious disorders, though they are often tall and may have some learning disabilities (Figure 8-8).

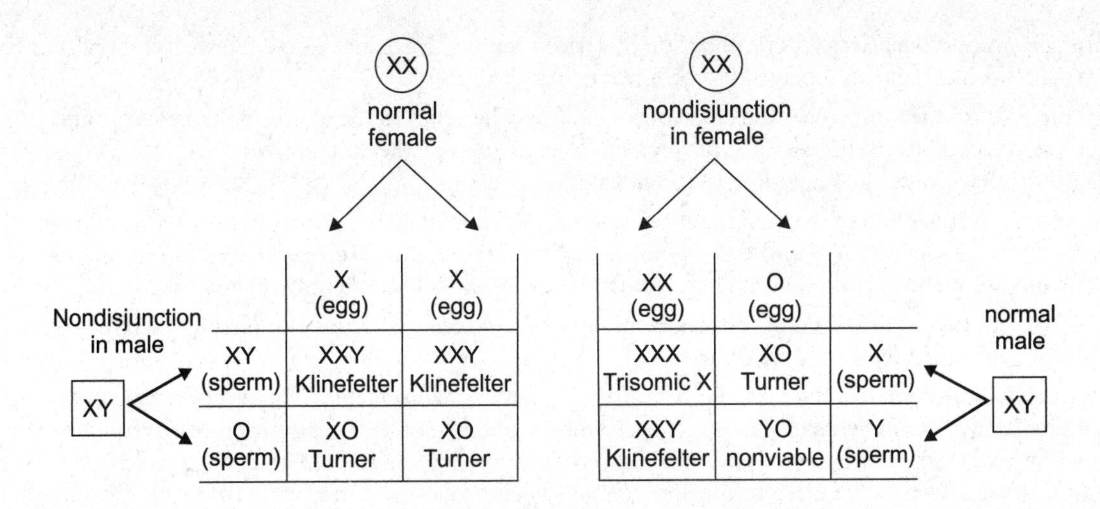

Aneuploidy of Sex Chromosomes

Figure 8-8

3. **Chromosomal aberrations** are caused when chromosome segments are changed.
   - **Duplications** occur when a chromosome segment is repeated on the same chromosome. For example, **Huntington's disease** is caused by the insertion of multiple repeats of three nucleotides. The mutant gene codes for a defective enzyme, which results in the death of nerve cells in the brain.
   - **Inversions** occur when chromosome segments are rearranged in reverse orientation on the same chromosome. Individuals with inversions usually do not express any abnormalities as long as the inversion does not introduce any duplications or deletions.
   - **Translocations** occur when a segment of a chromosome is moved to another chromosome. For example, Down syndrome ordinarily occurs when an individual inherits an extra chromosome 21 (as a result of nondisjunction). However, Down syndrome can also occur after a translocation of a chromosome segment from chromosome 21 to chromosome 14. An individual inheriting a chromosome with a 14/21 translocation has a normal number of chromosomes but inherits three copies of a segment from chromosome 21 (two chromosomes 21 and a chromosome 14 with a segment of chromosome 21). The result has the same phenotypic effect as trisomy 21.

Environmental Influences on Phenotypic Expression

If you have ever compared "identical" twins, you know that, though very similar, they are not identical. Although their DNA may be identical, the expression of their genes is influenced by environmental factors. Before identical twins are born, they may experience slightly difference environments in the uterus, and these differences lead to different phenotypic expressions of height, weight, and fingerprints. Some examples of environmental factors that influence gene expression follow:

- Nutrition can heavily influence physical development in both animals and plants. Insufficient calcium or other dietary deficiencies may result in short stature, while plants growing in soils that lack adequate amounts of nitrogen may not flower or may flower and produce smaller-than-normal fruits.
- Nutrition may also influence the expression of genetic disorders. For example, individuals with phenylketonuria cannot metabolize the amino acid phenylalanine. As a result, phenylalanine accumulates, brain cells die, and death follows. However, if dietary phenylalanine is reduced to minimal levels, quantities of phenylalanine in the body remain safe. Similarly, individuals who are lactose intolerant and unable to break down lactose can avoid symptoms of nausea, gas, and diarrhea by omitting dairy products containing lactose from their diet.

- Temperature influences sex determination in various reptiles. Eggs incubated at lower temperatures become males; those at higher temperatures become females.
- Temperature often influences the color of animal fur. The genes for fur color in Siamese cats and Himalayan rabbits produce a dark fur pigment in cold areas of the animal (feet, ears, face, and tail) and a light fur pigment in the remaining warm areas.
- Seasonal changes in daylight length influence the expression of hair color from brown (in summer) to white (in winter, for camouflage) in the snowshoe hare. Similarly, an increased exposure to UV radiation in humans stimulates a corresponding increase in melanin, the skin-darkening pigment.
- Soil pH influences flower color in certain species of *Hydrangea.* Flower color is blue when the soil pH is acidic, pink when the soil pH is basic.
- Expression of one individual's genes is sometimes dependent upon chemicals in the environment that are produced by other individuals. This kind of chemical signaling between organisms is often required to elicit mating. For example, when nutrient availability is scarce, slime bacteria secrete signaling molecules that stimulate nearby bacteria to aggregate, form a multicellular collective, grow into a fruiting body, and produce spores. Similarly, certain yeast cells only mate with yeast cells of an opposite mating type. To signal its presence, a yeast cell secretes a pheromone (a signaling molecule), to which only yeasts of the opposite mating type respond.

Non-Nuclear Inheritance

Mitochondria and chloroplasts carry within their own DNA genes that are responsible for some of their metabolic processes. Mitochondria assort randomly during cell division and, in most animals, are inherited only from the mother, as the male gamete (sperm) delivers very little cytoplasm. Similarly, in many plants, both mitochondria and chloroplasts are inherited through the female gamete (ovules), as the male gamete (pollen) does not usually carry these organelles. Where it occurs, this **maternal inheritance** can be used to trace a specific genome from progeny back through multiple generations to its original mother. One example of maternal inheritance in humans is a genetic mutation that impairs oxidative phosphorylation, reduces ATP output, and results in nerve deterioration. Because the mutation occurs on a mitochondrial gene, the trait can only be inherited from mothers who possess the defective mitochondria.

Review Questions

Multiple-Choice Questions

The questions that follow provide a review of the material presented in this chapter. Use them to evaluate how well you understand the terms, concepts, and processes presented. Actual AP multiple-choice questions are often more general, covering a broad range of concepts, and often more lengthy. For multiple-choice questions typical of the exam, take the two practice exams in this book.

Directions: Each of the following questions or statements is followed by four possible answers or sentence completions. Choose the one best answer or sentence completion.

1. If you roll a pair of dice, what is the probability that they will both turn up a three?

A. $\frac{1}{2}$
B. $\frac{1}{8}$
C. $\frac{1}{16}$
D. $\frac{1}{36}$

2. Which of the following best expresses the concept of the word *allele*?

A. genes for wrinkled and yellow
B. genes for wrinkled and round
C. the expression of a gene
D. phenotypes

3. The ability to taste a chemical called PTC is inherited as an autosomal dominant allele. What is the probability that children can taste PTC if they are descended from parents who are both heterozygous for this trait?

A. 0
B. $\frac{1}{4}$
C. $\frac{1}{2}$
D. $\frac{3}{4}$

4. In fruit flies, dumpy wings are shorter and broader than normal wings. The allele for normal wings (*D*) is dominant to the allele for dumpy wings (*d*). Two normal-winged flies mated and produced 300 normal-winged and 100 dumpy-winged flies. The parents were probably

A. *DD* and *DD*
B. *DD* and *Dd*
C. *Dd* and *Dd*
D. *Dd* and *dd*

5. Which of the following is true of the gametes produced by an individual with genotype *Dd*?

A. $\frac{1}{4}$ *D* + $\frac{1}{2}$ *Dd* + $\frac{1}{4}$ *D*
B. $\frac{1}{2}$ *D* and $\frac{1}{2}$ *d*
C. $\frac{1}{2}$ *Dd* and $\frac{1}{2}$ *dD*
D. all *Dd*

6. Suppose that in sheep, a dominant allele (*B*) produces black hair and a recessive allele (*b*) produces white hair. If you saw a black sheep, you would be able to identify

A. its phenotype for hair color
B. its genotype for hair color
C. the genotypes for only one of its parents
D. the phenotypes for both of its parents

Questions 7–9 refer to the following key and description of fruit fly traits. Each answer in the key may be used once, more than once, or not at all.

A. 0
B. $\frac{1}{16}$
C. $\frac{9}{16}$
D. 1

In fruit flies, the gene for curved wings (*c*) and the gene for spineless bristles (*s*) are on different chromosomes. The respective wild-type alleles for each of these genes produce normal wings and normal bristles.

7. From the cross *CCSS* × *ccss,* what is the probability of having an offspring that is *CcSs*?

8. From the cross *CcSs* × *CcSs,* what is the probability of having an offspring that is *ccss*?

9. From the cross *CcSs* × *CcSs,* what is the probability of having an offspring that is normal for both traits?

Questions 10–11 refer to the following.

In snapdragons, the allele for tall plants (*T*) is dominant to the allele for dwarf plants (*t*), and the allele for red flowers (*R*) is codominant with the allele for white flowers (*R′*). The heterozygous condition for flower color is pink (*RR′*).

10. If a dwarf red snapdragon is crossed with a white snapdragon homozygous for tall, what are the probable genotypes and phenotypes of the F1 generation?

A. all *TtRR′* (tall and pink)
B. all *TtRR* (tall and red)
C. all *TtR′R′* (tall and white)
D. all *ttRR* (dwarf and red)

11. If *ttRR′* is crossed with *TtRR,* what would be the probable frequency for offspring that are dwarf and white?

A. 0
B. ¼
C. ½
D. ¾

12. For the cross *AABBCCDd* × *AAbbCcDd,* what is the probability that an offspring will be *AABbCcDd*?

A. 1/16
B. ⅛
C. ¼
D. ½

13. The inheritance of skin color in humans is an example of which of the following?

A. X-linked inheritance
B. codominance
C. polygenic inheritance
D. gene linkage

14. Red-headed people frequently have freckles. This is best explained by which of the following?

A. The genes for these two traits are linked on the same chromosome.
B. The genes for these two traits are sex-linked.
C. Alleles for these two traits are codominant.
D. Both parents have red hair and freckles.

15. Let *A* and *a* represent two alleles for one gene and *B* and *b* represent two alleles for a second gene. If for a particular individual, *A* and *B* were on one chromosome and *a* and *b* were on a second chromosome, then all of the following are true EXCEPT:

A. The two genes are linked.
B. The two chromosomes are homologous.
C. All gametes would be either *AB* or *ab.*
D. The genotype of this individual is *AaBb.*

16. Four genes, *J, K, L,* and *M,* reside on the same chromosome. Given that the crossover frequency between *K* and *J* is 3, between *K* and *L* is 8, between *J* and *M* is 12, and between *L* and *M* is 7, what is the order of the genes on the chromosome?

A. *J K L M*
B. *J K M L*
C. *K J L M*
D. *K J M L*

Questions 17–19 refer to the following pedigree. Circles indicate females, and squares indicate males. A horizontal line connecting a male and a female indicates that these two individuals produced offspring. Offspring are indicated by a descending vertical line that branches to the offspring. A filled circle or filled square indicates that the individual has a particular trait, in this case, red-green color blindness. Color blindness is inherited as a sex-linked, recessive allele.

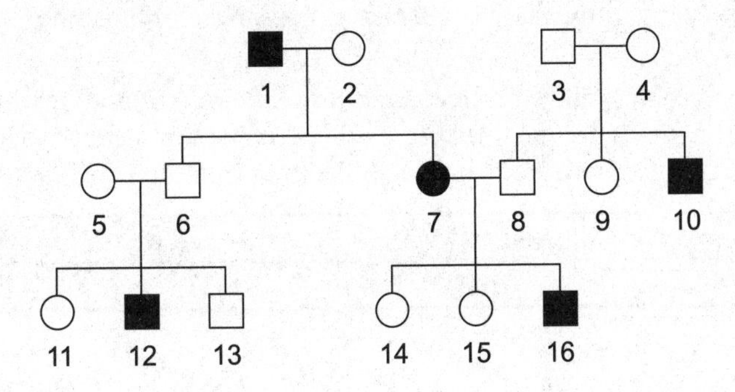

Use the following key for the next three questions. Each answer in the key may be used once, more than once, or not at all.

A. $X^N X^N$
B. $X^N X^n$
C. $X^N Y$
D. $X^n Y$

17. Identify the genotype for individual 10.

18. Identify the genotype for individual 5.

19. Identify the genotype for individual 14.

20. In domestic cats, two alleles of a sex-linked (X-linked) gene code for hair color. One allele codes for orange hair, and the other allele codes for black hair. Cats can be all orange or all black, or they can be calico, a coat characterized by randomly arranged patches of orange and black hair. With respect to this gene, all of the following are true EXCEPT:

A. A black female and an orange male can produce a black male cat.
B. A black female and an orange male can produce a female calico cat.
C. A calico female and an orange male can produce a female calico cat.
D. A calico female and an orange male can produce a male calico cat.

21. From which parent(s) did a male with red-green color blindness inherit the defective allele?

A. *only* the mother
B. *only* the father
C. the mother or the father, but not both
D. *both* the mother and the father

22. A human genetic disorder that is caused by nondisjunction of the sex chromosomes is

A. sickle-cell disease
B. hemophilia
C. Down syndrome
D. Turner syndrome

23. Two genes, *A* and *B*, are linked. An individual who is *AaBb* produces equal numbers of four gametes: *AB*, *Ab*, *aB*, and *ab*. The best explanation for this would be that

A. nondisjunction occurred
B. the genes are on nonhomologous chromosomes
C. the two genes are close together on the same chromosome
D. the two genes are separated by a large distance on the same chromosome

24. In peas, the gene for seed color (yellow or green) and flower color (colored or white) are located on the same chromosome. Find the crossover frequency if a cross between a plant heterozygous for both traits and a plant homozygous recessive for both traits produces the following progeny.

yellow and colored	green and white	yellow and white	green and colored
652	683	77	88

A. 5.5%
B. 11%
C. 15%
D. 22%

25. For the cross *AaBbCCddEE* × *AaBbCcDdEE*, what is the probability that an offspring will be *aaBBCCddEE*?

A. $1/16$
B. $1/32$
C. $1/64$
D. $1/128$

Free-Response Questions

The AP exam has long and short free-response questions. The long questions have considerable descriptive information that may include tables, graphs, or figures. The short questions are brief but may also include figures. Both kinds of questions have four parts and generally require that you bring together concepts from multiple areas of biology.

The questions that follow are designed to further your understanding of the concepts presented in this chapter. Unlike the free-response questions on the exam, they are narrowly focused on the material in this chapter. For free-response questions typical of the exam, take the two practice exams in this book.

Directions: The best way to prepare for the AP exam is to write out your answers as if you were taking the exam. Use complete sentences for all your answers. But do *not* use outline form or bullets. You may use diagrams to supplement your answers, but be sure to describe the importance or relevance of your diagrams.

1. Color blindness is inherited as an X-linked recessive allele. Explain why males have a greater chance of expressing the color-blindness trait.

2. A form of vitamin D resistance is inherited as an X-linked dominant allele. Explain why females have a greater chance of inheriting and expressing the vitamin D–resistant trait.

3. The height of an individual can be any value over a wide range, from short to tall. Describe a mechanism that can produce such continuous variation.

4. Describe Mendel's laws of segregation and independent assortment with respect to

- **a.** genes that are not linked
- **b.** genes that are linked
- **c.** crossing over
- **d.** sex-linkage
- **e.** Down syndrome
- **f.** Turner syndrome

Answers and Explanations

Multiple-Choice Questions

1. D. The chance that one die will turn up a three is 1 in 6, or $\frac{1}{6}$. For both dice to turn up a three, the probability is determined by multiplying the probability of each event happening independently, or $\frac{1}{6} \times \frac{1}{6} = \frac{1}{36}$.

2. B. Alleles refer to the various forms of a gene, or the various forms in which a particular gene can be expressed. Wrinkled and round are alleles that refer to two forms of a single gene (at a particular locus) that code for seed texture. Answer choice A is incorrect because wrinkled and yellow refer to two different genes (at different loci), one for seed texture and the other for seed color.

3. D. If you let T represent the dominant allele for the ability to taste PTC, then the cross would be $Tt \times Tt$. The Punnett square that follows shows that $\frac{3}{4}$ of the offspring have ability to taste PTC ($\frac{1}{4}$ TT + $\frac{1}{2}$ Tt).

	T	*t*
T	*TT*	*Tt*
t	*Tt*	*tt*

4. C. If both parents have normal wings (*DD* or *Dd*), there are three possible parent crosses: *DD* × *DD*, *DD* × *Dd*, or *Dd* × *Dd*. All of the progeny of *DD* × *DD* and of *DD* × *Dd* would have normal wings. Only the progeny of *Dd* × *Dd* would consist of $\frac{3}{4}$ normal-winged flies.

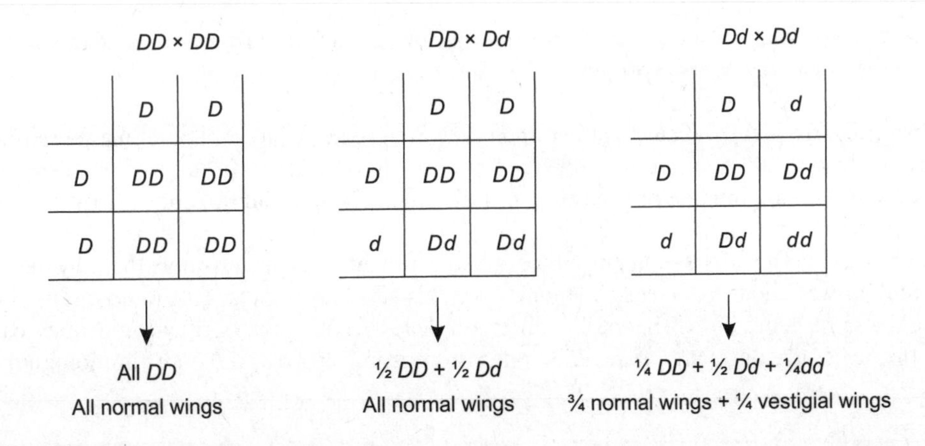

5. **B.** At the end of meiosis I, the two homologous chromosomes, one with *D* and one with *d*, would separate and migrate to opposite poles, which will form separate cells. During meiosis II, each chromosome separates into two chromatids (both of which will have exactly the same allele, assuming no crossing over), which migrate to opposite poles and become separate gametes. Thus, the cell containing the *D* chromosome will produce two gametes, each with a *D* chromosome (previously a chromatid). Similarly, the cell containing the *d* chromosome will produce two gametes, each with a *d* chromosome. At the end of meiosis II, then, there will be two gametes with a *D* chromosome and two gametes with a *d* chromosome.

6. **A.** Black is the phenotype of the sheep. That is given to you in the question. Without further information, you cannot identify the genotype of a black sheep because it could be either *BB* or *Bb*. The possible genotypes of the parents of a black sheep could be *BB* × *BB*, *BB* × *Bb*, *BB* × *bb*, or *Bb* × *Bb*. Thus, there is no one single genotype for either parent. Answer choice D is incorrect because although one parent would always be black, you cannot be certain whether the second parent is black or white.

7. **D.** In the *CCSS* × *ccss* cross, *CCSS* produces only *CS* gametes, and *ccss* produces only *cs* gametes. Thus, all offspring are *CcSs*.

8. **B.** The cross of *CcSs* × *CcSs* is the same kind of cross illustrated in Figure 8-5. Among the 16 genotypes in the Punnett square, only one would be *ccss*.

9. **C.** The cross of *CcSs* × *CcSs* is the same kind of cross illustrated in Figure 8-5. Among the 16 genotypes in the Punnett square, 9 would be normal for both traits (1 of *CCSS*, 2 of *CCSs*, 2 of *CcSS*, and 4 of *CcSs*).

10. **A.** The question involves the progeny of the cross *ttRR* × *TTR′R′*. Since *ttRR* produces only *tR* gametes, and *TTR′R′* produces only *TR′* gametes, all progeny will be *TtRR′*. The *Tt* genotype codes for tall, and *RR′* codes for pink.

11. **A.** This is a trick question because there's a long way and a very short way to solve this problem. The long way would be to construct a Punnett square and sort and count all the offspring. The short, easy way is to recognize that a white flower has the genotype *R′R′*. Looking at only the color gene for each parent, the cross is *RR′* × *RR*. In order for a cross to produce a white-flowered offspring (*R′R′*), both parents must contribute *R′*. Since this is not the case, no offspring will be white flowered.

12. **C.** It is not usually practical to make a Punnett square for genotypes involving more than two genes. In this problem, you are asked about the frequency of one specific offspring, *AABbCcDd*. To solve this problem, look at each gene separately. Looking at the first gene, the parents are *AA* × *AA* and all offspring will be *AA* (frequency of 1). For the second gene, *BB* × *bb*, all offspring will be *Bb* (1). For the third gene, the parents are *CC* × *Cc*, which produces ½ *CC* and ½ *Cc* (do a Punnett square to confirm this). Finally, a cross of the fourth gene, *Dd* × *Dd*, produces ¼ *DD*, ½ *Dd*, and ¼ *dd*. To find the probability of *AABbCcDd*, first find the frequency of each gene separately. The probability of *AA* is 1, of *Bb* is 1, of *Cc* is ½, and of *Dd* is ½. Then find the product of these frequencies. For *AABbCcDd*, the product is 1 × 1 × ½ × ½ = ¼.

13. **C.** Since the range of skin colors in humans shows continuous variation from very pale to very dark, it is most likely coded by many genes (polygenic inheritance).

14. **A.** When two traits frequently occur together, then they are probably linked. Sometimes, a red-headed person may not have freckles, or a freckled person may not have red hair. In these cases, there was probably a crossover event, exchanging one of the two genes with an allele that did not code for freckles or red hair.

15. **C.** Answer choice C is the false statement. Since *A* and *B* are on one chromosome, then, by definition, they are linked and answer choice A is true. Answer choice B is also true because the homologous chromosome would have the same genes but with possibly different alleles. In this case, both chromosomes carry the same genes, but the alleles are different (*A* and *B* on one chromosome and *a* and *b* on its homologue). Taking

both chromosomes together, you get the genotype *AaBb,* and, thus, answer choice D is true. Answer choice C is false because even though the chromosomes separate during meiosis to produce gametes that are *AB* and *ab,* crossing over can take place, producing some gametes that are *Ab* and some that are *aB.*

16. **C.** Begin by drawing a horizontal line with about 30 tick marks. Since *K* and *J* have a crossover frequency of 3, write the letters *K* and *J* on two marks 3 ticks apart, near the middle. Next, add the letter *L* in two positions, 8 units to the right of *K* and 8 units to the left of *K.* At this point of the solution, both positions are possible. Next, add the letter *M* 12 units to the right of *J* and, again, 12 units to the left of *J.* Both are possible. Last, use the *L-M* frequency to determine which configuration of *M* and *L* is correct. Since the *L-M* frequency is 7, only the *M* and *L* positions at the right are correct. That leaves only one possible sequence, *K-J-L-M.*

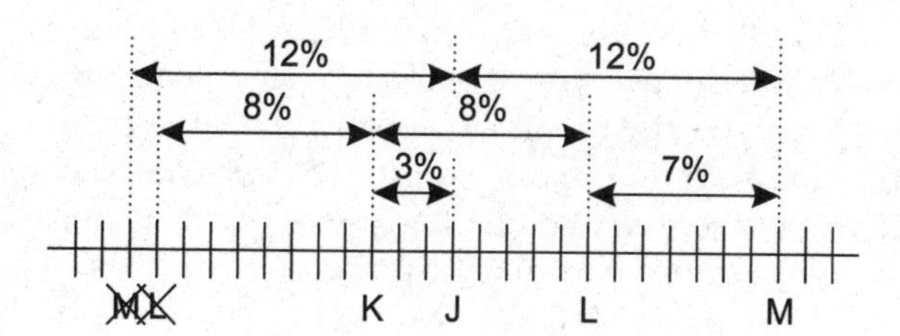

17. **D.** In any pedigree problem, you should begin by first identifying genotypes for which there is only one possibility. For a sex-linked recessive pattern of inheritance, you can identify the genotypes of all males and of all females that express the trait that the pedigree is describing. In this case, color-blind males are X^nY (filled boxes, 1, 10, 12, 16), normal males are X^NY (open boxes, 3, 6, 8, 13), and color-blind females are X^nX^n (filled circle, 7).

18. **B.** The next step in this pedigree problem is to identify the normal females (open circles). Are they X^NX^N or X^NX^n? Note that box 12 is a color-blind son (X^nY). Since a son can inherit only his Y chromosome from his father (box 6), box 12 must have inherited his X^n gene from his mother (circle 5). Thus, you can conclude that the mother, circle 5, is X^NX^n.

19. **B.** There are two possibilities for female 14, X^NX^N or X^NX^n. The color-blind mother, 7, is X^nX^n. The normal father, 8, is X^NY. A cross between these two individuals can produce only one kind of daughter, X^NX^n (confirm this with a Punnett square). For practice, you should try to identify every female. Female 7 is color blind (X^nX^n), females 2, 4, 5, 14, and 15 are carriers (X^NX^n), and females 9 and 11 are either normal (X^NX^N) or carriers (X^NX^n). (There is not enough information to determine which genotype is correct.)

20. **D.** To be a calico, a cat must have two X chromosomes, one with the orange allele and one with the black allele. Since a male cat has only one X chromosome, it normally can be only orange or black.

21. **A.** Since red-green color blindness is inherited as a sex-linked recessive allele, a color-blind male must be X^nY. Because he is a male, he received the *Y* from his father. Therefore, he inherited the X^n from his mother.

22. **D.** Turner syndrome is caused by the nondisjunction of the sex chromosomes. The result is a sperm or egg that is missing a sex chromosome. The formation of a zygote from the union of one of these sperm or eggs with a normal egg or sperm (with an X chromosome) results in Turner syndrome (XO). It is also possible to form an OY zygote, but because this zygote is missing an X chromosome, a chromosome with many essential genes, the zygote is nonviable. Answer choice C, Down syndrome, is also caused by nondisjunction, but of chromosome 21, not the sex chromosomes. Sickle-cell disease is inherited as an autosomal recessive allele, while hemophilia is inherited as a sex-linked recessive allele.

23. **D.** Since the genes *A* and *B* are on the same chromosome (linked), there are two possible allele arrangements for an *AaBb* individual. The first is that *AB* is on one chromosome and *ab* is on the homologous chromosome. The second possibility is that *Ab* is on one chromosome and *aB* is on the homologous chromosome. Using the first possibility as an example, one would expect only two kinds of gametes in the absence of crossing over—*AB* and *ab* in equal quantities. Crossing over would produce gametes that are *Ab* and *aB*. If the two genes are very close together, there would be very few crossovers because there are few places between the genes where the chromosomes can break and cross over. If the genes are far apart, there would be many crossovers because there are many places for chromosome breaks. When the genes are very far apart, they cross over so frequently that by observing the allele frequencies of the gametes the genes seem to assort independently, as if they were on different chromosomes (not linked). That is exactly what has happened in this question. The observed frequencies are those that would have been expected had the genes been on different chromosomes. Since the question states that the genes are linked, they must be far apart to allow so large a number of crossovers.

24. B. Without crossing over, the homozygous recessive parent can only produce one gamete genotype and the heterozygous parent can only produce two gamete genotypes (see Figure 8-6). Without crossing over, then, only two kinds of progeny phenotypes are produced, and they would be produced in equal numbers. These would be the progeny in the table with the largest numbers, 652 and 683. The smaller numbers, then, represent the progeny that are produced as a result of crossing over. The total number of crossover progeny produced is 77 + 88, or 165. Since the total number of progeny is 77 + 88 + 652 + 683, or 1,500, the crossover frequency is 165 ÷ 1,500, or 0.11 (11%).

25. C. First, find the probabilities for each gene:

$Aa \times Aa \rightarrow \frac{1}{4} AA + \frac{1}{2} Aa + \frac{1}{4} aa$

$Bb \times Bb \rightarrow \frac{1}{4} BB + \frac{1}{2} Bb + \frac{1}{4} bb$

$CC \times Cc \rightarrow \frac{1}{2} CC + \frac{1}{2} Cc$

$dd \times Dd \rightarrow \frac{1}{2} Dd + \frac{1}{2} dd$

$EE \times EE \rightarrow 1\ EE$

Then multiply the probabilities for each gene in the offspring: $aa \times BB \times CC \times dd \times EE = \frac{1}{4} \times \frac{1}{4} \times \frac{1}{2} \times \frac{1}{2} \times 1 = \frac{1}{64}$.

Free-Response Questions

1. Because males inherit only one X chromosome, they express all the genes on that chromosome, regardless of whether the alleles are dominant or recessive. In contrast, females inherit two X chromosomes and both alleles for a gene must be recessive in order for a female to express a recessive trait.

2. Because females have two X chromosomes, they have two chances of inheriting a trait on that chromosome and they need only one copy of a dominant allele to express the trait.

3. A trait that is produced by the expression of many interacting genes (polygenic inheritance) generates continuous variation in a population. Also, multiple alleles of a single gene can recombine to produce many genotypic and phenotypic combinations.

4. a. When chromosomes align (on the metaphase plate) during meiosis I, homologous chromosomes are paired. Each homologue migrates to a separate pole and becomes a member of a separate gamete. The migration to separate poles is random—that is, either chromosome of a homologous pair can migrate to either pole (Mendel's law of segregation).

Different homologous pairs of chromosomes act independently of other homologous chromosome pairs. Thus, genes that are on different chromosomes (unlinked) migrate independently of genes on other chromosomes (Mendel's law of independent assortment).

b. When two genes are linked, they are on the same chromosome. If they are on the same chromosome, they migrate together to either pole (unless crossing over occurs). Thus, they violate Mendel's law of independent assortment and are inherited together as if they were a single gene in a monohybrid cross. If the dominant alleles *A* and *B* are on one chromosome and the recessive alleles *a* and *b* are on the homologous chromosome, they produce only two kinds of gametes, *AB* and *ab.* Then, the dihybrid cross *AaBb* × *AaBb* would produce a 1:2:1 genotypic ratio for *AABB, AaBb,* and *aabb* with a phenotypic ratio of 3:1, not the typical 9:3:3:1 phenotypic ratio that Mendel found when using unlinked genes. The 1:2:1 and 3:1 genotypic and phenotypic ratios are those expected from a typical monohybrid cross.

c. Crossing over occurs between linked genes. Instead of producing only two kinds of gametes—say, *AB* and *ab*—exchanges occur between homologous chromosomes, producing some *Ab* and *aB* gametes, quantities of which depend on the frequency of crossing over. The frequency of crossing over increases as the distance between the gene loci increases.

d. Sex-linkage occurs when a gene is located on one of the sex chromosomes, usually the X chromosome. For example, in humans, hemophilia is inherited as a recessive allele on the X chromosome. Females receive two copies of the gene, one on each of their X chromosomes. If they receive two recessive alleles, they are hemophiliacs. If they receive two normal alleles, they are normal, but if they inherit one normal and one hemophilia allele, they will have normal clotting abilities (because the normal allele is dominant) but will be carriers of the disease. Males, on the other hand, inherit only one X chromosome and, thus, only one copy of the allele. If they receive the normal allele, they will have normal clotting; if they receive the hemophilia allele, they will be hemophiliacs. Because they need only one copy of the allele to express the trait, sex-linked diseases are more common in males than in females.

e. Down syndrome occurs as a result of the nondisjunction of the two number 21 chromosomes. As a result, the homologous pair does not separate and move to opposite poles (as the law of segregation implies), but rather both chromosomes end up at the same pole and in the same gamete. Two kinds of gametes are formed, one with two copies of chromosome 21 and one with no chromosome 21. Only the gamete with two copies of the chromosome is viable. The zygote formed between this gamete and a normal gamete will have three copies of chromosome 21, and the infant will express the Down syndrome phenotype, which consists of physical abnormalities and mental retardation.

f. Turner syndrome results from a nondisjunction of the sex chromosomes. This results in a gamete that has either two sex chromosomes or no sex chromosomes. If a gamete with no sex chromosomes (O) fuses with a normal gamete bearing the X chromosome, the resulting zygote will have only a single X chromosome (XO) and express the Turner syndrome phenotype. Turner syndrome individuals are female and exhibit physical abnormalities, including sterility.

Chapter 9

Gene Expression and Regulation

Review

Chromosomes bear the genetic information that is passed from parents to offspring. The genetic information is stored in molecules of DNA. The DNA is used to make RNA, and RNA assembles amino acids into proteins. Some proteins form the basic structure and appearance of cells, while other proteins, those that function as enzymes, regulate chemical reactions that direct metabolism for cell development, growth, and maintenance. Following is the summary of this process that begins with DNA to create a living, functioning organism.

DNA → RNA → proteins/enzymes → traits, metabolism, homeostasis

This flow of information is described as the **central dogma** of molecular biology. The underlying molecular mechanisms for this process are the subject of this chapter.

The structure of DNA and RNA was presented earlier in Chapter 2, "Chemistry of Life." As a review, both DNA and RNA are polymers of **nucleotides.** The nucleotide monomer consists of three covalently bonded parts—a nitrogen base, a sugar, and a phosphate. Figure 9-1 reviews the differences in the structures of DNA and RNA and summarizes their functions. Details of the functions of these molecules will be presented in this chapter.

	Nucleotide Components		Function	Structure
	Sugar	Nitrogen Bases		
DNA	deoxyribose	adenine, *thymine*, guanine cytosine	contains hereditary information (genes) of the cell	double helix
RNA (involved in protein synthesis)	ribose	adenine, *uracil*, guanine cytosine	mRNA (messenger RNA) - provides the instructions for assembling amino acids into a polypeptide chain	linear
			tRNA (transfer RNA) - delivers amino acids to a ribosome for their addition into a growing polypeptide chain	upside-down "L" shape
			rRNA (ribosomal RNA) - combines with proteins to form ribosomes	globular
RNA (involved in RNA processing)	ribose	adenine, *uracil*, guanine cytosine	snRNA (small nuclear RNA) - combines with proteins to form small nuclear ribonucleoproteins (snRNPs) which process RNAs before they leave the nucleus	globular
RNA (involved in regulating gene expression)	ribose	adenine, *uracil*, guanine cytosine	miRNA (microRNA) - regulates gene expression by blocking or degrading mRNA	linear
			siRNA (short interfering RNA) - regulates gene expression by blocking or degrading mRNA	linear

Comparison of DNA and RNA

Figure 9-1

Except for asexually reproducing organisms and identical twins, the DNA of every individual is different. These differences, resulting from variations in the sequence of nucleotides, generate different RNA, which produces different proteins.

Early Experiments

Experiments during the first half of the twentieth century lead to the identification of DNA as the hereditary material and that the three-dimensional shape of a DNA molecule is a double helix. Four important research efforts are summarized here:

1. **Griffith discovers that genetic information can be transferred from dead bacteria to living bacteria.** Microbiologist Frederick Griffith experimented with two strains of a bacterium—one that produces a polysaccharide coat and causes pneumonia, and a mutant strain, without the coat, that does not cause pneumonia. Griffith killed the disease-causing bacteria with heat and showed that they could no longer cause pneumonia in mice. He then injected into mice both the dead bacteria that once could cause disease and live bacteria that could not cause disease. These mice died, and Griffith found live bacteria in the mice that had polysaccharide coats. The descendants of these bacteria also had polysaccharide coats and would cause disease. Griffith concluded that genetic information from the dead bacteria with polysaccharide coats *transformed* the bacteria without coats, giving them the ability to make coats and cause disease. Today, the term **transformation** is used to describe the ability of bacteria to absorb and express genetic information (now known to be DNA) obtained from their surroundings.
2. **Avery, MacLeod, and McCarty identify DNA as the heredity information of a cell.** Using the same bacteria used by Griffith, bacteriologists Oswald Avery, Colin MacLeod, and Maclyn McCarty removed the proteins and polysaccharide coats from the dead, disease-causing bacteria. They found that the remaining material was still able to transform bacteria, giving previously harmless bacteria the ability to cause disease. Further tests confirmed that the transforming material was not RNA, but a substance with the same properties as DNA.
3. **The Hershey and Chase experiments establish that DNA was the genetic material of phages.** At the time of these experiments, it was known that **phages,** viruses that infect bacteria, consisted of DNA and protein. In the first part of their experiment, geneticists Alfred Hershey and Martha Chase substituted radioactive sulfur for the sulfur in the amino acids of the phage proteins and mixed these phages with *E. coli* bacteria. After separating the bacteria from the growing medium using a centrifuge, they found the culture media, not the bacteria, were radioactive, indicating that the phage proteins did not enter the bacteria. In the second part of the experiment, they substituted radioactive phosphorous for the phosphorous in the phage DNA. Following the same procedure as in the first part of the experiment, they found that the bacteria, not the growing media, were radioactive, indicating that the phage DNA had entered the bacteria. In a follow-up experiment, the researchers found that infected radioactive bacteria released new phages that were also radioactive. Hershey and Chase concluded that the radioactive DNA from the phages provided the genetic information needed to make new viruses.
4. **Watson, Crick, Wilkins, and Franklin determine the structure of DNA.** Using DNA prepared in the lab of biophysicist Maurice Wilkins, chemist Rosalind Franklin produced an X-ray diffraction photograph of DNA. X-ray diffraction creates a black-and-white pattern of spots that reveal certain structural characteristics of crystals. For DNA, the pattern revealed that the molecule consisted of two strands wrapped around each other (a double helix). Franklin also proposed that sugar-phosphate material formed the outside of the double helix because of its hydrophilic properties, while the hydrophobic nitrogenous bases were located on the inside of the molecule. Using that information, molecular biologists James Watson and Francis Crick proposed a model of DNA resembling a twisted ladder, where the vertical sides of the ladder are sugar-phosphate molecules and its horizontal rungs are pairs of nitrogen bases in which adenine pairs with thymine and guanine pairs with cytosine.

Later research elucidated the flow of information from genes, the hereditary units of DNA, to traits. In the **one-gene-one-enzyme hypothesis,** the gene was defined as the segment of DNA that codes for a particular enzyme. But because many genes code for polypeptides that are not enzymes (such as structural proteins, regulatory proteins, or individual components of enzymes), the gene has since been redefined as the DNA segment that codes for a particular polypeptide **(one-gene-one-polypeptide hypothesis).** In Chapter 8, "Heredity," the terms **gene** and **genotype** are used to represent the genetic information for a particular trait. From the molecular viewpoint presented in this chapter, *traits are the end products of metabolic processes regulated by enzymes.*

DNA Replication

During interphase of the cell cycle, a second chromatid containing a copy of the DNA molecule is assembled. The process, called **DNA replication,** involves separating (unzipping) the double-stranded DNA molecule into two strands, each of which serves as a template to assemble a new, complementary strand. The result is two identical double-stranded molecules of DNA. Because each of these double-stranded molecules of DNA consists of a single strand of old DNA (the template strand) and a single strand of new, replicated DNA (the complementary strand), the process is called **semiconservative replication.**

During DNA replication, the enzyme **helicase** unwinds the DNA helix, forming a Y-shaped **replication fork. Single-strand binding proteins** attach to each strand of the uncoiled DNA to keep them separate. As helicase unwinds the DNA, it forces the double-helix in front of it to twist. A group of enzymes, called **topoisomerases,** break and rejoin the double helix, allowing the twists to unravel and preventing the formation of knots.

Since a DNA double-helix molecule consists of two opposing DNA strands, the uncoiled DNA consists of a $3' \rightarrow 5'$ template strand and a $5' \rightarrow 3'$ template strand. The enzyme that assembles the new DNA strand, DNA polymerase, moves in the $3' \rightarrow 5'$ direction along each template strand. A new (complement) strand grows in the **antiparallel,** $5' \rightarrow 3'$ direction.

For the $3' \rightarrow 5'$ template strand, replication occurs continuously as the DNA polymerase follows the replication fork, assembling a $5' \rightarrow 3'$ complementary strand. This complementary strand is called the **leading strand.**

For the $5' \rightarrow 3'$ template strand, however, the DNA polymerase moves away from the uncoiling replication fork. This is because it can assemble nucleotides only as it travels in the $3' \rightarrow 5'$ direction. As the helix is uncoiled, DNA polymerase assembles short segments of nucleotides along the template strand in the direction away from the replication fork. After each complement segment is assembled, the DNA polymerase must return back to the replication fork to begin assembling the next segment. These short segments of complementary DNA are called **Okazaki fragments.** The Okazaki fragments are connected by **DNA ligase,** producing a single complementary strand. Because this complementary strand requires more time to assemble than the leading strand, it is called the **lagging strand.**

DNA polymerase is able to attach nucleotides only to an already existing complementary strand. Therefore, to initiate a new complementary strand, another enzyme, **primase,** begins replication with a short segment of RNA (not DNA) nucleotides, called an **RNA primer.** The leading strand and every Okazaki fragment on the lagging strand must begin with an RNA primer. When the primer is in place, DNA polymerase can attach succeeding DNA nucleotides to the primer. The RNA nucleotides of the RNA primer are later replaced with DNA nucleotides by DNA polymerase.

Figure 9-2 illustrates the growth of leading and lagging DNA complements. In the figure, the RNA primer that initiated the leading strand is not shown because it was replaced with DNA nucleotides earlier in its synthesis. The growing Okazaki fragment on the lagging strand, however, still has its RNA primer attached, because a primer must initiate each new fragment.

The details of DNA replication are summarized here. Numbers correspond to events illustrated in Figure 9-2:

1. **Helicase** unwinds the DNA, producing a **replication fork** (1A, 1B). **Single-strand binding proteins** prevent the single strands of DNA from recombining (1C). **Topoisomerase** removes twists and knots that form in the double-stranded template as a result of the unwinding induced by helicase (1D).
2. **Primase** (2A) initiates DNA replication at special nucleotide sequences, called **origins of replication,** with short segments of RNA nucleotides, called **RNA primers** (2B).
3. **DNA polymerase** attaches to the RNA primers and begins **elongation,** the adding of DNA nucleotides to the complementary strand.
4. The **leading complementary strand** is assembled continuously toward the replication fork as the double-helix DNA uncoils.

5. The **lagging complementary strand** (5A) is assembled away from the replication fork in multiple, short **Okazaki fragments** (5B). Each new Okazaki fragment begins when DNA polymerase attaches to an RNA primer (5C).
6. The Okazaki fragments are joined by **DNA ligase.**
7. The RNA primers are replaced with DNA nucleotides.

Energy for elongation is provided by two additional phosphates that are attached to each new nucleotide (making a total of three phosphates attached to the nitrogen base). Breaking the bonds holding the two extra phosphates provides the chemical energy for the process.

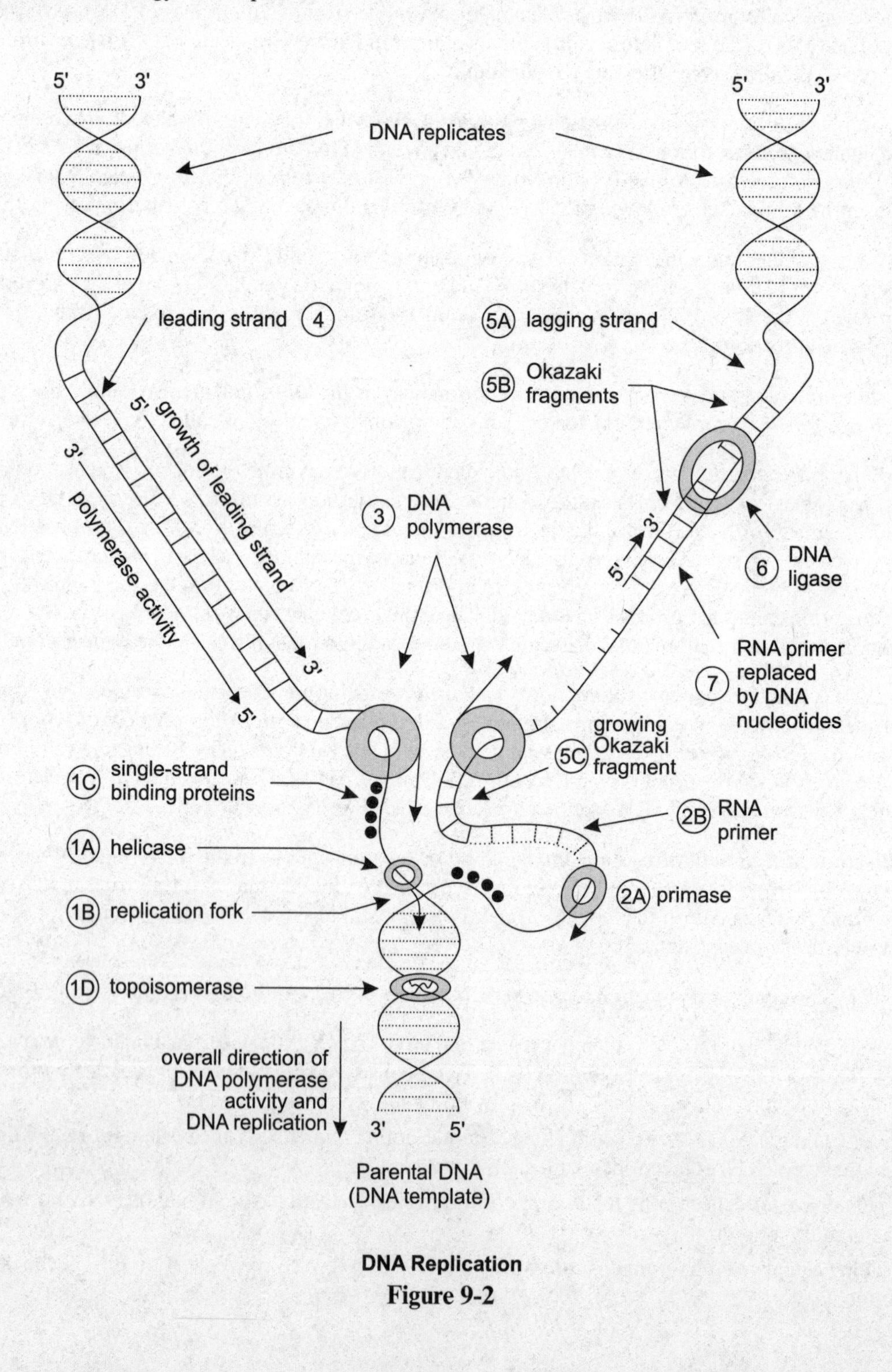

DNA Replication

Figure 9-2

DNA replication in prokaryotic and eukaryotic organisms is basically the same, with the following differences:

1. **Chromosome structure.** A prokaryotic chromosome is circular. Eukaryotic chromosomes are linear with ends called **telomeres.**
2. **Origins of replications.** A prokaryotic chromosome has one unique origin of replication. Eukaryotes have multiple origins to accommodate the much larger size of their chromosomes.

DNA Repair

DNA replication is not perfect. Errors happen. Various mechanisms, however, are present to repair the errors, including the following:

1. **Proofreading** of a newly attached base to the growing replicate strand is carried out by DNA polymerase. DNA polymerase checks to make sure that each newly added nucleotide correctly base-pairs with the template strand. If it does not, the nucleotide is removed and replaced with the correct nucleotide.
2. **Mismatch repair proteins** repair errors that escape the proofreading ability of DNA polymerase.
3. **Excision repair proteins** identify and remove damaged nucleotides caused by environmental factors, such as toxins or radiation (UV, X-rays). A polymerase then uses the undamaged complementary strand as a template to repair the damage.

Protein Synthesis

The DNA in chromosomes contains genetic instructions that regulate development, growth, and the metabolic activities of cells. The DNA instructions determine whether a cell will be that of a pea plant, a human, or some other organism, as well as establish specific characteristics of the cell in that organism. For example, the DNA in a cell may establish that it is a human cell. If, during development, it becomes a cell in the iris of an eye, the DNA will direct other information appropriate for its location in the organism, such as the concentration of melanin pigmentation (which influences the appearance of different colors). DNA controls the cell in this manner because it contains codes for polypeptides. Many polypeptides are enzymes that regulate chemical reactions, and these chemical reactions influence the resulting characteristics of the cell.

The process that describes how enzymes and other proteins are made from DNA is called **protein synthesis.** The three steps in protein synthesis are **transcription, RNA processing,** and **translation.** In transcription, RNA molecules are created by using one strand of the DNA molecule as a template. After transcription, RNA processing modifies the RNA molecule with deletions and additions. In translation, the processed RNA molecules are used to assemble amino acids into a polypeptide.

There are three kinds of RNA molecules necessary for protein synthesis that are produced during transcription, as follows:

1. **Messenger RNA (mRNA)** is a single strand of RNA that provides the template used for sequencing amino acids into a polypeptide. A triplet group of three adjacent nucleotides on the mRNA, called a **codon,** codes for one specific amino acid. Since there are 64 possible ways that four nucleotides can be arranged in triplet combinations ($4 \times 4 \times 4 = 64$), there are 64 possible codons. However, there are only 20 amino acids, and thus, some codons code for the same amino acid. The **genetic code,** given in Figure 9-3, provides the decoding for each codon. That is, it identifies the amino acid specified by each of the possible 64 codon combinations. For example, the codon composed of the three nucleotides cytosine-guanine-adenine (CGA) codes for the amino acid arginine. This can be found in Figure 9-3 by aligning the C found in the first column with the G in the center part of the table and the A in the column at the far right. Note that three of the codons in the genetic code are stop codons. They signal an end to translation rather than code for an amino acid. Therefore, only 61 of the codons actually code for amino acids. The codon that codes for the amino acid methionine is also the codon that signals the beginning of translation.

   Note: *You do not need to memorize the genetic code (Figure 9-3) for the AP exam.*

2. **Transfer RNA (tRNA)** is a short RNA molecule (consisting of about 80 nucleotides) that is used for transporting amino acids to their proper place on the mRNA. Interactions among various parts of the tRNA molecule result in base-pairings between nucleotides, folding the tRNA in such a way that it forms a three-dimensional molecule. (In two dimensions, a tRNA resembles the three leaflets of a clover leaf; in three dimensions, it resembles an upside-down L.) One end of the tRNA attaches to an amino acid. Another portion of the tRNA, specified by a triplet combination of nucleotides, is the **anticodon.** During translation, the anticodon of the tRNA base-pairs with the codon of the mRNA. Exact base-pairing between the third nucleotide of the tRNA anticodon and the third nucleotide of the mRNA codon is often not required. This relaxed base-pairing requirement, called **wobble pairing,** allows the anticodon of some tRNAs to base-pair with more than one kind of codon. As a result, about 45 different tRNAs base-pair with the 61 codons that code for amino acids.
3. **Ribosomal RNA (rRNA)** molecules combine with various proteins to form ribosomes. The rRNA molecules are transcribed in the nucleolus and assembled with proteins imported from the cytoplasm to form a large and a small ribosome subunit. In the cytoplasm, the two subunits join to form a ribosome that coordinates the activities of the mRNA and tRNA during translation.

First Letter ↓	Second Letter: U	C	A	G	Third Letter ↓
U	phenylalanine	serine	tyrosine	cysteine	U
	phenylalanine	serine	tyrosine	cysteine	C
	leucine	serine	STOP	STOP	A
	leucine	serine	STOP	tryptophan	G
C	leucine	proline	histidine	arginine	U
	leucine	proline	histidine	arginine	C
	leucine	proline	glutamine	arginine	A
	leucine	proline	glutamine	arginine	G
A	isoleucine	threonine	asparagine	serine	U
	isoleucine	threonine	asparagine	serine	C
	isoleucine	threonine	lysine	arginine	A
	methionine and START	threonine	lysine	arginine	G
G	valine	alanine	aspartate	glycine	U
	valine	alanine	aspartate	glycine	C
	valine	alanine	glutamate	glycine	A
	valine	alanine	glutamate	glycine	G

The Genetic Code

Figure 9-3

Transcription

Transcription begins with **initiation,** continues with **elongation,** and ends with **termination.** The details follow, with numbers corresponding to events illustrated in Figure 9-4:

1. In **initiation,** the RNA polymerase attaches to a **promoter region** on the DNA and begins to unzip the DNA into two strands. A promoter region for mRNA transcriptions often contains the sequence T–A–T–A (called the TATA box).
2. **Elongation** occurs as the RNA polymerase unzips the DNA and assembles RNA nucleotides using one strand of the DNA as a template. As in DNA replication, elongation of the RNA molecule occurs in the $5' \rightarrow 3'$ direction. In contrast to DNA replication, new nucleotides are RNA nucleotides (rather than DNA nucleotides), only one DNA strand is transcribed, and primers are not required.
3. **Termination** occurs when the RNA polymerase reaches a special sequence of nucleotides that serve as a termination point.

mRNA Processing

Before the mRNA molecule leaves the nucleus, it undergoes the following alterations:

1. A **5′ GTP cap** (–P–P–P–G-5′) is added to the 5′ end of the mRNA (see 5 in Figure 9-4). The 5′ GTP cap is a guanine nucleotide with two additional phosphate groups, forming GTP (in the same way that ATP is an adenine nucleotide with two additional phosphates). Capping provides stability to the mRNA and a point of attachment for the small subunit of the ribosome.

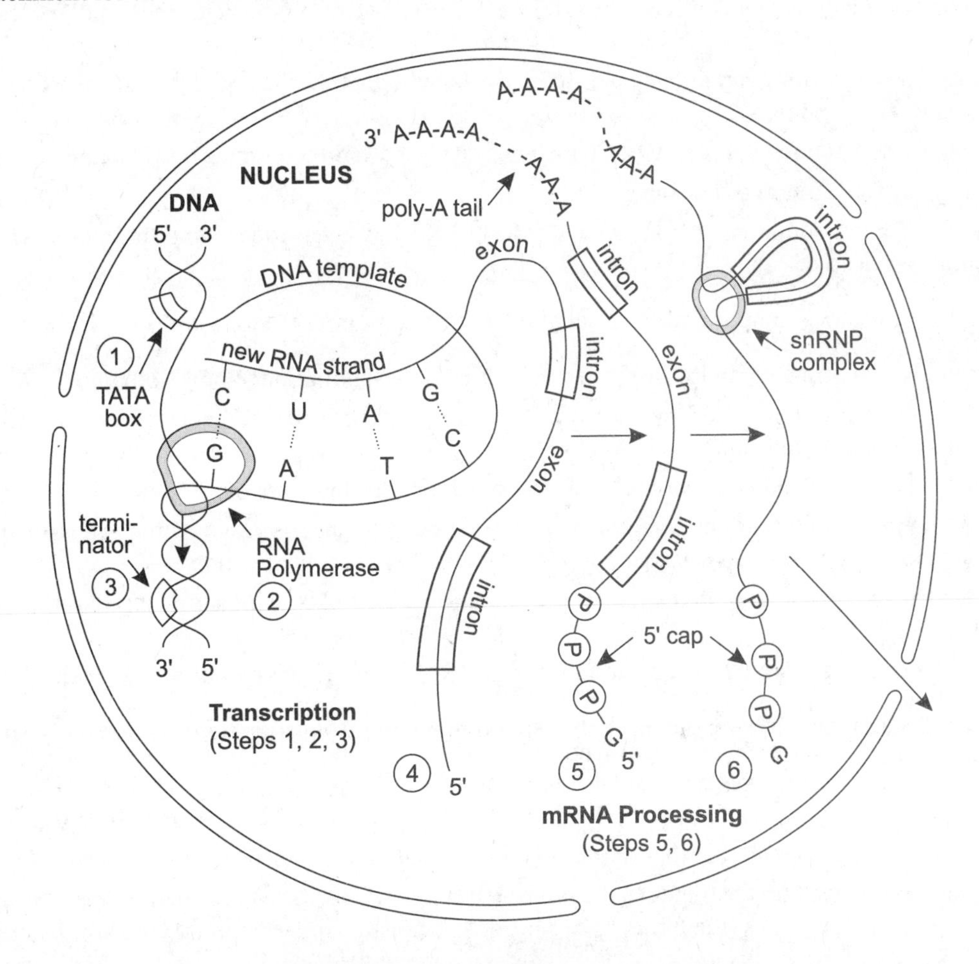

Transcription and mRNA Processing

Figure 9-4

2. A **poly-A tail** (–A–A–A . . . A–A-3′) is attached to the 3′ end of the mRNA (see 5 in Figure 9-4). The tail consists of about 200 adenine nucleotides. It provides stability to the mRNA and also appears to control the movement of the mRNA across the nuclear envelope.
3. **RNA splicing** removes nucleotide segments from mRNA. A transcribed DNA segment contains two kinds of sequences—**exons,** which are sequences that *ex*press a code for a polypeptide, and **introns,** *in*tervening sequences that are noncoding. The original unprocessed mRNA transcript contains both the coding and the noncoding sequences (see 4 in Figure 9-4). Before the mRNA moves to the cytoplasm, **small nuclear ribonucleoproteins,** or **snRNPs** (pronounced "snurps"), delete the introns and splice the exons (see 6 in Figure 9-4). A snRNP consists of **small nuclear RNAs (snRNAs)** and various proteins.
4. **Alternative splicing** allows different mRNAs to be generated from the same RNA transcript. By selectively removing different parts of an RNA transcript, different mRNAs can be produced, each coding for a different protein product.

Translation

After transcription, the mRNA, tRNA, and ribosomal subunits are transported across the nuclear envelope and into the cytoplasm. Once in the cytoplasm, a specific amino acid attaches to each of the tRNAs using energy from ATP.

As an introduction, this is what happens during translation:

1. The mRNA attaches to the ribosome.
2. The sequence of codons on the mRNA determines the sequence of amino acids in the polypeptide to be synthesized.
3. One by one, a tRNA brings an amino acid to the ribosome such that the anticodon of the tRNA base-pairs with the codon of the mRNA.
4. The newly arrived amino acid is attached with a peptide bond to other amino acids already present.
5. A tRNA is released from the ribosome.
6. The process is repeated until a "stop" codon on the mRNA has been reached and the completed polypeptide is released.

The details are slightly more complicated. As a prelude to the details, note the following:

1. As in transcription, translation is categorized into three steps—initiation, elongation, and termination.
2. Energy for translation is provided by several GTP (guanosine triphosphate) molecules. GTP acts as an energy supplier in the same manner as ATP.
3. A ribosome has three binding sites for tRNA molecules arranged in a series:
   - The A site (for *a*mino acid or "*a*cceptor"), in the *first* position, accepts an incoming tRNA carrying an amino acid. The amino acid is passed on to the tRNA in the second position.
   - The P site (for *p*olypeptide), in the *second* position, holds the tRNA with a growing chain of amino acids (polypeptide).
   - The E site (for *e*xit), in the *third* position, holds the tRNA after it gives up its amino acid.

The details of translation follow, with numbers corresponding to events illustrated in Figure 9-5:

1. **Initiation** begins when the small ribosomal subunit attaches near the beginning of the mRNA.
2. A tRNA (with anticodon UAC) carrying the amino acid methionine attaches to the mRNA at the start codon AUG. (You can remember that the start codon is AUG because school often starts in *Aug*ust.)
3. The large ribosomal subunit attaches to the mRNA with the tRNA (bearing a methionine), occupying the middle of three binding sites (the P site). The ribosome is now completely assembled with the mRNA and one tRNA.

4. **Elongation** occurs as additional tRNAs arrive bearing their amino acids. A newly arriving tRNA attaches to the first binding site (the A site), always with the anticodon of the tRNA appropriately base-pairing with the codon of the mRNA.
5. The amino acid on the tRNA in the second binding site is transferred to the amino acid on the newly arrived tRNA in the first binding site. In Figure 9-5, several amino acids of the growing polypeptide chain are transferred because the figure shows elongation after several tRNAs have delivered amino acids.

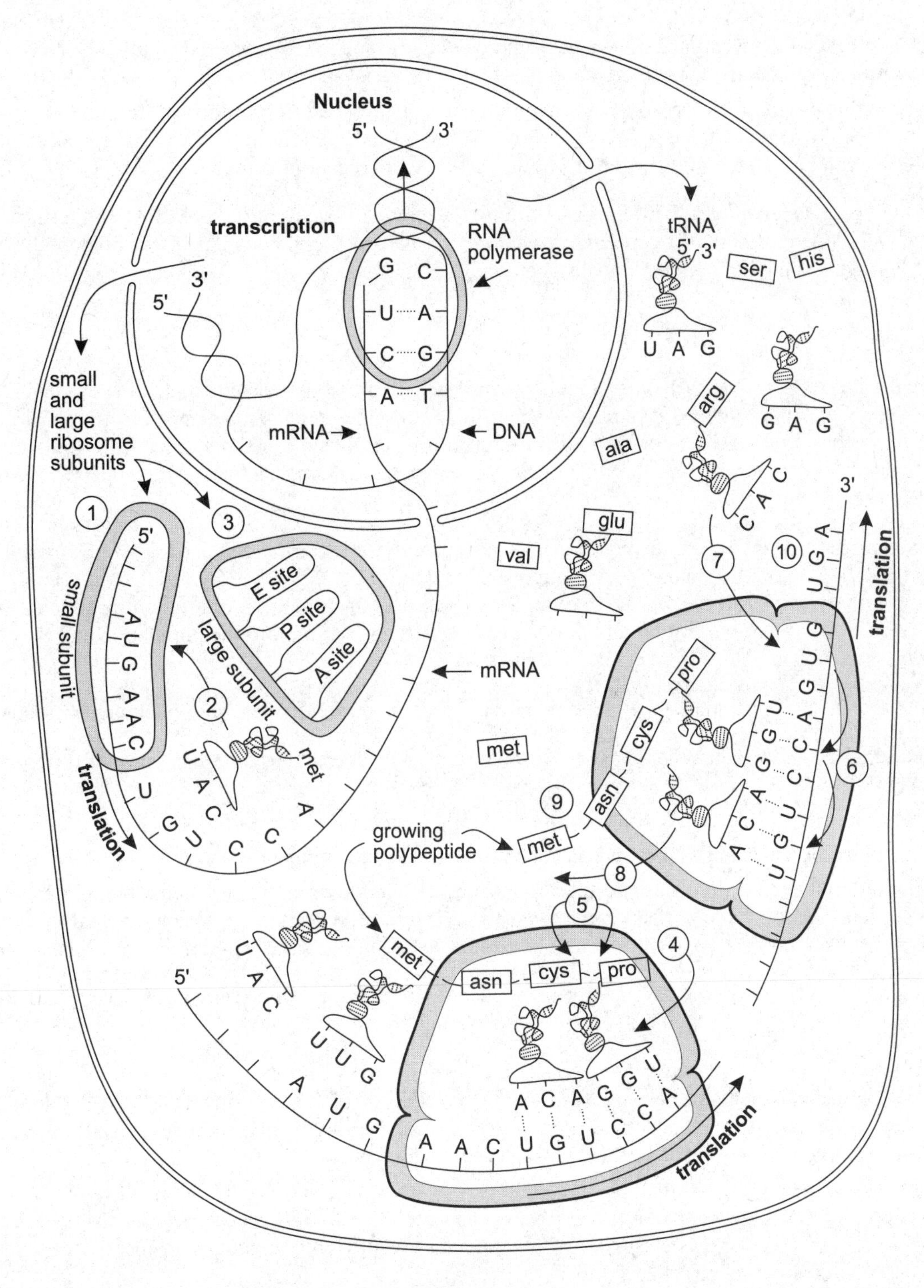

Protein Synthesis
Figure 9-5

6. Translocation occurs as the ribosome moves over one binding site: The tRNA in the middle moves to the third position (the E site), and the tRNA in the first position (bearing the amino acids) moves to the second position.
7. This leaves the first position binding site empty, allowing for the arrival of a new tRNA.
8. Meanwhile, the tRNA in the third position is released, now free to bind again with its specific amino acid and provide another delivery to the mRNA.
9. Elongation continues as each successive tRNA delivers an amino acid. As each new tRNA arrives, the polypeptide chain is elongated by one new amino acid, growing in sequence and length as prescribed by the sequence of codons on the mRNA.
10. **Termination** occurs when the ribosome encounters one of the three stop codons. At termination, the completed polypeptide, the last tRNA, and the two ribosomal subunits are released. The ribosomal subunits can now attach to the same or another mRNA and repeat the process.

Once the polypeptide is completed, interactions among the amino acids give it its secondary and tertiary structures. Subsequent processing by the endoplasmic reticulum or a Golgi body may make final modifications before the protein attains its final functional structure.

Mutations

A **mutation** is any sequence of nucleotides in a DNA molecule that does not exactly match the original DNA molecule from which it was copied. Mutations can occur as a result of replication errors, or they may result from environmental effects such as radiation (for example, ultraviolet or X-ray) or reactive chemicals. Radiation or chemicals that cause mutations are called **mutagens. Carcinogens** are mutagens that activate uncontrolled cell growth (cancer).

As reviewed earlier in this chapter, there are various mechanisms for repairing damaged DNA. Damaged DNA that is not repaired becomes a mutation. Although some mutations have no phenotypic effect, most are deleterious—that is, they lead to a partial or complete loss of cell functionality. A mutation that occurs in a sex cell is passed on to the next generation and introduces new allelic variation into the population and the potential for evolutionary change.

There are various kinds of mutations. A **point mutation** is a single nucleotide error and includes the following:

1. A **substitution** occurs when the DNA sequence contains an incorrect nucleotide in place of the correct nucleotide.
2. A **deletion** occurs when a nucleotide is omitted from the nucleotide sequence.
3. An **insertion** occurs when a nucleotide is added to the nucleotide sequence.
4. A **frameshift** mutation occurs as a result of a nucleotide deletion or insertion. Such mutations cause all subsequent nucleotides to be displaced one position. If a frameshift mutation occurs in a DNA segment whose transcription produces the mRNA, all codons following the transcribed mutation will change.

A point mutation may or may not have a significant phenotypic effect. If the mRNA is produced from a DNA segment that contains a point mutation, one of the following results:

1. A **silent mutation** occurs when the new codon still codes for the *same* amino acid. This occurs most often when the nucleotide substitution results in a change of the last of the three nucleotides in a codon. This relaxed requirement for the nucleotide in the third position is called **wobble pairing.** Consult Figure 9-3 for examples of codons that differ by their third nucleotide but code for the same amino acid.
2. A **missense mutation** occurs when the new codon codes for a *new* amino acid. The effect can be minor, or it may result in the production of a protein that is unable to fold into its proper three-dimensional shape and, therefore, is unable to carry out its normal function. The hemoglobin protein that causes sickle-cell disease is caused by a missense mutation.
3. A **nonsense mutation** occurs when the new codon codes for a stop codon. The hemoglobin protein that causes some forms of thalassemia is caused by a nonsense mutation.

Chromosomal aberrations are changes in the chromosome structure or in the makeup of the genome:

1. **Deletions** occur when segments of a chromosome are lost. Most chromosome deletions result in a significant loss of important DNA and are fatal.
2. **Duplications** occur when segments of a chromosome are repeated. If the duplication occurs within a gene segment, it is likely to cause a frameshift mutation with deleterious consequences. On the other hand, a duplicated gene can have beneficial effects by providing additional gene products for processes that are in high demand. For example, most species, including bacteria, have gene redundancy for the gene that codes for rRNA, a product in high demand. Also, extra copies of a gene provide the opportunity for subsequent mutations to create novel gene variations without interfering with the normal operation of the original gene. Some examples of this follow.
   - **Globin genes.** The great similarities among the various globin chains of hemoglobin suggest that they each evolved from a common gene. In humans, multiple variations of the gene occur on two separate chromosomes.
   - **Antifreeze genes.** In certain arctic fish, glycoproteins in the blood provide resistance to freezing. The genes for these glycoproteins appear to be the result of multiple duplications and divergence of the gene that codes for trypsinogen, an enzyme whose activated form, trypsin, digests proteins in the small intestine. This illustrates how novel genes can originate from gene duplication of genes originally used for a totally different purpose.
3. **Inversions** occur when a DNA segment is reversed. Depending upon where the chromosome breaks occur, the mutation may or may not have a significant effect.
4. **Translocations** occur when segments of the chromosome are deleted or copied and then inserted elsewhere, either within the same chromosome or in another chromosome. For example, one form of Down syndrome occurs when a piece of chromosome 21 is translocated to chromosome 14. Offspring who inherit this chromosome 14, along with the two normal copies of chromosome 21, effectively received three copies of the translocated segment.
5. **Transposons** (transposable elements, mobile genetic elements, or "jumping genes") are naturally occurring mutations. They are DNA segments that insert themselves throughout the genome after copying or deleting themselves from another area.
   - In corn (maize), transposons are responsible for mutant strains whose kernels lack pigmentation or have spotted pigmentation.
   - The human genome has as much as 50% of its DNA derived from transposons, although most transposons appear to be inactively sitting within introns.

Viruses

Viruses are parasites of cells. A typical virus penetrates a cell, commandeers its metabolic machinery, assembles hundreds of new viruses that are copies of itself, and then leaves the cell to infect other cells. In the process, the host cell is usually destroyed.

Viruses are specific for the kinds of cells they will parasitize. Some viruses attack only one kind of cell within a single host species, and others attack similar cells from a range of closely related species. **Bacteriophages,** or **phages,** for example, are viruses that attack only bacteria.

Viruses consist of the following structures:

1. A **nucleic acid,** either RNA or DNA (but not both), contains the hereditary information of the virus. The DNA or RNA may be double-stranded (**dsDNA** or **dsRNA**) or single-stranded (**ssDNA** or **ssRNA**).
2. A **capsid,** or **protein coat,** encloses the nucleic acid.
3. An **envelope** surrounds the capsid of some viruses. Envelopes incorporate phospholipids and proteins obtained from the cell membrane of the host cell.

Replication of viruses occurs by one of the following two cycles:

1. In the **lytic cycle,** a virus penetrates the cell membrane of the host and uses the enzymes of the host to produce viral nucleic acids and viral proteins. The nucleic acids and proteins are then assembled into new viruses, which subsequently erupt from the host cell, destroying the cell in the process. The new viruses then infect other cells, and the process repeats. There are variations of this theme, depending upon whether the nucleic acid of the virus is DNA or RNA, double-stranded or single-stranded. Some important variations follow.
   - For most DNA viruses, the DNA is replicated to generate new viral DNA and the DNA is transcribed to produce viral mRNA. Viral mRNA is then translated to produce viral proteins. The DNA and proteins are assembled into new viruses.
   - For some RNA viruses, the RNA serves as mRNA or as a template to make mRNA. The mRNA is translated to make proteins, and these proteins are assembled with RNA to make new viruses.
2. In the **lysogenic cycle,** the viral DNA is temporarily incorporated into the DNA of the host cell. A virus in this dormant state is called a **provirus** (or, if a bacteriophage, a **prophage**). The virus remains inactive until some trigger, often an external environmental stimulus (such as radiation or certain chemicals), causes the virus to begin the destructive lytic cycle.

Retroviruses are ssRNA viruses that use an enzyme called **reverse transcriptase** to make a DNA complement of their RNA. The DNA complement can then be transcribed immediately to manufacture mRNA (lytic cycle), or it can begin the lysogenic cycle by becoming incorporated into the DNA of the host. Because there is often little specificity to where the viral DNA is inserted into the host genome, retroviruses are a special kind of transposon. Human immunodeficiency virus (HIV, the cause of AIDS) is a retrovirus.

Because viruses produce many copies over very short intervals, their potential for rapid evolution is great. RNA viruses, in particular, have much higher rates of replication errors because RNA replication lacks the repair mechanisms associated with DNA replication. As a result, mutations in RNA viruses are more frequent than in DNA viruses.

Evolution of viruses is also augmented when the genetic material of several different, but related, viruses recombine when present together in a host cell. There is evidence that two strains of simian immunodeficiency virus (SIV), an HIV-like virus in monkeys, recombined to create HIV.

The high rates of mutation in viruses intensify their pathogenicity because host populations do not evolve immune-system defenses as fast as the viruses evolve. Note that HIV, influenza (flu) viruses, and most of the viruses that cause the common cold are all RNA viruses. These viruses mutate and evolve quickly, producing new strains on a regular (or seasonal) basis and, as a result, remain infectious throughout the human lifetime.

Prokaryotes

Archaea and bacteria are prokaryotes. They do not contain a nucleus, nor do they possess any of the specialized organelles of eukaryotes. The primary genetic material of prokaryotes is a chromosome consisting of a single, circular DNA molecule without the proteins associated with the DNA of eukaryotic chromosomes. A prokaryotic cell reproduces by **binary fission.** In binary fission, the chromosome replicates and the cell divides into two cells, each cell bearing one chromosome. The spindle apparatus, microtubules, and centrioles found in eukaryotic cell divisions are lacking, since in prokaryotes, there is no nucleus to divide.

Prokaryotes may also contain **plasmids,** short, circular dsDNA molecules outside the chromosome. Plasmids carry genes that are beneficial but not normally essential to the survival of the prokaryote. A group of plasmids, called **R plasmids,** provide bacteria with resistance against antibiotics. Plasmids replicate independently of the chromosome. Some plasmids, called **episomes,** can become incorporated into the prokaryotic chromosome.

DNA replication in prokaryotic and eukaryotic organisms is basically the same. But because prokaryotic chromosomes are circular, replication begins at a single, unique origin, progressing in both directions until they meet at the termination site. In contrast, eukaryotes, with much larger chromosomes, have multiple points of origin.

Transcription in prokaryotes is also similar to that of eukaryotes. However, prokaryotic genes do not have introns. But a single mRNA may contain transcripts of multiple genes whose enzyme products function sequentially as part of a metabolic pathway. Such gene clusters are called **operons.** Also, translation is coupled with transcription—that is, the ribosome attaches to one end of the mRNA transcript and begins translation while the mRNA transcript is still being produced at the other end.

Genetic variation is introduced into prokaryotes through **horizontal gene transfer.** Although it occurs among both archaea and bacteria, it is best understood among bacteria. It occurs in the following ways:

1. **Conjugation** is a process of DNA exchange between bacteria. A donor bacterium produces a tube, or **pilus** (plural, **pili**), that connects to a recipient bacterium. Through the pilus, the donor bacterium sends chromosomal or plasmid DNA to the recipient. In some cases, copies of large portions of a donor's chromosome are sent, allowing recombination with the recipient's chromosome. One plasmid, called the **F plasmid,** contains the genes that enable a bacterium to produce pili. When a recipient bacterium receives the F plasmid, it, too, can become a donor cell.
2. **Transduction** occurs when new DNA is introduced into the genome of a bacterium by a virus. When a virus is assembled during a lytic cycle, it is sometimes assembled with some bacterial DNA in place of some of the viral DNA. When this aberrant virus infects another cell, the bacterial DNA that it delivers can recombine with the resident DNA. Like conjugation, transduction can transfer bacterial resistance or other pathogenic traits to host cells.
3. **Transformation** occurs when bacteria absorb DNA from their surroundings and incorporate it into their genome. Specialized proteins on the cell membranes of some bacteria facilitate this kind of DNA uptake.

Regulation of Gene Expression

Although some genes, like those that code for ribosomal proteins, are always turned on, most are not. That is because a cell is constantly modifying biochemical activities to respond to varying internal and external conditions. In order to accomplish this feat, mechanisms exist to turn genes on and off at appropriate times.

Regulatory mechanisms in prokaryotes occur as part of **operons.** An operon is a unit of DNA that contains multiple genes whose products work together to direct a single metabolic pathway. It contains the following components:

1. The **promoter** region is a sequence of DNA to which the RNA polymerase attaches to begin transcription.
2. The **operator** region is engaged by a regulatory protein to either block or promote the action of the RNA polymerase.
3. The **structural genes** contain coding DNA, that is, DNA sequences that code for various related enzymes that direct the production of some particular end product.
4. A **regulatory gene,** lying outside the operon region, produces a regulatory **protein** that engages the operator region and governs whether RNA polymerase can attach to the promoter region and begin transcription. Regulatory proteins are allosteric, that is, they become active (or inactive) only when they bind to some specific substrate molecule. A regulatory protein can be one of two kinds:
   - A **repressor protein** *blocks* the attachment of RNA polymerase to the promoter region. Repressor proteins characterize **negative regulation** because they must be *inactive* in order for transcription to occur.
   - An **activator protein** *promotes* the attachment of RNA polymerase to the promoter region. Activator proteins characterize **positive regulation** because they must be *active* in order for transcription to occur.

Three examples of gene regulation in bacteria follow (summarized in Figure 9-6):

1. The ***trp* operon** in *E. coli* produces enzymes for the synthesis of the amino acid tryptophan. A regulatory gene produces an inactive repressor that does not bind to the operator. As a result, the RNA polymerase proceeds to transcribe the structural genes necessary to produce enzymes that synthesize tryptophan. When tryptophan is available to *E. coli* from the surrounding environment, the bacterium no longer needs to manufacture its own tryptophan. In this case, rising levels of tryptophan induce some tryptophan to react

with the inactive repressor and make it active. Here, tryptophan is acting as a **corepressor.** The active repressor now binds to the operator region, which, in turn, prevents the transcription of the structural genes. Since these structural genes stop producing enzymes only in the presence of an active repressor, they are called **repressible enzymes,** and the operon is a **repressible operon.** Because a repressor protein is involved, it is an example of **negative regulation.**

2. The ***lac* operon** in *E. coli* controls the breakdown of lactose. A regulatory gene produces an active repressor that binds to the operator region. When the operator region is occupied by the repressor, RNA polymerase is unable to transcribe several structural genes that code for enzymes that control the uptake and subsequent breakdown of lactose. When lactose is available, however, some of the lactose (in a converted form) combines with the repressor to make it inactive. When the repressor is inactivated, RNA polymerase is able to transcribe the genes that code for the enzymes that break down lactose. Since a substance (lactose, in this case) is required to induce (turn on) the operon, the enzymes that the operon produces are said to be **inducible enzymes,** and the operon is an **inducible operon.** Because a repressor protein is involved, it is an example of **negative regulation.**

Kind of Regulation	Operon Description	Regulatory Protein	Regulatory Protein Status	Regulatory Activity	RNA Polymerase Access to Operator	Action
Negative Regulation	Repressible Operon (trp operon)	repressor	repressor + corepressor (tryptophan)	repressor active	blocked	do not produce tryptophan
			repressor only	repressor inactive	promoted	produce tryptophan
	Inducible Operon (lac operon)	repressor	repressor + inducer (lactose)	repressor inactive	promoted	break down lactose
			repressor only	repressor active	blocked	do not break down lactose
Positive Regulation	Glucose Repression	activator	activator (CAP) only	activator inactive	blocked	do not break down lactose
			activator (CAP) + effector (lactose)	activator active	promoted	break down lactose

Summary of Operon Activity

Figure 9-6

3. **Glucose repression** is a second regulatory process that influences the *lac* operon. Glucose is a preferential source of energy when both glucose and lactose are present. But when only lactose is present, this process enhances the breakdown of lactose (already permitted by the inactive repressor of the *lac* operon). This is accomplished by an **activator** regulatory protein, CAP, that is activated by cyclic AMP (cAMP). When glucose levels are up, cAMP levels are down, and the CAP activator is inactive. But when glucose is absent, cAMP levels are up, CAP is activated and binds to the operator, promoting RNA polymerase transcription of the enzymes that break down lactose. Because an activator protein is involved (CAP), this is an example of **positive regulation.**

In general, repressible operons are associated with genes that regulate *anabolic* biochemical pathways (pathways that consume energy to synthesize new molecules). Inducible operons are associated with *catabolic* pathways (pathways that release energy when they break down molecules). It should also be noted that these regulatory processes are negative feedback mechanisms: They turn on remedial processes in response to changes in environmental conditions (too much or too little tryptophan or lactose, for example) and turn off the processes when suitable conditions return.

The regulation of gene expression in eukaryotic cells is more complicated than in prokaryotes. There are several reasons for this:

1. **Multicellularity.** Many eukaryotic organisms are multicellular. This requires different gene regulation programs for different cell types.
2. **Chromosome complexity.** The chromosomes of eukaryotic organisms are more complex than those of prokaryotes due to their much larger size and elaborate organization with histone proteins. Also, some metabolic processes require the activation of multiple genes, each of which is located on a different chromosome. In these cases, coordinated expression of these genes requires a more sophisticated system of regulation than that which is present in prokaryotes.
3. **Uncoupling of transcription and translation.** In prokaryotes, translation begins while transcription is still in progress. In contrast, eukaryotic transcription occurs in the nucleus isolated from translation, which takes place in the cytoplasm. This allows for a greater range of mechanisms to control gene expression.

In line with the complexity of eukaryotic cells, the mechanisms that regulate eukaryotic transcription are similarly complex. Every stage that contributes to the final protein product, from accessing the gene in the chromatin to the final folding of the translated protein, can be subject to some kind of regulation. Following are the more prominent gene-regulating mechanisms that occur:

1. **DNA methylation** occurs when methyl groups ($-CH_3$) attach to DNA bases. This makes it more difficult for transcription factors to access the DNA. DNA methylation seems to be associated with long-term inactivation of genes.
2. **Histone modification** refers to changes in the organization of histone proteins with DNA. The DNA double helix in chromatin is wrapped around a bundle of eight histone molecules to form tightly knit complexes called **nucleosomes.** Access to DNA for transcription can be increased or decreased by the following:
   - **Acetylation.** Histone molecules loosen their grip on the DNA molecule when they are acetylated— that is, when an acetyl group ($-COCH_3$) is attached. Acetylated histones are associated with *activated* transcription.
   - **Methylation.** Histones are methylated when a methyl group ($-CH_3$) is attached. In most cases, methylated histones are associated with *repressed* transcription.
3. **X inactivation** is a special case of chromosome inactivation in female mammals. A few days after fertilization, one of the two X sex chromosomes in each cell of female embryos is randomly inactivated. A gene on the chromosome that is silenced produces a noncoding RNA transcript that is associated with a loss of acetylation in the histone proteins of nucleosomes. The descendants of each cell maintain the same inactivated X chromosome. The purpose of X inactivation is to equalize the gene dosage that both males (who have only one X chromosome) and females express. (See "X-Inactivation" in Chapter 8, "Heredity.")
4. **Transcription initiation** is regulated by a **transcription complex,** a group of various proteins that are associated with RNA polymerase activity. The makeup of the transcription complex determines the degree to which transcription is activated or repressed. There are several components to the complex:
   - **General transcription factors** are proteins that are required by *all* transcription events to successfully initiate transcription by RNA polymerase. General transcription factors attach with the RNA

polymerase to the *promoter region* upstream and adjacent to the gene to be transcribed. Some general transcription factors target the TATA box sequence associated with the promoter region.

- **Specific transcription factors** are additional proteins associated with regulating *specific* transcription activities—specific to cell type, specific to the particular genes, or specific to the timing of the transcription. There are two kinds of specific transcription factors: activators and repressors. Specific transcription factors attach to **enhancers,** DNA binding sites that can be thousands of nucleotides upstream or downstream from the gene. There may be one or more enhancers that are unique to a particular gene, and each of those enhancers may be specific to a different timing of transcription or to a specific cell type. Because an enhancer may be quite a distance away from the gene it influences, the DNA segment containing the enhancer (and bearing its *specific* transcription factor) folds such that it can join the *general* transcription factors and RNA polymerase on the promoter.
- **Coactivators** and **mediators** are additional proteins that contribute to the binding of transcription complex components.

5. **RNA processing,** as discussed earlier in this chapter, can produce different mRNAs by slicing the primary RNA transcript in different ways. This allows a single gene to encode proteins that are specific to the cell type or to its developmental stage.
6. **RNA interference (RNAi)** refers to gene silencing caused by short RNA molecules. They can do this in three ways:
   - They bind to complementary sequences of mRNAs in the cytoplasm and block their translation.
   - They bind to, cleave, and degrade complementary sequences of mRNA.
   - They bind to chromatin in the nucleus, preventing transcription of genes.

   The two best understood kinds of these molecules are both single-stranded (ssRNA) and about 20 nucleotides long. They differ mostly by their origin as described here:

   - microRNAs (miRNAs) originate from mRNAs that have been transcribed from regulatory genes. The mRNAs are subsequently truncated in the cytoplasm to form miRNAs.
   - Short interfering RNAs (siRNAs) originate from double-stranded RNAs (dsRNAs) that have formed in the cytoplasm from ssRNAs or are dsRNAs that have been introduced into the cell experimentally. In either case, the dsRNAs are subsequently truncated in the cytoplasm to form siRNAs.
7. **mRNA degradation** occurs as a result of RNAi (above), but also because mRNAs are unstable molecules. The poly-A tail and the 5′ GTP cap help maintain mRNA stability on a scale of hours. But degradation of the tail occurs as the mRNA ages and degrading enzymes, targeting the tail and cap, quickly follow. Meanwhile, sequences rich in adenine and uracil in untranslated regions of the mRNA are recognition sites for other degrading enzymes.
8. **Protein degradation** is the final stage in the life of proteins. As proteins age, they lose functionality as their three-dimensional shape changes. Nonfunctional proteins are marked for destruction with the protein **ubiquitin** (so called because it is ubiquitous, present in all eukaryotic cells).

Cells in the early stages of embryonic development are **stem cells.** When stem cells divide, daughter cells have the potential to become any kind of fetal or adult cell. But as development continues, cells differentiate, become specialized, and subsequent cell divisions produce cells that are similarly specialized. Cells become specialized because transcription factors activate some genes while repressing others. The process is often self-reinforcing—once certain genes are turned on or off, they remain so. Genes that are permanently turned off are often associated with DNA methylation and histone modification. The process that fixes a cell's fate is called **cell determination.**

Reproductive cloning is the process of making an individual with the same nuclear DNA as another animal. The first mammal to be cloned was Dolly, a sheep. The process for cloning Dolly and many mammals since was "somatic cell nuclear transfer." In this process, the nucleus of an unfertilized egg cell is replaced by the nucleus from a fully differentiated adult cell (for Dolly, a cell taken from her udder). This demonstrated that cell determination can be reversed, that is, a nucleus from a cell whose fate was fixed was unfixed by implanting it in the cytoplasm of a stem cell, where transcription factors and other regulatory factors were absent.

Biotechnology

Recombinant DNA contains DNA segments or genes from different sources. DNA transferred from one part of a DNA molecule to another, from one chromosome to another chromosome, or from one organism to another, all constitute recombinant DNA. The transfer of DNA segments can occur naturally through viral transduction, bacterial conjugation, or transposons. It is also a regular event in eukaryotes as a result of crossing over during prophase of meiosis. Recombinant DNA can also be produced artificially with biotechnology. **Biotechnology** is the use of biological systems to modify organisms or produce desired products. This definition addresses a range of activities, from selected breeding of plants or animals to manipulating bacteria to produce human insulin.

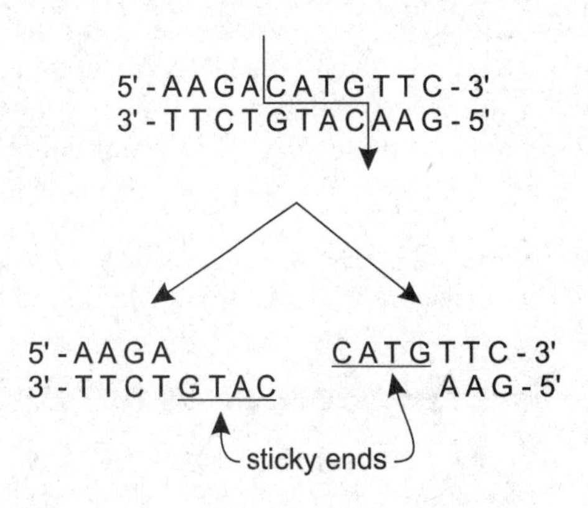

Restriction Enzyme *PciI* Activity

Figure 9-7

Recombinant DNA technology uses **restriction enzymes (restriction endonucleases)** to cut up DNA. Restriction enzymes are obtained from bacteria that manufacture these enzymes to combat invading viruses. Restriction enzymes are very specific, cutting DNA at specific recognition sequences of nucleotides called **restriction sites** (Figure 9-7). The cut across a double-stranded DNA is usually staggered, producing fragments that have one strand of the DNA extending beyond the complementary strand. The unpaired extension is called a **sticky end.**

DNA Cloning

DNA cloning is a procedure that allows DNA fragments or genes to be copied. The major features of the process follow:

1. **Use a restriction enzyme to cut up the foreign DNA that contains a gene to be copied.** The restriction enzyme produces multiple fragments of the foreign DNA with sticky ends. Some of these fragments will contain the gene to be copied.
2. **Use the same restriction enzyme to cut up the DNA of a cloning vector.** A cloning vector is a vehicle used to transfer DNA between cells. Plasmids are commonly used as vectors because they can subsequently be introduced into bacteria by transformation. Using the same restriction enzyme produces the same sticky ends as that produced in the foreign DNA. To help in the identification of the copied gene, use a plasmid that has one restriction site for the restriction enzyme and is engineered to contain the following:
   - ***amp^R* gene.** This gene gives a bacterium resistance against the antibiotic ampicillin.
   - ***GFP* gene.** This gene, originally obtained from jellyfish, presents a bright green fluorescence.
   - ***lac*Z gene.** This gene codes for an enzyme that naturally breaks down lactose. Conveniently, the enzyme also breaks down a related, but artificially made, molecule called **X-gal.** X-gal is colorless, but it forms a blue product when broken down by the *lac*Z enzyme. *The one restriction site for the restriction enzyme occurs within the* lacZ *gene.*
3. **Mix cut foreign DNA with cut plasmids.** As fragments reattach by base-pairing at their sticky ends, foreign fragments, some of which contain the gene to be copied, will fuse with plasmid fragments.
4. **Apply DNA ligase to stabilize attachments.** Some of the plasmids, by chance, are now recombinant plasmids containing the foreign gene. Other plasmids will not contain the foreign gene.
5. **Mix plasmids with bacteria to allow transformation.** Some of the bacteria will absorb the plasmids (transformation). Some bacteria will absorb recombinant plasmids. Others will absorb nonrecombinant plasmids.
6. **Grow the transformed bacteria in the presence of ampicillin and X-gal.** Only bacteria that have absorbed a plasmid (transformed bacteria) will grow in the presence of ampicillin because only the introduced plasmids contain the *amp^R* gene. Of those bacteria, only bacteria that have absorbed a *recombinant* plasmid will be white because they lack a functioning *lac*Z gene. For these bacteria, the foreign DNA was inserted within

the *lac*Z gene of the plasmid, making the gene dysfunctional. Bacteria that absorbed a *non*recombinant plasmid have a functioning *lac*Z gene and turn X-gal blue.

The end product of this process is a **genomic library,** a collection of bacteria, each of which contains a fragment of the genome of the foreign DNA but together contain the entire genome of the foreign DNA.

When foreign genes are inserted into the genome of a bacterium, introns often prevent their transcription. To avoid this problem, the DNA fragment bearing the required gene is obtained directly from the mRNA that codes for the desired polypeptide. Reverse transcriptase (obtained from retroviruses) is used to make a DNA molecule directly from the mRNA. DNA obtained in this manner is called **complementary DNA (cDNA)** and comprises cDNA libraries that lack the introns that suppress transcription.

As a result of DNA cloning, the human gene for insulin was inserted into *E. coli.* The transformed *E. coli* produces insulin, which is isolated and used to treat diabetes.

The Polymerase Chain Reaction (PCR)

The **polymerase chain reaction (PCR)** is a technique that makes large numbers of DNA copies faster than the DNA cloning process previously described. Steps of the process are outlined here:

1. **DNA is heated.** Heating denatures (separates) the hydrogen bonding holding the double-stranded DNA (dsDNA) together and forms two single-stranded DNA (ssDNA) molecules.
2. **DNA is cooled and single-stranded DNA primers are added.** Two primers are added, each complementary to the 3′ end of the single strands of DNA. These single-stranded DNA primers serve the same purpose as RNA primers during normal DNA replication.
3. **DNA polymerase is added.** A special heat-tolerant DNA polymerase derived from bacteria adapted to living in hot springs is added. The DNA polymerase attaches to the primers at each end of a single strand of DNA and synthesizes a complementary DNA strand. At the end of this step, the one initial dsDNA molecule becomes two dsDNA molecules.
4. **Repeat the above steps.** Each repetition of the sequence doubles the number of DNA molecules, and the total number of molecules increases exponentially. PCR can amplify DNA to billions of copies in hours.

Gel Electrophoresis and DNA Fingerprinting

Gel electrophoresis is a procedure that separates restriction fragments. In this procedure, DNA fragments of different lengths are separated as they diffuse through a gelatinous material under the influence of an electric field. Since DNA is negatively charged (because of the phosphate groups), it moves toward the positive electrode. Shorter fragments migrate further through the gel than longer, heavier fragments. Gel electrophoresis is often used to compare DNA fragments of closely related species in an effort to determine evolutionary relationships.

When restriction fragments between individuals of the same species are compared, the fragments differ in length because of polymorphisms, which are slight differences in DNA sequences. These fragments are called **restriction fragment length polymorphisms (RFLPs).** In **DNA fingerprinting,** RFLPs produced from DNA left at a crime scene are compared to RFLPs from the DNA of suspects. Areas of the human genome that are particularly polymorphic contain **short tandem repeats (STRs).** STRs are short sequences of nucleotides (two to five base pairs) that repeat multiple times, with the number of repeats varying markedly among individuals.

Concerns about Biotechnology

The advent of biotechnology has introduced new capabilities to human civilization. Improvements in the identification and treatment of disease and advances in forensic science are clear benefits. But some benefits can also be abused while the benefits of others are debated. In some cases, social and ethical questions arise. Here are some of the issues that biotechnology presents:

1. **Pharmaceuticals.** DNA cloning allows quick and inexpensive production of many pharmaceuticals. For example, human insulin (previously isolated from animals) and human growth hormone, or HGH

(previously obtained from human cadavers), are now readily available as products of DNA cloning. Insulin is used to treat diabetes, and HGH is used to treat certain forms of dwarfism. But HGH is also used by some athletes to increase athletic performance and by celebrities to supposedly reduce the effects of aging.

2. **Human disease profiles.** Some diseases are inherited and can be identified before symptoms appear by evaluating the genes of the individual. Sometimes, treatments can ensue and symptoms, including early death, can be circumvented. Other times, there are no available treatments. In this case, it is not clear whether affected individuals should (or want to) know that they will suffer from a debilitating disease in the near future, whether medical insurance has a right to know that they are high-risk patients, and whether society should control the reproductive potential of the individual. These are issues of individual self-determination, medical privacy, and reproductive rights.
3. **Transgenic** organisms are those that possess genes that, through genetic engineering, have been taken from other organisms, including other species.
   - **Genetic engineering in plants.** Genes have been inserted into agricultural plants that provide resistance to pests and herbicides and tolerance to drought and other extreme environmental conditions. In other cases, genetic engineering has changed fruit color, increased crop yields, or extended shelf life. For example, several genetically modified (GM) crops, including corn (Bt corn), carry the Bt gene that gives plants insecticide properties. The gene, whose origin is the plasmid of a soil bacterium (*Bacillus thuringiensis,* or Bt), produces a chemical that is toxic only to certain insects. This reduces the need for pesticides, which often kill insects indiscriminately as well as their natural predators. Often, however, other insects not killed by the Bt toxin increase in numbers, requiring, once again, the application of insecticides. In addition, insects once susceptible to Bt toxin evolve resistance to the toxin. Also, in some genetically engineered crops, there are uncertain consequences, should the transgenic gene, through gene flow via pollen, spread to wild plant populations.
   - **Genetic engineering in animals.** Genes have been inserted into domestic animals to produce desirable products or to produce animals that are more vigorous or convenient to rear. Salmon have been modified with a growth hormone gene (from a different salmon species) to make them grow faster. Like plants, a major concern for the transgenic salmon is gene flow into wild populations. Goats have been genetically modified to produce milk that contains spider fibers (for use as sutures or industrial products). Predictably, there are often unexpected results with genetically modified organisms (GMOs). A GM breed of featherless chickens, developed to simplify marketing preparation, was more prone to insect bites and more sensitive to UV radiation. Also, mating success was impaired because courtship requires feather displays.
   - **GMOs in the food chain.** Whether animal or plant, GMOs in the food chain are controversial. A major concern is one of health: Unidentified genes, accidently and unknowingly inserted into the GM plant or animal, may produce products that create allergies.
4. **Reproductive cloning.** Traditional selective breeding of animals, where two animals with the desired traits are bred, is a slow process. Each new generation must reach reproductive age before another round of selective breeding can occur. Reproductive cloning, however, promises to produce, effectively within a single generation, copies of any desirable individual. Many identical copies of exceptional individuals could be created using the best available individuals—a cow that produces the most milk, a prize-winning racehorse, or your favorite pet. So far, however, reproductive cloning has had mediocre success. Problems include organ failure, a high susceptibility to disease, shorter-than-normal life spans, and low success rates. (Hundreds of trials are sometimes required before one successful clone is produced.)

Review Questions

Multiple-Choice Questions

The questions that follow provide a review of the material presented in this chapter. Use them to evaluate how well you understand the terms, concepts, and processes presented. Actual AP multiple-choice questions are often more general, covering a broad range of concepts, and often more lengthy. For multiple-choice questions typical of the exam, take the two practice exams in this book.

Directions: Each of the following questions or statements is followed by four possible answers or sentence completions. Choose the one best answer or sentence completion.

1. The two strands of a DNA molecule are connected by

A. hydrogen bonds between the codons and anticodons
B. hydrogen bonds between the bases of one strand and the bases of the second strand
C. covalent bonds between phosphate groups
D. covalent bonds between the nitrogen bases

2. All of the following combinations of nucleotides are examples of normal base-pairing EXCEPT:

A. an adenine DNA nucleotide to a thymine DNA nucleotide
B. a guanine DNA nucleotide to a cytosine DNA nucleotide
C. a thymine RNA nucleotide to an adenine DNA nucleotide
D. a uracil RNA nucleotide to an adenine DNA nucleotide

3. Which of the following is true?

A. Ribosomes contain RNA nucleotides and amino acids.
B. The uracil nucleotide consists of the uracil nitrogen base, a deoxyribose sugar, and a phosphate group.
C. When tRNA attaches to mRNA during translation, cytosine nucleotides base-pair with guanine nucleotides and adenine nucleotides base-pair with thymine nucleotides.
D. In eukaryotes, DNA is manufactured in the nucleus and RNA is manufactured in the cytoplasm.

4. All of the following enzymes are involved in DNA replication EXCEPT:

A. helicase
B. DNA ligase
C. RNA polymerases
D. primase

5. ATP, the common energy-carrying molecule, most resembles the

A. adenine DNA nucleotide
B. adenine RNA nucleotide
C. adenine DNA nucleotide with two extra phosphates
D. adenine RNA nucleotide with two extra phosphates

6. The end products of translation are

A. amino acids
B. polypeptides
C. lipids
D. RNA molecules

7. Which of the following contains a code for a protein?

A. DNA polymerase
B. RNA polymerase
C. rRNA
D. mRNA

8. Which of the following changes following the start codon in the mRNA would most likely have the greatest deleterious effect?

A. a deletion of a single nucleotide
B. a deletion of a nucleotide triplet
C. a single nucleotide substitution of the nucleotide occupying the first codon position
D. a single nucleotide substitution of the nucleotide occupying the third codon position

Questions 9–15 refer to the following diagram of DNA and RNA segments. Boxes represent nucleotides. The letters A, G, and C refer to the names of the nucleotides that occupy a particular position.

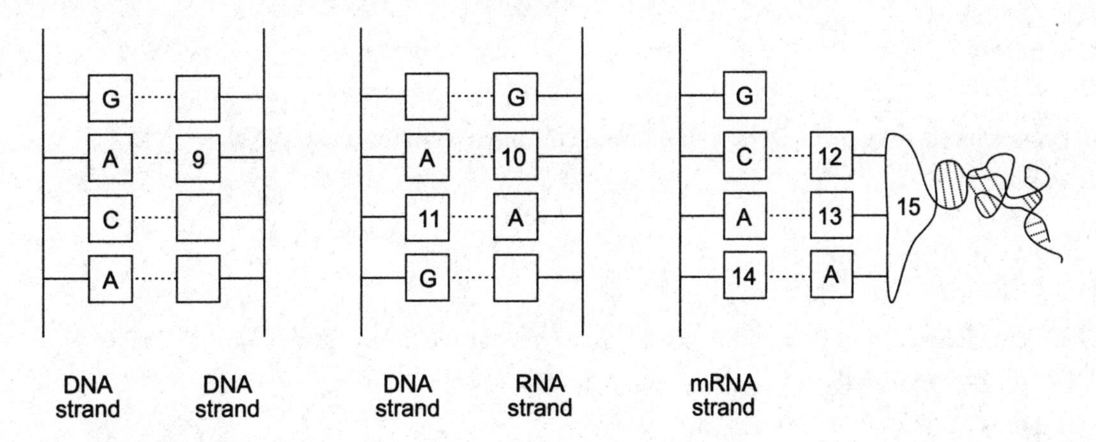

Use the following key for questions 9–14. Each answer in the key may be used once, more than once, or not at all.

A. cytosine nucleotide
B. guanine nucleotide
C. thymine nucleotide
D. uracil nucleotide

9. Which nucleotide would occupy box 9?

10. Which nucleotide would occupy box 10?

11. Which nucleotide would occupy box 11?

12. Which nucleotide would occupy box 12?

13. Which nucleotide would occupy box 13?

14. Which nucleotide would occupy box 14?

15. The nucleotide strand represented by 15 is

A. tRNA
B. rRNA
C. mRNA
D. ATP

16. The DNA of an elephant and the DNA of a cherry tree will most likely differ in all of the following respects EXCEPT:

- **A.** the kinds of genes for which the DNA codes
- **B.** the kinds of nucleotides utilized in forming DNA
- **C.** the number of DNA molecules
- **D.** the length of DNA molecules

17. Protein synthesis consists of all of the following steps EXCEPT:

- **A.** replication
- **B.** transcription
- **C.** translation
- **D.** elongation

18. The genetic instructions for forming a polypeptide chain are carried to the ribosome by the

- **A.** tRNA
- **B.** rRNA
- **C.** mRNA
- **D.** DNA

19. In bacteria, a small circle of DNA found outside the main chromosome is called a

- **A.** plasmid
- **B.** cDNA
- **C.** RFLP
- **D.** PCR

20. Genetic variation can be introduced into bacteria by all of the following methods EXCEPT:

- **A.** transfer of DNA between bacteria through pili
- **B.** DNA amplification
- **C.** mutation
- **D.** transformation

21. All viruses consist of

- **A.** DNA and a protein coat
- **B.** RNA and a protein coat
- **C.** a nucleic acid and a protein coat
- **D.** proteins and polysaccharides

22. The mRNA actively being translated in the cytoplasm would have all of the following EXCEPT:

- **A.** a poly-A tail
- **B.** a 5′ GTP cap
- **C.** exons
- **D.** introns

23. The *lac* operon in *E. coli* is involved in

- **A.** regulating the expression of a gene
- **B.** regulating the translation of mRNA
- **C.** controlling the formation of ribosomes
- **D.** controlling DNA replication

Free-Response Questions

The AP exam has long and short free-response questions. The long questions have considerable descriptive information that may include tables, graphs, or figures. The short questions are brief but may also include figures. Both kinds of questions have four parts and generally require that you bring together concepts from multiple areas of biology.

The questions that follow are designed to further your understanding of the concepts presented in this chapter. Unlike the free-response questions on the exam, they are narrowly focused on the material in this chapter. For free-response questions typical of the exam, take the two practice exams in this book.

Directions: The best way to prepare for the AP exam is to write out your answers as if you were taking the exam. Use complete sentences for all your answers and do *not* use outline form or bullets. You may use diagrams to supplement your answers, but be sure to describe the importance or relevance of your diagrams.

---

1. Gene regulation in eukaryotic cells is considerably more complicated than gene regulation in prokaryotes. Describe two reasons eukaryotic organisms require a more complex approach to gene regulation.

2. In some cases, a single nucleotide mutation does not lead to the creation of a different protein. Explain how this can happen.

3. Traditional fingerprints and DNA fingerprints are both used to identify suspected criminals. Can either of these two techniques distinguish identical twins? Justify your answer.

4. Genes store the hereditary information of the cell. Describe how each of the following contributes to the process of transforming the information stored in a gene into the expression of a physical trait.

- **a.** transcription
- **b.** RNA processing
- **c.** translation
- **d.** protein folding

5. Both of the following cause disease. Describe how their disease-causing activities differ.

- **a.** viruses
- **b.** bacteria

6. Describe how gene regulation occurs in

- **a.** bacterial cells
- **b.** eukaryotic cells

Answers and Explanations

Multiple-Choice Questions

1. **B.** Weak hydrogen bonds form between bases of the two strands. In particular, a pyrimidine (a base with one nitrogen ring) in one strand bonds to a purine (a base with two nitrogen rings) in the second strand.

2. **C.** Thymine is not used as a base in any RNA nucleotide. Instead, RNA uses the uracil base, which base-pairs with the adenine DNA nucleotide during transcription.

3. **A.** Ribosome is a complex of rRNA molecules (RNA nucleotides) and proteins (amino acids). The remaining answer choices are incorrect because the uracil nucleotide contains a ribose sugar (not a deoxyribose sugar), adenine nucleotides base-pair with uracil nucleotides, and both DNA and RNA are produced in the nucleus.

4. **C.** RNA polymerases are involved in the transcription of DNA into RNA, not the replication of DNA. The major enzyme for DNA replication is DNA polymerase (not one of the answer choices). Primase directs the attachment of RNA nucleotides to a DNA template, but those RNA primers must be in place before DNA polymerase can attach and begin attaching DNA nucleotides.

5. **D.** Since ATP contains the adenine nitrogen base, the sugar ribose (not deoxyribose), and three phosphate groups, it is equivalent to the adenine RNA nucleotide with two extra phosphate groups. In contrast, an adenine DNA nucleotide with two extra phosphates contains a deoxyribose sugar and is written as dATP.

6. **B.** Translation is the process in which ribosomes conduct the matching of tRNA with mRNA, producing an amino acid chain, or polypeptide.

7. **D.** The mRNA is a sequence of nucleotides. Each triplet of nucleotides codes for a particular amino acid. The sequence of triplets on the mRNA corresponds to the sequence of amino acids in an entire polypeptide, or protein.

8. **A.** A deletion of a nucleotide in the mRNA produces a change in the triplet-codon reading frame and creates a frameshift mutation. For example, if the first nucleotide of a codon is deleted, then its second nucleotide becomes its first, its third nucleotide becomes its second, and the first nucleotide of the next codon becomes its third. This, then, is repeated in every subsequent codon. Such an arrangement is likely to change many of the amino acids in the sequence (depending upon where in the sequence the frameshift begins) and, thus, affect the final sequence of the polypeptide considerably. Answer choice B will result in a missing amino acid, and answer choice C will very likely change one amino acid to a different amino acid. These changes may alter the effectiveness of the polypeptide, but not as severely as changing many amino acids, as would occur in answer choice A. Answer choice D may have no effect at all because a change in the third position of a codon will often code for the same amino acid. (This results from the wobble of the third position of the tRNA anticodon.) The inherited disorder sickle-cell disease is caused by the replacement of one amino acid by another in two chains of the hemoglobin protein, severely reducing the effectiveness of hemoglobin in carrying oxygen. However, a frameshift in the mRNA coding for hemoglobin would certainly make it entirely ineffective.

9. **C.** In a DNA double helix, thymine base-pairs with adenine.

10. **D.** During transcription of DNA, the uracil RNA nucleotide base-pairs with the adenine DNA nucleotide.

11. **C.** During transcription of DNA, the adenine RNA nucleotide base-pairs with the thymine DNA nucleotide.

12. **B.** Guanine base-pairs with cytosine.

13. **D.** When base-pairing occurs between the anticodon nucleotides of the tRNA and the codon nucleotides of the mRNA, a uracil nucleotide base-pairs with an adenine nucleotide.

14. **D.** This is the same base-pair as in question 13.

15. A. The process illustrated here is translation. In particular, the anticodon of a tRNA is shown base-pairing with the codon of the mRNA.

16. B. The DNA of all cells uses the same DNA nucleotides—adenine, cytosine, guanine, and thymine nucleotides. On the other hand, the DNA of two unrelated species is likely to differ considerably in the genes produced, as well as in the number of DNA molecules (that is, the number of chromosomes), the DNA lengths, and the DNA nucleotide sequences.

17. A. Replication is the process of copying DNA. Protein synthesis involves transcription and translation. Translation begins with initiation, continues with elongation, and ends with termination.

18. C. The genetic instructions originate on the DNA and are carried to the cytoplasm by the mRNA. The rRNA and tRNA operate together to translate the code on the mRNA into a polypeptide.

19. A. Plasmids are small circular DNA molecules that a bacterium contains in addition to its primary chromosome.

20. B. DNA amplification refers to a laboratory process, such as PCR, that generates multiple copies of DNA. Answer choice A is a description of conjugation, which, together with mutation, transformation, and transduction, is a naturally occurring process that can introduce genetic variation into the genome of bacteria.

21. C. Viruses consist of a nucleic acid (DNA or RNA) surrounded by a protein coat. Some viruses contain an envelope made from lipids or glycoproteins obtained from the membranes of their hosts, but they do not have the phospholipid bilayer membrane typical of cells.

22. D. Introns are intervening sequences in the mRNA that are cleaved from the mRNA by snRNPs before export to the cytoplasm and subsequent translation.

23. A. Operons are DNA segments that include a promoter region, an operator region, and a series of structural genes. Together with a regulatory gene lying outside the operon, the three parts of an operon work collectively to control transcription, which results in the regulation of gene expression.

Free-Response Questions

1. Eukaryotic organisms are often multicellular. As a result, gene regulation must be able to regulate gene expression specific to cell type. Eukaryotes also have more than one chromosome, requiring simultaneous regulation of multiple genes on different chromosomes when they all contribute to a metabolic pathway.

2. In many cases, no new amino acid is designated if the mutation is a substitution in the third position of the mRNA codon. More than one codon often codes for the same amino acid, and the third position is the most variable (wobble pairing).

3. Yes. The traditional fingerprint can distinguish identical twins because the expression of dermal ridges (the ridges on the skin that make fingerprints) are the product of genes whose expression has been modified by gene regulation. Gene regulation is often influenced by environmental factors that differ between identical twins during fetal development. In contrast, a DNA fingerprint is a snapshot of the DNA, the "raw" genetic material, uninfluenced by gene regulation; it cannot distinguish identical twins. *(Rare exceptions to this occur as a result of somatic mutations—mutations in cells other than those that produce gametes. Somatic mutations originate from mitotic errors during development and create cell lines within an individual that are genetically different.)*

4. a. Genes are segments of the DNA that contain instructions for producing a specific polypeptide. Many polypeptides are enzymes that regulate cellular reactions, which, in turn, produce chemical end products that appear as traits. The process by which information is transferred from gene to enzyme is called protein synthesis. Transcription, the first step in the process, describes how RNA is synthesized from DNA. RNA polymerase, in association with various regulating transcription factors, attaches to DNA at a promoter region on the DNA. RNA polymerase directs RNA nucleotides to base-pair with the DNA fragment that represents the gene. If a fragment of DNA contained the nucleotide sequence adenine, cytosine, guanine, and thymine, the RNA nucleotides that would base-pair with it are uracil, guanine, cytosine, and adenine, respectively. The products of transcription are three kinds of RNA—mRNA, tRNA, and rRNA.

b. The mRNA contains the code for the polypeptide. Noncoding intervening sequences, called introns, are removed and the mRNA is stabilized with a 5′ GTP cap and a poly-A tail. The mRNA then moves to the cytoplasm.

c. Translation occurs in the cytoplasm and describes the actual assembly of amino acids into proteins. It requires all three kinds of RNA—rRNA, tRNA, and mRNA.

First, ribosomes, consisting of rRNA and proteins, attach to the mRNA. Then a tRNA attaches to the ribosome. There are various kinds of tRNA molecules. Each kind is specific for a particular amino acid that attaches to one end of the tRNA. Each tRNA has a special region of three nucleotides, called an anticodon.

During the process of translation, ribosomes direct the pairing of the anticodons of tRNAs with appropriate triplet regions of the mRNA, called codons. Each mRNA codon specifies a particular amino acid. The genetic code describes which amino acid is indicated by each of the 64 different mRNA codons. Some codons indicate a stop code, signaling that translation of the mRNA is complete. Another codon indicates methionine, the start amino acid.

During translation, the ribosome provides binding sites to incoming tRNAs. Each tRNA brings the appropriate amino acid as dictated by the codon sequence on the mRNA. As each new tRNA arrives, the growing polypeptide chain that is attached to the previous tRNA is transferred to the new tRNA. The old tRNA is released; the ribosome moves over one binding site; and the process repeats until the stop codon is encountered. At this point the ribosome separates into two subunits, and the polypeptide is released.

d. Once released, the amino acids in the polypeptide may interact with one another, giving the polypeptide secondary and tertiary protein structures. Secondary structure results from hydrogen bonding and produces two kinds of structures—a helix or a pleated sheet. Tertiary structure results from additional bonding between R groups of the amino acids and disulfide bonds between cysteine amino acids. In its final form, the polypeptide may be an enzyme that can regulate a reaction that will produce some end product, or trait.

In the limited time you have available to answer this free-response question, your goal should not be to describe a single facet of protein synthesis in extreme detail. Rather, you should try to describe each step in the entire process, pursuing detail only when you have available time. When your essay is evaluated, points are given for each step in the process. There is a maximum number of points given for each step, so pursuing excess detail in one step does not improve your score. Providing a few pieces of information for every part optimizes your time and maximizes your score. For example, you would not improve your score by describing, in detail, the structure of the poly-A tail or 5' GTP cap of the mRNA. Also, details about the regulation of transcription would provide more than the question asks. However, if you omit an explanation of the genetic code, you will probably lose 1 or 2 points out of a maximum of 10.

5. **a.** Viruses cause disease by destroying cells. Viruses consist of a nucleic acid core (either DNA or RNA) and a protein coat. In the lytic cycle of reproduction, a DNA virus enters a cell and uses the metabolic machinery and raw materials of the cell to manufacture more viral DNA and viral protein coats. The viral DNA and protein assemble into hundreds of new viruses that burst from the cell, killing the cell in the process. In the lysogenic cycle, the virus may temporarily remain dormant as part of the host's genome, to become active only when exposed to radiation or other environmental disturbance. When activated, the viral DNA begins the lytic cycle.

Some viruses are RNA retroviruses. These RNA viruses produce the enzyme reverse transcriptase to first manufacture DNA, which, in turn, enters a lytic or lysogenic cycle.

b. Bacteria, unlike viruses, do not usually cause disease by direct destruction of host cells. Rather, most bacteria cause disease by producing toxins, usually waste products of their normal metabolism. When the toxins affect the normal metabolism of the host, disease results.

Some bacteria also cause disease by competing for the same resources as do the host cells. In other cases, the symptoms of a disease are the result of the host's response to invasion by foreign bodies. For example, in pneumonia, the mucus that accumulates in the lungs is produced by the lung cells in response to the presence of the bacteria.

6. **a.** *To answer this question, describe one (or more, if time permits) of these operons: lac operon, trp operon, or how CAP regulates lactose metabolism.*

b. *For this part of the question, you should describe how transcription factors work. If time permits, describe other mechanisms such as DNA methylation, histone modification, RNA interference, and mRNA degradation.*

Chapter 10

Evolution

Review

In general, **evolution** is about changes in populations, species, or groups of species. More specifically, evolution is the process by which the frequency of heritable traits in a population changes from one generation to the next. For example, in one research study of Darwin finches, it was found that the average size of a bird beak (a heritable trait) increased by 10% over a period of about one year. That was evolution.

An allele is one of several (or many) varieties of a gene. Individuals inherit alleles that code for traits that establish morphology (form or structure), physiology, or behavior. Evolution is when the frequency of those alleles in the population changes over time.

One of the earliest advocates for evolution was naturalist **Jean-Baptiste Lamarck.** His theory included the following two ideas:

1. **Use and disuse** described how body parts of organisms can develop with increased usage, while unused parts weaken. This idea was correct, as is commonly observed among athletes who train for competitions.
2. **Inheritance of acquired characteristics** described how body features acquired during the lifetime of an organism (such as muscle bulk) could be passed on to offspring. This, however, was incorrect. Only changes in the genetic material of cells can be passed on to offspring.

Fifty years after Lamarck published his ideas, naturalist **Charles Darwin** published *The Origin of Species.* Darwin's theory, discussed later in this chapter, was that **natural selection,** or "survival of the fittest," was the driving force of evolution.

There is abundant evidence that evolution occurs—that some species change over time, that other species diverge and become one or more new species, and that still other species become extinct. The question that evolutionary biologists try to answer is *how* evolution occurs. For this they propose theories. Lamarck theorized, incorrectly, that evolution occurs through the inheritance of acquired characteristics. Darwin's theory was that evolution progresses through natural selection. These theories, together with others discussed in this chapter, propose mechanisms responsible for the evolutionary patterns unequivocally observed in nature.

Evidence for Evolution

Evidence for evolution is provided by the following five scientific disciplines:

1. **Paleontology** provides fossils that reveal the prehistoric existence of extinct species. As a result, changes in species and the formation of new species can be studied.
   - Fossil deposits are often found among sediment layers, where the deepest fossils represent the oldest specimens. For example, fossil oysters removed from successive layers of sediment show gradual changes in the size of the oyster shell alternating with rapid changes in shell size. Large, rapid changes produced new species.
   - The age of many fossils can be determined using C-14 dating. In this procedure, the natural decay rate of a radioactive isotope of carbon (C-14) is used to determine the age of the fossil.

2. **Biogeography** uses geography to describe the distribution of species. This information has revealed that unrelated species in different regions of the world look alike when found in similar environments. This provides strong evidence for the role of natural selection in evolution.
   - Rabbits did not exist in Australia until they were introduced by humans. A native Australian hare wallaby resembles a rabbit both in structure and habit. As similar as these two animals appear, they are not that closely related. The rabbit is a placental mammal, while the wallaby is a marsupial mammal. The fetus of

a placental mammal develops in the female uterus, obtaining nourishment from the mother through the placenta. The fetus of a marsupial leaves the mother's uterus at an early stage of development and completes the remaining development while attached to a teat in the abdominal pouch. The great similarity of the rabbit and the wallaby is the result of natural selection.

3. **Embryology** reveals similar stages in development **(ontogeny)** among related species. The similarities help establish evolutionary relationships **(phylogeny).**
   - Gill slits and tails are found in the embryos of fish, chickens, pigs, and humans.
4. **Comparative anatomy** describes two kinds of structures that contribute to the identification of evolutionary relationships among species.
   - **Homologous structures (homologies)** are body parts that resemble one another in different species because they have evolved from a common ancestor. Because anatomy may be modified for survival in specific environments, homologous structures may look different but will resemble one another in pattern (how they are put together). The forelimbs of cats, bats, whales, and humans are homologous because they have all evolved from a common ancestral mammal. In some species, homologous structures have become **vestigial,** that is, they no longer serve any function. The remnants of limbs in snakes, hind limbs in whales, and the wings of flightless birds are examples of vestigial structures, structures that provide evidence of evolutionary heritage.
   - **Analogous structures (analogies)** are body parts that resemble one another in different species, not because they have evolved from a common ancestor, but because they evolved independently as adaptations to their environments. Some species of plants in Africa resemble cacti of North America because both have photosynthetic green stems with spines. But the similar-looking plants differ markedly in their flower structures and their DNA. The similarities in appearance result from adaptations to a hot, dry environment.
5. **Molecular biology** examines the nucleotide and amino acid sequences of DNA and proteins from different species. Closely related species share higher percentages of sequences than species distantly related. In addition, all living things share the same genetic code and, with minor variations, the same basic biochemical pathways, including those for replication, protein synthesis, respiration, and photosynthesis. This data strongly favors evolution of different species through modification of ancestral genetic information.
   - More than 98% of the nucleotide sequences in humans and chimpanzees are identical.

Natural Selection

Natural selection is the differences in survival and reproduction among individuals in a population as a result of their interaction with the environment. In other words, some individuals possess alleles (genotypes) that generate traits (phenotypes) that enable them to cope more successfully in their environment than other individuals do. The more successful individuals produce more offspring. Superior inherited traits are **adaptations** to the environment and increase an individual's **fitness,** or relative ability to survive and leave offspring. When the environment favors a trait, that is, when a trait increases the survival of its bearer, selection is said to act for that trait. In contrast, selection is said to act against unfavorable traits. Favorable traits are adaptive, while unfavorable traits are maladaptive.

Darwin presented his theory of evolution by natural selection using the following arguments:

1. **Populations possess an enormous reproductive potential.** Darwin calculated that two elephants would produce a population of 19 million individuals after 750 years if all offspring survived to reproductive maturity and fostered their normal number of offspring.
2. **Population sizes remain stable.** Darwin observed that populations generally fluctuate around a constant size.
3. **Resources are limited.** Resources, such as food, water, or light, do not increase as populations grow larger.
4. **Individuals compete for survival.** Eventually, the needs of a growing population will exceed the available resources. As a result, individuals must compete for resources.

5. **There is variation among individuals in a population.** Most traits reveal considerable variety in their form. In humans, for example, skin, hair, and eye color occur as continuous variation from very dark to very light.
6. **Much variation is heritable.** Although Darwin was unaware of the mechanism for heredity, he recognized that traits were passed from parents to offspring. This contrasts with characteristics acquired during the life of an organism as a result of environmental influences. The amputation of a limb or characteristics acquired as the result of exposure to pathogens or radiation, for example, are not heritable. We now know that most traits are produced by the action of enzymes that are coded by DNA, the hereditary information that is passed from generation to generation.
7. **Only the most fit individuals survive.** "Survival of the fittest" occurs because individuals with traits best adapted for survival and reproduction are able to outcompete other individuals for resources and mates.
8. **Evolution occurs as favorable traits accumulate in the population.** The best adapted individuals survive and leave offspring who inherit the traits of their parents. In turn, the best adapted of these offspring leave the most offspring. Over time, traits best adapted for survival and reproduction, and the alleles that generate them, accumulate in the population.

Natural selection may affect populations in a variety of ways. These are illustrated in Figure 10-1 and discussed below. Note that in all cases, natural selection acts on individual phenotypes already present in the population. Alleles may be expressed with other alleles in new combinations (as a result of genetic recombination) to create novel phenotypes, or new phenotypes may appear in the population as a result of mutations. It is important to note that natural selection does *not* cause mutations or create new phenotypes. It only "selects" phenotypes—already present in the population—that maximize fitness.

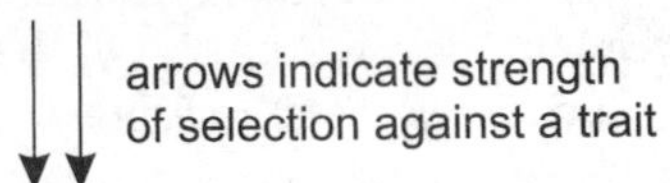

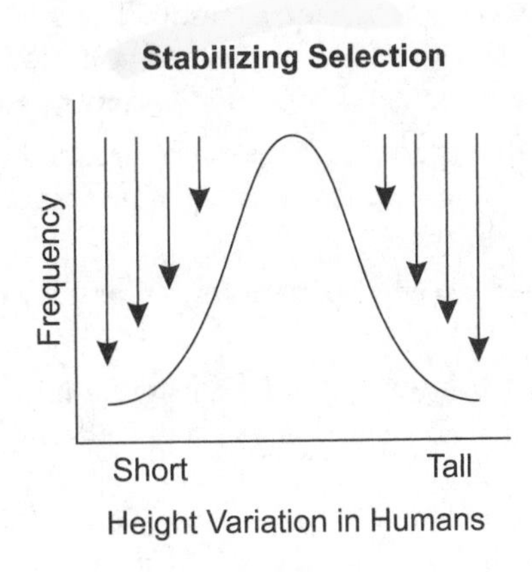

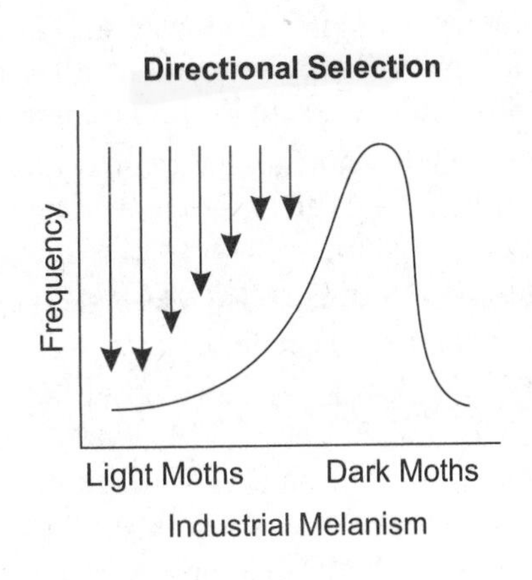

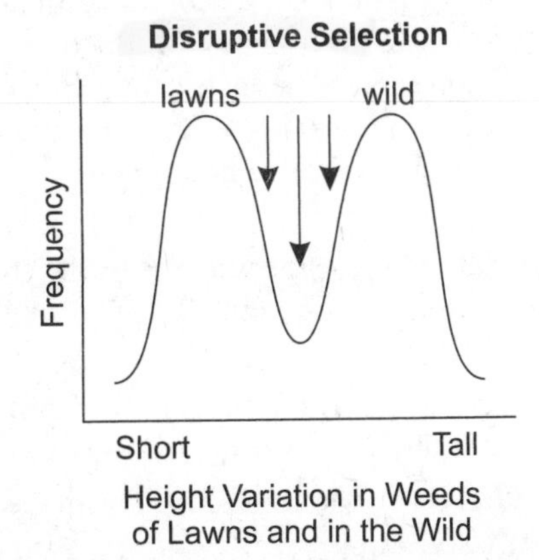

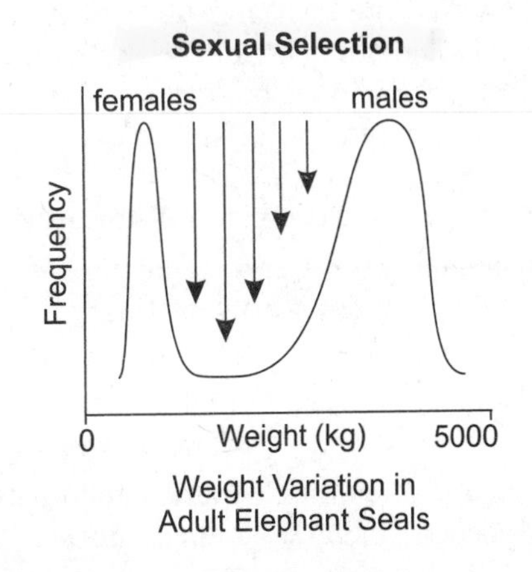

Kinds of Selection

Figure 10-1

1. **Stabilizing selection** eliminates individuals that have extreme or unusual traits. Under this condition, individuals with the most common form of a trait are the best adapted, while individuals who differ from the common form are poorly adapted. As a result, stabilizing selection maintains the existing population frequencies of common traits while selecting against all other trait variations.
2. **Directional selection** favors traits that are at one extreme of a range of traits. Traits at the opposite extreme are selected against. If directional selection continues for many generations, favored traits become more and more extreme, leading to distinct changes in the allele frequencies of the population.
   - The Darwin finches described at the beginning of this chapter were an example of directional selection. Beak size increased because only large seeds with hard seed coats were available due to a drought. However, when rains returned, directional selection reversed direction toward smaller beaks as smaller seeds with softer seed coats dominated.
   - **Insecticide resistance** occurs as a result of directional selection. Because traits of individuals vary in a population, some individuals may possess some degree of resistance to the insecticide. These few individuals survive and produce offspring, most of whom will inherit the insecticide-resistance trait. After several generations of directional selection, the population will consist of nearly all insecticide-resistant individuals.
   - The **peppered moth** provides an example of directional selection of moth color from a light to a dark color. Before the industrial revolution, the light form of the moth was well camouflaged among the light-colored lichens that grew on tree barks around London. Since color variation is known to exist in other moths, the dark form of the moth probably existed but was never observed because it was so easily spotted and eaten by predator birds. With the advent of the industrial revolution, soot killed the pollution-sensitive lichens, exposing the dark tree bark below. As a result, the dark form of the moth became the better camouflaged of the two forms, and increased in frequency. A hundred years after the first dark moth was discovered in 1848, 90% of the moths were dark colored. Meanwhile, the light form of the moth continued to dominate populations in unpolluted areas outside London. The selection of dark-colored (melanic) varieties in various species of moths as a result of industrial pollution is called **industrial melanism.** It is an example of how changes in environmental conditions promote evolution.
   - As a result of global climate change, many habitats are experiencing **season creep,** the shortening of winters and earlier arrivals of spring. In response, there is selection for plants that germinate and flower earlier. In general, invasive plant species appear to be favored.
3. **Disruptive selection** (or **diversifying selection**) occurs when the environment favors extreme or unusual traits, while selecting against the common traits.
   - In the wild, many species of weeds occur in a range of heights, but tall forms predominate. Because of disruptive selection, however, only very short forms of these same weeds occur in lawns. On lawns, short weeds are selectively advantageous because they escape mowing, allowing them to flower and produce seeds. Weeds in the wild are primarily tall because tallness makes them better competitors for sunlight.
4. **Sexual selection** is the differential mating of males (sometimes females) in a population. Since females usually make a greater energy investment in producing offspring than males do, they can increase their fitness by increasing the *quality* of their offspring by choosing superior males. Males, on the other hand, contribute little energy to the production of offspring and, thus, increase their fitness by maximizing the *quantity* of offspring produced. Thus, traits (physical qualities or behaviors) that allow males to increase their mating frequency have a selective advantage and, as a result, increase in frequency within the population. This leads to two kinds of sexual selection, as follows:
   - **Male competition** leads to contests of strength that award mating opportunities to the strongest males. The evolution of antlers, horns, and large stature or musculature are examples of this kind of sexual selection.
   - **Female choice** leads to traits or behaviors in males that are attractive to females. Colorful bird plumage (the peacock's tail is an extreme example) or elaborate mating behaviors are examples.

   Sexual selection often leads to **sexual dimorphism,** differences in the appearance of males and females. When this occurs, sexual selection is a form of disruptive selection.

5. **Artificial selection** is a form of directional selection carried out by humans when they sow seeds or breed animals that possess desirable traits. Since it is carried out by humans, it is not "natural" selection, but it is given here for comparison.
   - The various breeds of dogs have originated as a result of humans breeding animals with specific desirable traits.
   - Brussels sprouts, broccoli, cabbage, and cauliflower are all varieties of a single species of wild mustard after artificial selection of offspring possessing specific traits.

Sources of Variation

In order for natural selection to operate, there must be variation among individuals in a population. Indeed, considerable variation exists in nearly all populations. The variation arises from or is maintained by the following mechanisms:

1. **Mutations** provide the raw material for new variation. All other contributions to variation, listed here, occur by rearranging existing alleles into new combinations. Mutations, however, can invent alleles that never before existed in the gene pool.
   - Antibiotic and pesticide resistance alleles can be introduced into populations by mutation. However, these alleles may already exist as part of the genetic variation of the population. The application of antibiotics or pesticides eliminates those susceptible individuals, allowing the nonsusceptible individuals to reproduce rapidly without competition.
2. **Sexual reproduction** creates individuals with new combinations of alleles. These rearrangements, or **genetic recombinations,** originate from three events during the sexual reproductive process, as follows:
   - **Crossing over,** or exchanges of DNA between nonsister chromatids of homologous chromosomes, occurs during prophase I of meiosis.
   - **Independent assortment of homologues** during metaphase I creates daughter cells with random combinations of maternal and paternal chromosomes.
   - **Random joining of gametes** during fertilization contributes to the diversity of gene combinations in the fertilized egg (zygote).
3. **Diploidy** is the presence of two copies of each chromosome in a cell. In the heterozygous condition (when two different alleles for a single gene locus are present), the recessive allele is hidden from natural selection, allowing variation to be "stored" for future generations. As a result, more variation is maintained in the gene pool.
4. **Outbreeding,** or mating with unrelated partners, increases the possibility of mixing different alleles and creating new allele combinations.
5. **Balanced polymorphism** is the maintenance of different phenotypes in a population. Often, a single phenotype provides the best adaptation, while other phenotypes are less advantageous. In these cases, the alleles for the advantageous trait increase in frequency, while the remaining alleles decrease. However, examples of polymorphism (the coexistence of two or more different phenotypes) are observed in many populations. These polymorphisms can be maintained in the following ways:
   - **Heterozygote advantage** occurs when the heterozygous condition bears a greater selective advantage than either homozygous condition. As a result, both alleles and all three phenotypes are maintained in the population by selection. For example, the alleles for normal and sickle-cell hemoglobin proteins (Hb^A and Hb^S, respectively) produce three genotypes: Hb^AHb^A, Hb^AHb^S, and Hb^SHb^S. Hb^AHb^A individuals are normal, while Hb^SHb^S individuals suffer from sickle-cell disease, because the sickle-cell allele produces hemoglobin with an impaired oxygen-carrying ability. Without medical intervention, most Hb^SHb^S individuals die early in life. The heterozygote Hb^AHb^S individuals are generally healthy, but their oxygen-carrying ability may be significantly reduced during strenuous exercise or exposure to low oxygen concentrations (such as at high altitudes). Despite fatal effects to homozygote Hb^SHb^S individuals and

reduced viability of heterozygote individuals, the frequency of the Hb^AHb^S condition exceeds 14% in parts of Africa, an unusually high value for a deleterious phenotype. However, heterozygote Hb^AHb^S individuals have a selective advantage (in Africa) because the Hb^AHb^S trait also provides resistance to malaria. When Hb^AHb^S phenotypes are selected, both Hb^A and Hb^S alleles are preserved in the gene pool, and all three phenotypes are maintained.

- **Hybrid vigor** (or **heterosis**) describes the superior quality of offspring resulting from crosses between two different inbred strains of plants. The superior hybrid quality results from a reduction of loci with deleterious homozygous recessive conditions and an increase in loci with heterozygote advantage. For example, a hybrid of corn, developed by crossing two different corn strains that were highly inbred, is more resistant to disease and produces larger corn ears than either of the inbred strains.
- **Frequency-dependent selection** (or **minority advantage**) occurs when the least common phenotypes have a selective advantage. Common phenotypes are selected against. However, since rare phenotypes have a selective advantage, they soon increase in frequency and become common. Once they become common, they lose their selective advantage and are selected against. With this type of selection, then, phenotypes alternate between low and high frequencies, thus maintaining multiple phenotypes (polymorphism). For example, some predators form a "search image," or standard representation of their prey. By standardizing on the most common form of its prey, the predator optimizes its search effort. The prey that is rare, however, escapes predation.

Not all variation has selective value. Instead, much of the variation observed, especially at the molecular level in DNA and proteins, is **neutral variation.** For example, the differences in fingerprint patterns among humans represent neutral variation. In many cases, however, the environment to which the variation is exposed determines whether a variation is neutral or has selective value.

Humans impact the evolutionary potential of many species by reducing the size of their populations and decreasing genetic variation. When genetic variation decreases, populations lack the variation necessary to respond to selection pressures imposed by changing environments.

- **Monocultures** in agriculture reduce genetic variation because only a few varieties (sometimes only one) of the many wild varieties of a plant are used. Meanwhile, wild varieties in their native habitats are lost due to habitat destruction or other human impacts. In addition, a monoculture, by definition, has no genetic variation and is extremely susceptible to changing environmental conditions. For example, potato crops infected with potato blight, a fungal disease, resulted in widespread crop failures and famine in the middle of the nineteenth century in Ireland.
- The **overuse of antibiotics** reduces variation in bacterial populations by eliminating those individuals that are susceptible to the antibiotic. In the absence of the susceptible individuals, however, nonsusceptible bacteria increase in number and dominate the population, causing new outbreaks of disease.

Causes of Changes in Allele Frequencies

Natural selection was the mechanism that Darwin proposed for evolution. With the understanding of genetics, it became evident that factors other than natural selection can change allele frequencies and, thus, cause evolution. These factors, together with natural selection, are given here:

1. **Natural selection** is the increase or decrease in allele frequencies due to the impact of the environment.
2. **Mutations** introduce new alleles that may provide a selective advantage. In general, however, most mutations are **deleterious,** or harmful.
3. **Gene flow** describes the movement of individuals between populations resulting in the removal of alleles from a population when they leave (emigration) or the introduction of alleles when they enter (immigration).
4. **Genetic drift** is a *random* increase or decrease of alleles. In other words, some alleles may increase or decrease for no other reason than by chance. When populations are small (usually fewer than 100 individuals), the effect of genetic drift can be very strong and can dramatically influence evolution.

- An analogy of genetic drift can be made with the chances associated with flipping a coin. If a coin is flipped 100 times, the number of heads obtained would approach the expected probability of ½. However, if the coin is flipped only 5 times (analogous to a small population), one may obtain, by chance, all tails. Similarly, gene frequencies, especially in small populations, may change by chance.

Two special kinds of genetic drift are commonly observed, as follows:

- The **founder effect** occurs when allele frequencies in a group of migrating individuals are, by chance, not the same as that of their population of origin. For example, one of the founding members of the small group of Germans who began the Amish community in Pennsylvania possessed an allele for polydactylism (more than five fingers or toes on a limb). After 200 years of reproductive isolation, the number of cases of this trait among the 8,000 Amish exceeded the number of cases occurring in the remaining world's population.
- A **bottleneck** occurs when a population undergoes a dramatic decrease in size. Regardless of the cause of the bottleneck (natural catastrophe, predation, or disease, for example), the small population that results becomes severely vulnerable to genetic drift. In addition, when bottlenecks are caused by forces that strike individuals randomly (such as natural catastrophes), gene frequencies may change due to chance. Which individuals survive such a catastrophe is random—being in the wrong place at the wrong time is a random event. The remaining allele frequencies in the population after the event can be very much different than those before the event. Destructive geological or meteorological events such as floods, volcanic eruptions, and ice ages have created bottlenecks and generated genetic drift for many populations of plants and animals.

5. **Nonrandom mating** occurs when individuals choose mates based upon their particular traits. For example, they may always choose mates with traits similar to their own or traits different from their own. Nonrandom mating also occurs when mates choose only nearby individuals. In all of these cases, mate selection is not random, and only the alleles possessed by the mating individuals are passed to the next generation. The following two kinds of nonrandom mating are commonly observed:
   - **Inbreeding** occurs when individuals mate with relatives.
   - **Sexual selection** occurs when females choose to mate with males based upon their attractive appearance or behavior or mate only with males who defeat other males in contests.

Hardy-Weinberg (Genetic) Equilibrium

When the allele frequencies in a population remain constant from generation to generation, the population is said to be in **Hardy-Weinberg equilibrium** or **genetic equilibrium.** *At Hardy-Weinberg equilibrium, there is no evolution.* In order for equilibrium to occur, the factors that normally change gene frequencies do not occur. Thus, the following conditions hold:

1. **All traits are selectively neutral (no natural selection).**
2. **Mutations do not occur.**
3. **The population must be isolated from other populations (no gene flow).**
4. **The population is large (no genetic drift).**
5. **Mating is random.**

Hardy-Weinberg equilibrium is determined by evaluating the following values:

1. Allele frequencies for each allele (p, q)
2. Frequency of homozygotes (p^2, q^2)
3. Frequency of heterozygotes ($pq + qp = 2pq$)

Also, the following two equations hold:

1. $p + q = 1$ (all alleles sum to 100%)
2. $p^2 + 2pq + q^2 = 1$ (all individuals sum to 100%)

As an example, suppose a plant population consists of 84% plants with red flowers and 16% with white flowers. Assume the red allele (R) is dominant and the white allele (r) is recessive. Using the above notation and converting percentages to decimals:

$$q^2 = 0.16 = \text{white-flowered plants } (rr \text{ trait})$$
$$p^2 + 2pq = 0.84 = \text{red-flowered plants } (RR \text{ and } Rr \text{ trait})$$

To determine the allele frequency of the white-flower allele, calculate q by finding the square root of q^2.

$$q = \sqrt{0.16} = 0.4$$

Since $p + q = 1$, p must equal 0.6.

You can also determine the frequency (or percentages) of individuals with the homozygous dominant and heterozygous condition.

$$2pq = (2)(0.6)(.4) = 0.48 \text{ or } 48\% = \text{heterozygotes}$$
$$p^2 = (0.6)(0.6) = 0.36 \text{ or } 36\% = \text{homozygotes dominant}$$

In most natural populations, the conditions of Hardy-Weinberg equilibrium are not obeyed. However, the Hardy-Weinberg calculations serve as a starting point that reveal how allele frequencies are changing, which equilibrium conditions are being violated, and what mechanisms are driving the evolution of a population.

Speciation

A **species** is usually defined as a group of individuals capable of interbreeding. **Speciation,** the formation of new species, occurs by the following processes, as illustrated in Figure 10-2:

1. **Allopatric speciation** begins when a population is divided by a geographic barrier so that interbreeding between the two resulting populations is prevented. Common barriers include mountain ranges or rivers, but any region that excludes vital resources, such as a region devoid of water, a burned area devoid of food, or an area covered with volcanic lava, can act as a barrier because individuals cannot survive its crossing. Once the two populations are reproductively isolated by the barrier, gene frequencies in the two populations can diverge due to natural selection (the environments may be slightly different), mutation, or genetic drift. If the gene pools sufficiently diverge, then interbreeding between the populations will not occur if the geographic barrier is removed. Instead, differential evolution creates reproductive barriers that prevent interbreeding. As a result, new species have formed. In summary,

   geographic barrier → reproductive isolation → differential evolution → reproductive barriers → new species

2. **Sympatric speciation** is the formation of new species without the presence of a geographic barrier. This may happen in several different ways, as follows:
   - **Balanced polymorphism** among subpopulations may lead to speciation. Suppose, for example, a population of insects possesses a polymorphism for color. Each color provides a camouflage to a different substrate, and if not camouflaged, the insect is eaten by a predator. Under these circumstances, only insects with the same color can associate and mate. Thus, similarly colored insects are reproductively isolated from other subpopulations, and their gene pools diverge as in allopatric speciation.
   - **Polyploidy** is the possession of more than the normal two sets of chromosomes found in diploid ($2n$) cells. Polyploidy often occurs in plants (and occasionally in animals) where triploid ($3n$), tetraploid ($4n$), and higher ploidy chromosome numbers are found. Polyploidy occurs as a result of nondisjunction of all chromosomes during meiosis, producing two viable diploid gametes and two sterile gametes with no chromosomes. A tetraploid zygote can be established when a diploid sperm fertilizes a diploid egg. Since normal meiosis in the tetraploid individual will continue to produce diploid gametes, reproductive isolation with other individuals in the population (and, thus, speciation) occurs immediately in a single generation.

Allopatric Speciation

Species A
Species B
geographic
barrier
Time
Ancestral
Species
Variation in Phenotype

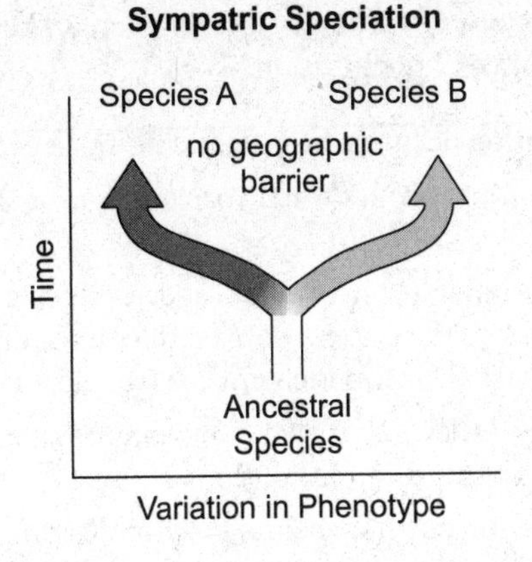

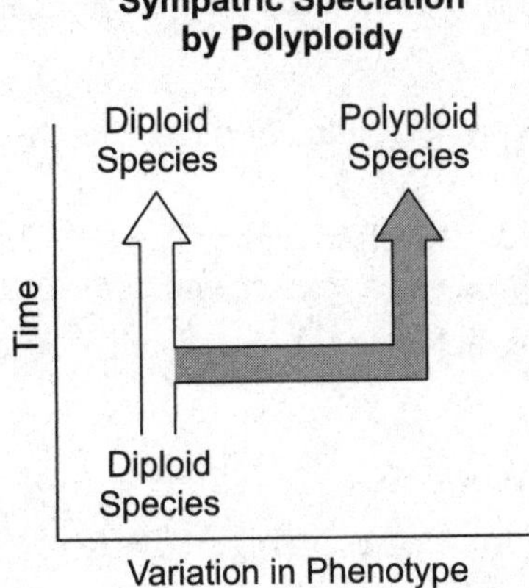

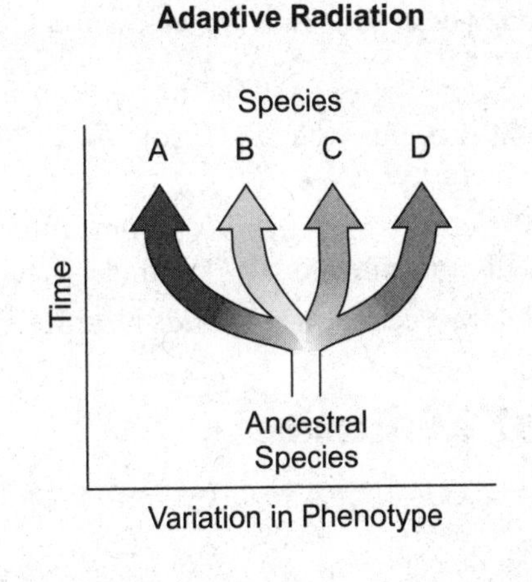

Processes of Speciation

Figure 10-2

- **Hybridization** occurs when two distinctly different forms of a species (or closely related species that are normally reproductively isolated) mate and produce progeny along a geographic boundary called a **hybrid zone.** In some cases, the genetic variation of the hybrids is greater than that of either parent and permits the population of hybrids to evolve adaptations to environmental conditions in the hybrid zone beyond the range of either parent. Exposed to different selection pressures, the hybrids eventually diverge from both parent populations.

3. **Adaptive radiation** is the relatively rapid evolution of many species from a single ancestor. It occurs when the ancestral species is introduced to an area where diverse geographic or ecological conditions are available for colonization. Variants of the ancestral species diverge as populations specialize for each set of conditions.
   - The marsupials of Australia began with the colonization and subsequent adaptive radiation of a single ancestral species.
   - The 14 species of Darwin's finches on the Galápagos Islands evolved from a single ancestral South American mainland species.
   - Adaptive radiations occurred after each of the five big mass extinctions. With up to 90% of species going extinct, the periods following extinctions provided numerous ecological opportunities for species to colonize. Colonization was followed by competition, which, in turn, promoted speciation.

Maintaining Reproductive Isolation

If species are not physically separated by a geographic barrier, various mechanisms exist to maintain reproductive isolation and prevent gene flow. These mechanisms may appear randomly **(genetic drift),** may occur from mutations, or may be the result of natural selection.

There are two categories of isolating mechanisms. The first category, **prezygotic isolating mechanisms,** consists of mechanisms that prevent fertilization:

1. **Habitat isolation** occurs when species do not encounter one another.
2. **Temporal isolation** occurs when species mate or flower during different seasons or at different times of the day.
3. **Behavioral isolation** occurs when a species does not recognize another species as a mating partner because it does not perform the correct courtship rituals, display the proper visual signals, sing the correct mating songs, or release the proper chemicals (scents, or pheromones).
4. **Mechanical isolation** occurs when male and female genitalia are structurally incompatible or when flower structures select for different pollinators.
5. **Gametic isolation** occurs when male gametes do not survive in the environment of the female gamete (such as in internal fertilization) or when female gametes do not recognize male gametes.

The second category, **postzygotic isolating mechanisms,** consists of mechanisms that prevent the formation of fertile progeny:

6. **Hybrid inviability** occurs when the zygote fails to develop properly and aborts, or dies, before reaching reproductive maturity.
7. **Hybrid sterility** occurs when hybrids become functional adults but are reproductively sterile (eggs or sperm are nonexistent or dysfunctional). The mule, a sterile offspring of a donkey and a horse, is a sterile hybrid.
8. **Hybrid breakdown** occurs when hybrids produce offspring that have reduced viability or fertility.

Patterns of Evolution

The evolution of species is often categorized into the following four patterns (Figure 10-3):

1. **Divergent evolution** describes two or more species that originate from a common ancestor and become increasingly different over time. This may happen as a result of allopatric or sympatric speciation or by adaptive radiation.
2. **Convergent evolution** describes two unrelated species that share similar traits. The similarities arise, not because the species share a common ancestor, but because each species has independently adapted to similar ecological conditions or lifestyles. The traits that resemble one another are called **analogous** traits.
   - Sharks, dolphins, and penguins have torpedo-shaped bodies with peripheral fins. These traits arise as a result of adaptations to aquatic life and not because these animals inherited the traits from a recent, common ancestor.
   - The eyes of squids and vertebrates are physically and functionally similar. However, these animals do not share a recent common ancestor. The fact that the eyes in these two groups of animals originate from different tissues during embryological development confirms that they have evolved independently.
3. **Parallel evolution** describes two related species or two related lineages that have made similar evolutionary changes after their divergence from a common ancestor.
   - Species from two groups of mammals, the marsupial mammals and the placental mammals, have independently evolved similar adaptations when ancestors encountered comparable environments.
4. **Coevolution** is the tit-for-tat evolution of one species in response to new adaptations that appear in another species. Suppose a prey species gains an adaptation that allows it to escape its predator. Although most of the predators will fail to catch prey, some variants in the predator population will be successful. Selection favors these successful variants and subsequent evolution results in new adaptations in the predator species.
   - Coevolution occurs between predator and prey, plants and plant-eating insects, pollinators and flowering plants, and pathogens and the immune systems of animals.

Divergent Evolution

African Elephant
Indian Elephant
Time
Ancestral Elephant
Variation in Morphology

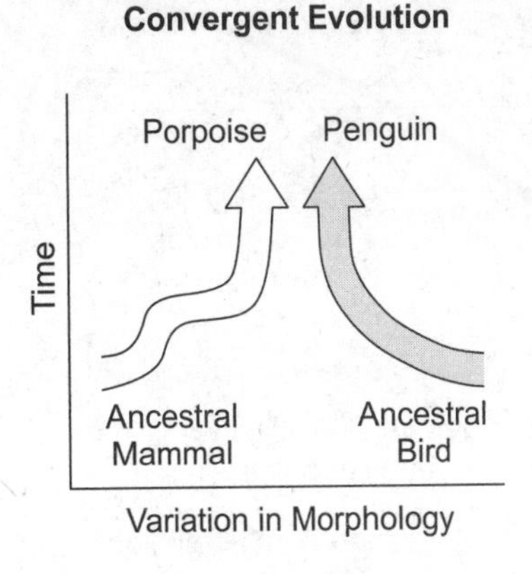

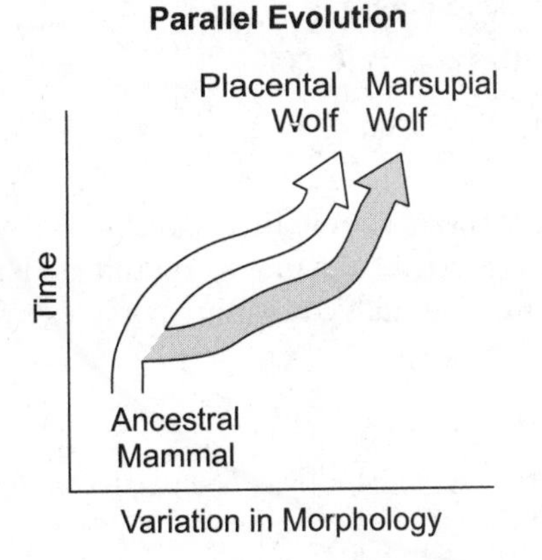

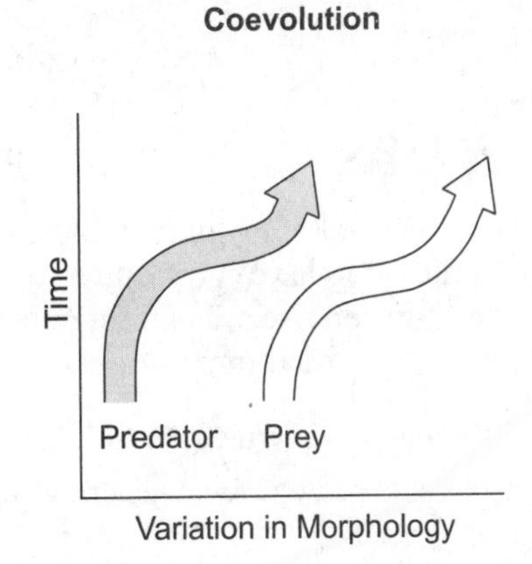

Patterns of Evolution

Figure 10-3

Microevolution vs. Macroevolution

There are two areas of evolutionary study:

1. **Microevolution** describes the details of how *populations* of organisms change *from generation to generation* and how new species originate. Microevolution is the subject of the previous sections in this chapter.
2. **Macroevolution** describes general patterns of change in *groups of related species* that have occurred over *broad periods of geologic time.* The patterns determine **phylogeny,** the evolutionary relationships among species and groups of species. Different interpretations of the fossil evidence have led to the development of two contrasting theories for the pace of macroevolution and the development of evolutionary history (Figure 10-4).
   - **Phyletic gradualism** argues that evolution occurs by the gradual accumulation of small changes. Individual speciation events or major changes in lineages occur over long periods of geologic time, from hundreds of thousands to millions of years. Fossil evidence provides snapshots of the evolutionary process, revealing only major changes in groups of organisms. That intermediate stages of evolution are not represented by fossils merely testifies to the incompleteness of the available fossil record.
   - **Punctuated equilibrium** argues that evolutionary history consists of geologically long periods of stasis with little or no evolution, interrupted, or "punctuated," by geologically short periods of rapid evolution ranging over tens of thousands of years. The fossil history, then, should consist of fossils mostly from the extended periods of stasis with few, if any, fossils available from the short bursts of evolution. Thus, in this theory, the absence of fossils revealing intermediate stages of evolution is considered data that confirm rapid evolutionary events.

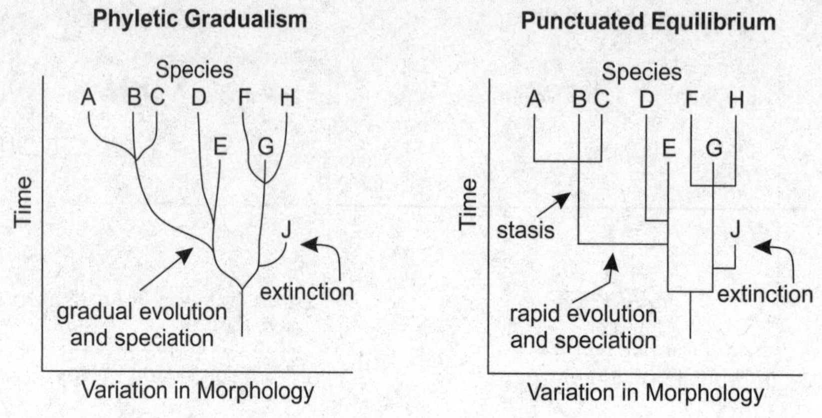

Patterns of Macroevolution

Figure 10-4

The Origin of Life

A topic related to evolution is the study of how life began, or **chemical evolution.** This kind of evolution describes the processes that are believed to have contributed to the formation of the first living things. The steps hypothesized to have led to the first primitive cell and the subsequent steps that led to more complex living cells are outlined below with supporting information:

1. **The earth and its atmosphere formed 4.6 billion years ago (bya).**
   - There is considerable geologic evidence that the earth formed 4.6 bya. The earth remained inhospitable to life for billions of years.
   - The primordial atmosphere originated from outgassing of the molten interior of the planet (through volcanos) and consisted primarily of CO_2 and N_2, but little or no O_2.
2. **The primordial seas formed.**
   - As the earth cooled, gases condensed to produce primordial seas consisting of water and minerals.
3. **Organic molecules were synthesized.**
   - Energy catalyzed the formation of organic molecules from inorganic molecules. An organic "soup" formed.
   - Energy was provided mostly by ultraviolet (UV) light, but also lightning, radioactivity, and heat.
   - Complex molecules, such as amino acids, formed. These kinds of molecules would later serve as monomers, or unit building blocks, for the synthesis of polymers.
   - Simple molecules were able to form only because oxygen was absent. As a very reactive molecule, oxygen, had it been present, would have prevented the formation of organic molecules by supplanting most reactants in chemical reactions.
   - Chemist **Stanley Miller** simulated primordial conditions by applying electric sparks to a flask containing heated water and simple gases (but no oxygen). After one week, the water contained various organic molecules, including amino acids.
4. **Polymers and self-replicating molecules were synthesized.**
   - Monomers combined to form polymers. Some of these reactions may have occurred by dehydration reactions, in which polymers formed from monomers by the removal of water molecules.
   - Self-replicating molecules, like DNA or RNA, form. DNA may have been fashioned before RNA, but the recently proposed **RNA world hypothesis** argues the reverse. This is based upon recent discoveries of the many diverse functions of RNA molecules. In particular, RNA molecules can act both as enzymes **(ribozymes)** and as vehicles of information storage (genetic material). Thus, RNA can perform the functions of both proteins (enzymes) and DNA.

5. **Organic molecules were concentrated and isolated into protobionts.**
   - **Protobionts** were the precursors of cells. They were able to carry out chemical reactions enclosed within a border across which materials could be exchanged. Borders formed in the same manner as hydrophobic molecules aggregate to form membranes (as phospholipids form plasma membranes).
   - **Liposomes** and **coacervates** are experimentally (and abiotically) produced protobionts that have some selectively permeable qualities. Both have borders that form when molecules with similar chemical properties (like hydrophilic molecules, such as lipids) separate from other molecules with different chemical properties (such as water).

6. **Primitive heterotrophic prokaryotes formed about 3.2 bya.**
   - **Heterotrophs** are living organisms that obtain energy by consuming organic substances. Pathogenic bacteria, for example, are heterotrophic prokaryotes.
   - The organic "soup" was a source of organic material for heterotrophic cells. As these cells reproduced, competition for organic material increased. Natural selection would favor those heterotrophs most successful at obtaining food.
   - The earliest fossils date to between 3.2 and 3.5 bya.

7. **Primitive autotrophic prokaryotes were formed.**
   - As a result of mutation, a heterotroph gained the ability to produce its own food. Now, as an **autotroph,** this cell would be highly successful.
   - Autotrophs manufacture their own organic compounds using light energy or energy from inorganic substances. Cyanobacteria (photosynthetic bacteria), for example, are autotrophic prokaryotes that obtain energy and manufacture organic compounds by photosynthesis.

8. **Oxygen and the ozone layer formed and abiotic chemical evolution ended.**
   - As a by-product of the photosynthetic activity of autotrophs, oxygen was released and accumulated in the atmosphere. The interaction of UV light and oxygen produced the ozone layer.
   - As a result of the formation of the ozone layer, incoming UV light was absorbed, preventing most of it from reaching the surface of the earth. Thus, the major source of energy for the abiotic synthesis of organic molecules and primitive cells was no longer available.

9. **Eukaryotes formed.**

   Eukaryotes differ from prokaryotes by the presence of mitochondria, chloroplasts, the nucleus, and various other organelles. These bodies may have formed by either (or both) of the following mechanisms.
   - **Invagination** describes how the membranes of some organelles arose by the folding in (invagination) of the plasma membrane. After folding in and separating from the plasma membrane, a new membrane forms inside the cell enclosing and isolating specialized processes.
   - **Endosymbiotic theory** describes how eukaryotic cells originated from a mutually beneficial association (symbiosis) among various kinds of prokaryotes. Specifically, mitochondria, chloroplasts, and other organelles established residence inside another prokaryote, producing a eukaryote.

     There is considerable evidence for the endosymbiotic theory. A sample of that evidence follows:
     - Mitochondria and chloroplasts possess their own DNA. The DNA is circular and without histone proteins, as is the DNA of bacteria and cyanobacteria.
     - Ribosomes of mitochondria and chloroplasts resemble those of bacteria and cyanobacteria, with respect to size and nucleotide sequence.
     - Mitochondria and chloroplasts reproduce independently of their eukaryotic host cell by a process similar to the binary fission of bacteria.
     - Mitochondria and chloroplasts have two membranes (both phospholipid bilayers). The second membrane could have been acquired when the introduced prokaryote is surrounded, in endocytosis fashion, by a vesicle produced by the host prokaryote.
     - The thylakoid membranes of chloroplasts resemble the photosynthetic membranes of cyanobacteria.

Review Questions

Multiple-Choice Questions

The questions that follow provide a review of the material presented in this chapter. Use them to evaluate how well you understand the terms, concepts, and processes presented. Actual AP multiple-choice questions are often more general, covering a broad range of concepts, and often more lengthy. For multiple-choice questions typical of the exam, take the two practice exams in this book.

Directions: Each of the following questions or statements is followed by four possible answers or sentence completions. Choose the one best answer or sentence completion.

1. Which of the following was most responsible for ending chemical evolution?

A. natural selection
B. heterotrophic prokaryotes
C. photosynthesis
D. the absence of oxygen in the atmosphere

2. Which of the following generates the formation of adaptations?

A. genetic drift
B. mutations
C. sexual reproduction
D. natural selection

3. The B blood-type allele probably originated in Asia and subsequently spread to Europe and other regions of the world. This is an example of

A. natural selection
B. genetic drift
C. gene flow
D. sexual reproduction

4. The appearance of a new mutation is

A. a random event
B. the result of natural selection
C. the result of sexual reproduction
D. usually a beneficial event

5. Which of the following is an example of sexual selection?

A. dark-colored peppered moths in London at the beginning of the Industrial Revolution
B. the mane of a lion
C. insecticide resistance in insects
D. Darwin's finches in the Galápagos Islands

6. After test-cross experiments, it was determined that the frequencies of homozygous dominant, heterozygous, and homozygous recessive individuals for a particular trait were 32%, 64%, and 4%, respectively. The dominant and recessive allele frequencies

A. are 0.2 and 0.8, respectively
B. are 0.32 and 0.68, respectively
C. are 0.36 and 0.64, respectively
D. cannot be determined because the population is not in Hardy-Weinberg equilibrium

7. *Cepaea nemoralis* is a land snail. Individual snails have shells with zero to five dark bands on a yellow, pink, or dark brown background. The various shell patterns could have occurred by all of the following EXCEPT:

A. convergent evolution
B. natural selection
C. a balanced polymorphism
D. chance

8. All of the following are homologous structures EXCEPT:

A. a bird wing
B. a butterfly wing
C. a human arm
D. a penguin flipper

Use the following key for questions 9–12. Each answer in the key may be used once, more than once, or not at all.

A. bottleneck
B. adaptive radiation
C. directional selection
D. sexual reproduction

9. Because of human predation, the sizes of and genetic variation in populations of many whale species have dramatically declined.

10. Progeny possess new combinations of alleles every generation.

11. Many strains of *Mycobacterium tuberculosis,* the bacterium that causes tuberculosis, are resistant to standard drug therapy.

12. There are more than 750,000 named species of insects inhabiting a wide range of habitats.

13. All of the following are examples of evolution EXCEPT:

A. mutations in an individual
B. changes in an allele frequency in a population
C. changes in an allele frequency in a species
D. divergence of a species into two species

14. A population consists of 9% white sheep and 91% black sheep. What is the frequency of the black-wool allele if the black-wool allele is dominant and the white-wool allele is recessive?

A. 0.09
B. 0.3
C. 0.7
D. 0.91

15. A blood group consists of two alleles, *M* and *N*. Calculate the frequency of the *M* allele if the following data were obtained for a population:

Blood Types	Number of Individuals
M	80
MN	240
N	180

A. 0.16
B. 0.4
C. 0.6
D. 8.9

Free-Response Questions

The AP exam has long and short free-response questions. The long questions have considerable descriptive information that may include tables, graphs, or figures. The short questions are brief but may also include figures. Both kinds of questions have four parts and generally require that you bring together concepts from multiple areas of biology.

The questions that follow are designed to further your understanding of the concepts presented in this chapter. Unlike the free-response questions on the exam, they are narrowly focused on the material in this chapter. For free-response questions typical of the exam, take the two practice exams in this book.

Directions: The best way to prepare for the AP exam is to write out your answers as if you were taking the exam. Use complete sentences for all your answers and do *not* use outline form or bullets. You may use diagrams to supplement your answers, but be sure to describe the importance or relevance of your diagrams.

---

1. Although muscles attach the human ear to the skull, few people can actually use these muscles to move their ears. Explain why the muscles are present if they serve no purpose.

2. "Species evolve because they have to adapt to survive." Explain why this statement is false.

3. Fingerprints, created by dermal ridges on fingers, are an example of neutral variation. The patterns vary among individuals, but differences have no selective value. If variation in fingerprints represents neutral variation, does this mean that the dermal ridges on fingers have no selective value? Justify your answer.

4. Describe how evolution occurs as a result of each of the following.

a. mutations
b. genetic drift
c. gene flow
d. nonrandom mating

5. Describe the process of speciation for each of the following.

a. allopatric speciation
b. sympatric speciation
c. adaptive radiation

6. Describe mechanisms that maintain reproductive isolation for

a. prezygotic
b. postzygotic

7. Explain how each of the following relates to speciation.

a. geographic barriers
b. polyploidy
c. sexual selection

8. Explain how each of the following is important for Darwin's theory of natural selection.

a. variation among individuals
b. heritability of traits
c. competition for resources

9. Explain how each of the following influenced the origin of living organisms.

a. primordial atmosphere
b. photosynthesis
c. oxygen and the ozone layer
d. endosymbiotic theory

Answers and Explanations

Multiple-Choice Questions

1. **C.** Chemical evolution was able to occur because highly reactive oxygen was not present. The production of oxygen from photosynthesis ended abiotic synthesis because oxygen interfered with the abiotic chemical reactions. Also, the oxygen interacted with UV light to form the ozone layer, which absorbed most incoming UV, the major energy source for abiotic reactions.

2. **D.** Only natural selection generates adaptations. Changes in gene frequencies from other factors may contribute to increases in fitness, but not because they produce adaptations. For example, mutations may introduce a new allele, but the allele will lead to an adaptation only if it increases in the population as a result of natural selection.

3. **C.** Gene flow is the increase in allele frequencies due to transfer from other populations.

4. **A.** Mutations occur randomly and are usually harmful. Whether the mutation increases or decreases in frequency in the population is the result of natural selection, genetic drift, gene flow, or nonrandom mating.

5. **B.** Only male lions have a mane. Differences in appearance between males and females (sexual dimorphisms) not directly required for reproduction are usually the result of sexual selection.

6. **D.** This population is not in Hardy-Weinberg equilibrium. The values given for p^2, $2pq$, and q^2 correctly total 1.

Calculating the value of q from q^2 gives $q = \sqrt{0.04}$ or 0.2, and the value of p from p^2 gives $p = \sqrt{0.32}$, which is approximately 0.57. The sum of these *calculated* values for p and q gives 0.77. Since $p + q$ *must* equal 1 (there are only two alleles and their frequencies must total 1), the population cannot be in equilibrium. This can be caused by the nonrandom nature of a test cross, as a population in equilibrium must be mating randomly.

7. **A.** The maintenance of various patterned shells in the snail population is an example of a balanced polymorphism. It may be (and there is good evidence that it is) maintained by natural selection, genetic drift (chance), mutations, and other factors as well. Convergent evolution does not apply here because it refers to two or more species not of common ancestral origin that share similar traits. This question deals with phenotypic variation within a single species.

8. **B.** Structures in different species are homologous because they have been inherited from a common ancestor. Insects (butterflies) are not closely related to the other listed animals. Mammals (including bats) and birds (including penguins) are related by descent from an early reptile. Insect wings, instead, are analogous structures.

9. **A.** A bottleneck occurs when population size precipitously falls. Surviving individuals may possess only a limited amount of the total genetic variation present previously. In addition, the effect of genetic drift intensifies when populations are small.

10. **D.** A consequence of sexual reproduction is that crossing over during prophase I of meiosis, mixing of maternal and paternal chromosomes, and random union of gametes produce new combinations of alleles in every generation. Except for identical twins, no two individuals will ever have exactly the same genetic makeup.

11. **C.** As a result of genetic variation, there will be some bacteria that are resistant to antibiotics. Extensive use of antibiotics kills bacteria that are susceptible, but resistant variants survive and reproduce. After many generations of (directional) selection for resistant bacteria, most surviving bacteria are antibiotic resistant.

12. **B.** The variety of insects and their range of habitat and ecological influence is an example of adaptive radiation on a grand scale.

13. **A.** Evolution does not occur for an individual. Only groups of individuals (of the same species) evolve.

14. **C.** The information given in the question is summarized as follows:

    Let $q^2 = 0.09$ = white sheep (homozygous recessives). Then $p^2 + 2pq = 0.91$ = black sheep (homozygous dominants and heterozygotes).

    The question asks for the frequency of the black-wool allele, p. Calculate the square root of $q^2 = 0.09$, using your calculator if necessary:

    $$q = 0.3 \text{ or } 30\%$$

    Because $p + q = 1$, then

    $$p = 1 - q = 0.7$$

15. **B.** First, add up all the individuals to find the size of the population: 80 + 240 + 180 = 500. Letting p and q represent the allele frequencies of N and M, respectively, then $p^2 = 180 \div 500 = 0.36$ and $q^2 = 80 \div 500 = 0.16$. Take the square root of q^2 to find $q = 0.4$.

Free-Response Questions

1. The ear muscles are vestigial, inherited from ancestors in whom they served a function. For many mammals, these muscles still serve to rotate the ear to capture sound from different directions.

2. The statement is false because the words "have to" incorrectly imply that species or individuals in a population are actively contributing to changes that lead to adaptations and increase survival. In fact, adaptations are inherited. An individual either inherits an advantageous trait or does not. If he inherits an advantageous trait, then he survives and produces offspring with similar traits.

3. Dermal ridges do have selective value. The ridges provide friction that allows for a better grip. It is not the pattern of ridges that is important, but the presence (as opposed to the absence) of the ridges that provides the selective value.

4. **a.** Evolution occurs when allele frequencies change from generation to generation in a group of interbreeding organisms. When evolution occurs as a result of natural selection, alleles of those individuals with traits enabling them to survive and reproduce better than others get passed on to the next generation. Over time, the best alleles accumulate in the population. Natural selection acts upon the available traits in the population. Mutations add new alleles, increase variation, and may introduce traits that are more successful that others in the population. Variation can be introduced into the population by mixing up existing alleles through genetic recombination, but mutation is the raw material for variation. It is the only mechanism that can introduce new alleles that didn't previously exist.

b. Genetic drift is another mechanism that can cause evolution, that is, cause allele frequencies to change. Genetic drift describes random changes in allele frequencies. This is especially influential in evolution when populations are small. When a population bottleneck occurs as a result of some catastrophic event (flood, epidemic, ice age), the small surviving population may change due to genetic drift. A founder population may also be subject to the effects of genetic drift. For example, if a small group of individuals becomes separated from the mother population (perhaps by emigration), the allele frequencies of the founder group may differ from the mother population by chance.

c. Evolution may also occur when allele frequencies change due to gene flow—the movement of alleles between populations. Gene flow occurs when individuals immigrate, bringing alleles into the population, or when they emigrate, removing alleles from the population. When gene flow causes a change in the relative frequencies of alleles, evolution occurs.

d. Nonrandom mating may also contribute to changes in allele frequencies and, therefore, cause evolution. Nonrandom mating increases the frequencies of alleles for traits that occur among the mating individuals. For example, in sexual selection, allele frequencies increase if they produce traits that give individuals a better chance of obtaining a mate. Traits that help males win contests or traits that make them more attractive to females have a selective advantage. Inbreeding is another form of nonrandom mating.

This question is about how evolution occurs, so make sure that for each part of the question your answer makes it clear how the mechanism causes evolution. In other words, don't just define each of the mechanisms.

5. **a.** Allopatric speciation occurs when a geographic barrier, such as a river or mountain range, divides the existing population into two populations. Separated in this manner, the two populations are reproductively isolated and gene flow does not occur. As a result, changes in allele frequencies in one population may not occur in the other population. If the environmental conditions vary between the two populations, natural selection may favor different traits in the two populations. Genetic drift may also cause differences in allele frequencies, either because of the founder effect or because either (or both) new population is small. In these two cases, allele frequencies are strongly influenced by chance (genetic drift). Also, mutations in one population may introduce new alleles absent in the other population, thus providing new variation upon which natural selection can act.

b. The defining characteristic of sympatric speciation is that it occurs in the absence of a geographic barrier that would isolate one or more groups of individuals. Instead, reproductive isolation occurs as a result of one of several other causes. Polyploidy, for example, creates reproductive isolation in a single generation. As a result of nondisjunction during meiosis, gametes contain all of the chromosomes instead of half of them. If such a gamete is fertilized by a similar gamete, then the resulting zygote has twice the number of chromosomes and is immediately reproductively isolated from individuals with chromosome numbers like its parents. Though rare in animals, polyploidy is common in plants.

Another source of reproductive isolation can occur when the habitats of two different species meet. In the zone where the two populations meet and overlap, some individuals may mate because of incomplete prezygotic or postzygotic reproductive isolating mechanisms. If the hybrids that form are better adapted to the features of the hybrid zone than either of the parent populations, then successful mating among the hybrids may result in a population that is isolated from either parent population.

A third source of reproductive isolation can result when a population maintains a balanced polymorphism. A balanced polymorphism occurs when multiple forms of a trait are maintained in the population at frequencies higher than would be expected from random mating. In these cases, one or more of the forms possesses an adaptation that has a greater selective value to some specific feature of the environment than other forms. In some cases, the adaptation may also create an isolating mechanism. For example, in response to seed size, a population consists of birds with large and small beaks. If the beak size influences bird song expression and results in segregated mating behavior, reproductive isolation and speciation will result.

c. Adaptive radiation occurs when a population is introduced to an area where many geographic or ecological conditions are available. When the introduced species enters the various new habitats, selection pressures will vary with habitat. For example, in colder habitats, larger animals may be favored (for insulation). In a habitat with many fruit-producing plants, fruit-eating abilities among the animals may be favored. Adaptive radiation occurs among plants as well. For example, in a rain forest habitat, individual plants that have adaptations to wet conditions are favored, whereas in dry regions, plants with water conservation adaptations (thick wax on leaf surfaces, perhaps) are favored.

Darwin's finches are a model for adaptive radiation. The finches inhabit the Galápagos Islands, a group of isolated islands off the coast of South America. Descendants from a single mainland species spread to the various islands where different ecological regions were available for colonization. The bodies of water between the islands provided a barrier that maintained isolation and led to allopatric speciation. But on each individual island, sympatric speciation occurred as finches competed for common resources. The ability to obtain food, an essential characteristic for survival, led to specialization in sizes and shapes of beaks, and, eventually, speciation. Today, as a result of allopatric speciation of populations on separate islands and sympatric speciation on each individual island, there are 14 species of finches, each adapted for obtaining different kinds of food (seeds, fruit, nectar, insects) and different sizes of food.

6. *For this question, separate your answer into two parts, a and b. In part a, define a prezygotic isolating mechanism, list the different forms that it can take (habitat, temporal, behavioral, mechanical, and gametic), and provide an example of each. In part b of your answer, define postzygotic mechanism, follow with the different forms (hybrid inviability, hybrid sterility, and hybrid breakdown), and provide an example of each.*

7. a. A geographic barrier separates a population into groups between which gene flow cannot occur. Once geographically isolated, the evolution of the two new populations may differ. For example, natural selection in one group may be different from that of the other group because their habitats differ. Resources, such as food or water, or predation may differ. Also, mutations in each group may be different. Over time, the two groups may become so different that they cannot (or will not) reproduce with each other even if the barrier is removed. As a result, they are now reproductively isolated and each is a separate species.

b. Polyploidy is the possession of one or more extra sets of chromosomes. As a result of nondisjunction during meiosis, gametes (sperm and eggs) have double the normal number of chromosomes. When a sperm produced in this manner fertilizes a similarly produced egg, the resulting diploid zygote also contains twice the normal number of chromosomes. The result is a polyploid individual. When this new individual undergoes a normal meiosis, gametes will contain twice the number of chromosomes (like its parent) and will be able to fertilize only similarly produced gametes. Thus, the polyploid individual and its progeny are reproductively isolated from the original population. The result is a speciation event occurring in a single generation. Polyploidy is common among plants and rare in animals.

c. Sexual selection is the differential mating of males within a species. Only males that win contests with other males or possess features that are attractive to females are able to mate. As a result, traits that improve a male's success in these two areas carry a selective advantage. Sexual selection results in attributes that improve success in contests (such as horns, antlers, large size, or increased musculature) or traits that are attractive to females (such as good nest-building ability, large territories, or long or colorful feathers as in peacocks and birds of paradise). Although sexual selection may change allele frequencies over time and result in new traits, speciation (the formation of a new species) does not necessarily occur.

8. a. Natural selection favors individuals with traits that increase their fitness, that is, their ability to survive and leave fertile offspring. If all individuals in a population were identical, no one individual would be any more capable of leaving offspring than any other. Without variation there can be no natural selection.

b. If a trait is not heritable, it doesn't matter how much it may increase fitness because it cannot be passed on to the next generation and cannot accumulate in the population. Such "acquired" characteristics do not contribute to evolution. Even mutations that occur in somatic tissues do not count. In order for a trait to contribute to evolution, it must be encoded in alleles that will be incorporated into gametes.

c. If there are unlimited resources and unlimited availability of mates, differences among individuals won't have any effect on their ability to produce offspring. But, eventually, as the population continues to grow, resources will become limited. Then, individuals must compete for those resources. Individuals with the traits that increase their ability to obtain resources will produce more offspring, passing their genes to the next generation. Without competition, there is no natural selection and no evolution.

This question specifically asks you to address each part of the question as to how it is important to Darwin's theory of natural selection. Do not stray from this target. For example, you would not get any points for explaining how a lack of variation can cause genetic drift because genetic drift is not part of Darwin's theory. Also, no points would be awarded for describing how mutations contribute to variation because mutations are also not part of his theory.

9. *Each part of this question corresponds to a major step leading to the origin of life and to eukaryotic cells. See "The Origin of Life," earlier in this chapter, for specific information. Be sure to separate your answers into paragraphs corresponding to each part of the question, each labeled with the appropriate letter.*

Chapter 11

Ecology

Review

Ecology is the study of the distribution and abundance of organisms, their interactions with other organisms, and their interactions with their physical environment.

The following terms provide a foundation for the study of ecology:

1. A **population** is a group of individuals all of the same species living in the same area. Thus, there are populations of humans, populations of black oaks, and populations of the bacteria *Streptococcus pneumoniae.*
2. A **community** is a group of populations living in the same area.
3. An **ecosystem** describes the interrelationships between the organisms in a community and their physical environment.
4. The **biosphere** is composed of all the regions of the earth that contain living things. This generally includes the top few meters of soil, the oceans and other bodies of water, and the lower 10 kilometers of the atmosphere.
5. The **habitat** of an organism is the type of place where it usually lives. A description of the habitat may include other organisms that live there (often the dominant vegetation), as well as the physical and chemical characteristics of the environment (such as temperature, soil quality, or water salinity).
6. The **niche** of an organism describes all the biotic (living) and abiotic (nonliving) resources in the environment used by an organism. When an organism is said to occupy a particular niche, it means that certain resources are consumed or certain qualities of the environment are changed in some way by the presence of the organism.

Population Ecology

Population ecology is the study of the growth, abundance, and distribution of populations. Population abundance and distribution are described by the following terms:

1. The **size** of a population, symbolically represented by N, is the total number of individuals in the population.
2. The **density** of a population is the total number of individuals per area or volume occupied. There may be 100 buffalo/km^2 or 100 mosquitos/m^3.
3. **Dispersion** describes how individuals in a population are distributed. They may be clumped (like humans in cities), uniform (like trees in an orchard), or random (like trees in a forest).
4. **Age structure** is a description of the abundance of individuals of each age. It is often graphically expressed in an age structure diagram (Figure 11-1). Horizontal bars or tiers of the diagram represent the frequency of individuals in a particular age group. A vertical line down the center of each tier divides each age group into males and females. A rapidly growing population is indicated when a large proportion of the population is young. Therefore, age structure diagrams that are pyramid-shaped, with tiers larger at the base and narrower at the top, indicate rapidly growing populations. In contrast, age structure diagrams with tiers of equal width represent populations that are stable, with little or no population growth (**zero population growth,** or ZPG).

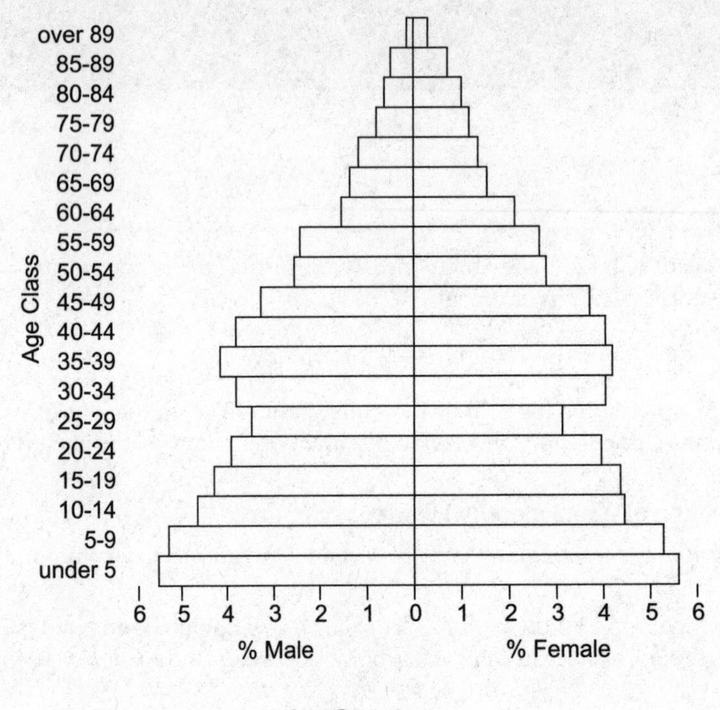

Age Structure

Figure 11-1

5. **Survivorship curves** describe how mortality of individuals in a species varies during their lifetimes (Figure 11-2):
   - **Type I** curves describe species in which most individuals survive to middle age. After that age, mortality is high. Humans exhibit type I survivorship.
   - **Type II** curves describe organisms in which the length of survivorship is random, that is, the likelihood of death is the same at any age. Many rodents and certain invertebrates (such as *Hydra*) are examples.
   - **Type III** curves describe species in which most individuals die young, with only a relative few surviving to reproductive age and beyond. Type III survivorship is typical of oysters and other species that produce free-swimming larvae that make up a component of marine plankton. Only those few larvae that survive being eaten become adults.

The following terms are used to describe population growth:

1. The **biotic potential** is the maximum growth rate of a population under ideal conditions, with unlimited resources and without any growth restrictions. For example, some bacteria can divide every 20 minutes. At that rate, one bacterium could give rise to over a trillion bacteria in 10 hours. In contrast, elephants require nearly two years for gestation of a single infant. Even at this rate, however, after 2,000 years, the weight of the descendants from two mating elephants would exceed that of the earth. The following factors contribute to the biotic potential of a species:
   - Age at reproductive maturity
   - **Clutch size** (number of offspring produced at each reproductive event)
   - Frequency of reproduction
   - Reproductive lifetime
   - Survivorship of offspring to reproductive maturity

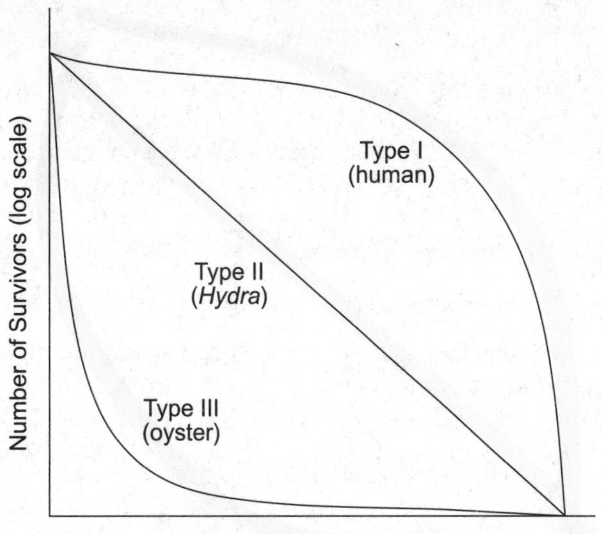

Survivorship Curves

Figure 11-2

2. The **carrying capacity** is the maximum number of individuals of a population that can be sustained by a particular habitat.
3. **Limiting factors** are those elements that prevent a population from attaining its biotic potential. Limiting factors are categorized into density-dependent and density-independent factors, as follows:
   - **Density-dependent** factors are those agents whose limiting effect becomes more intense as the population density increases. Examples include parasites and disease (transmission rates increase with population density), competition for resources (food, nesting materials, and space for growth or reproduction, including nesting sites and sunlight for photosynthesis), and the toxic effect of waste products. Also, predation is frequently density-dependent. In some animals, reproductive behavior may be abandoned when populations attain high densities. In such cases, stress may be a density-dependent limiting factor.
   - **Density-independent** factors occur independently of the density of the population. Natural disasters (fires, earthquakes, and volcanic eruptions) and extremes of climate (storms, floods, and frosts) are common examples.

The growth of a population can be described by the following equation:

$$r = \frac{\text{births} - \text{deaths}}{N}$$

In this equation, r is the **reproductive rate** (or **growth rate**) and N is the population size at the beginning of the interval for which the births and deaths are counted. The numerator of the equation is the net increase in individuals. If, for example, a population of size $N = 1{,}000$ had 60 births and 10 deaths over a one-year period, then r would equal $\frac{60-10}{1{,}000}$, or 0.05 per year.

If both sides of the equation are multiplied by N, the equation can be expressed as follows:

$$\frac{\Delta N}{\Delta t} = rN = \text{births} - \text{deaths}$$

The Greek letter delta (Δ) means "change in." Thus, $\frac{\Delta N}{\Delta t}$ means the change in the number of individuals in a given time interval. The expression can also be written in calculus terms, using $\frac{dN}{dt}$ for $\frac{\Delta N}{\Delta t}$. (Don't let the calculus expression intimidate you; for our purposes, the two expressions are essentially the same.)

$$\frac{dN}{dt} = r_{max}N$$

Here the reproductive rate, r_{max}, is maximum and so represents the biotic potential. It is called the **intrinsic rate** of growth. Note that when deaths exceed births, *r* will be negative and the population size will decrease. On the other hand, when births and deaths are equal, the growth rate is zero and the population size remains constant.

Population ecologists describe two general patterns of population growth, as follows:

1. **Exponential growth** occurs whenever the reproductive rate, as described by the equation $\frac{dN}{dt} = r_{max}N$, is greater than zero. On a graph where population size is plotted against time, a plot of exponential growth rises quickly, forming a **J-shaped** curve (Figure 11-3):

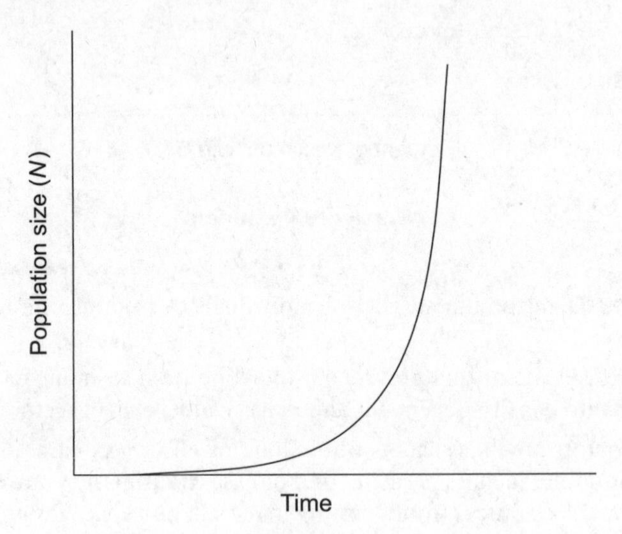

Exponential Population Growth

Figure 11-3

2. **Logistic growth** occurs when limiting factors restrict the size of the population to the carrying capacity of the habitat. In this case, the equation for reproductive rate given above is modified as follows:

$$\frac{dN}{dt} = r_{max}N\left(\frac{K-N}{K}\right)$$

K represents the carrying capacity. In logistic growth, when the size of the population increases, its reproductive rate decreases until, at carrying capacity (that is, when $N = K$), the reproductive rate is zero and the population size stabilizes. A plot of logistic growth forms an **S-shaped,** or **sigmoid,** curve (Figure 11-4):

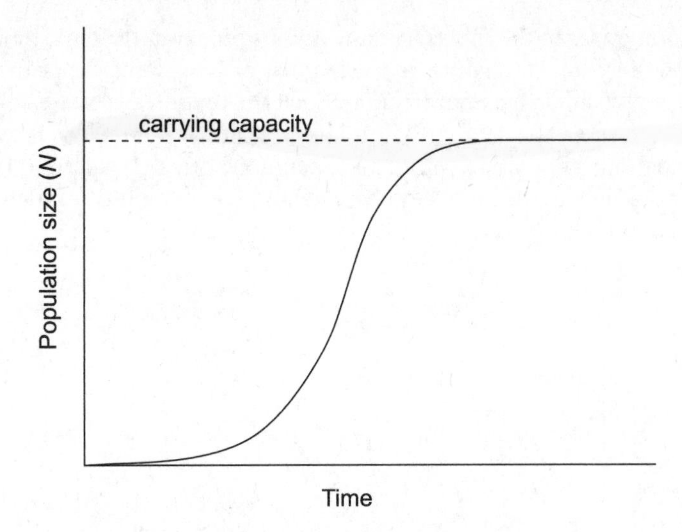

Logistic Population Growth

Figure 11-4

Population cycles are fluctuations in population size in response to varying effects of limiting factors. For example, since many limiting factors are density-dependent, they will have a greater effect when the population size is large as compared to when the population size is small. In addition, a newly introduced population may grow exponentially beyond the carrying capacity of the habitat before limiting factors inhibit growth (Figure 11-5). When limiting factors do bring the population under control, the population size may decline to levels lower than the carrying capacity (or it may even crash to extinction). Once reduced below carrying capacity, however, limiting factors may ease and population growth may renew. In some cases, a new carrying capacity, lower than the original, may be established (perhaps because the habitat was damaged by the excessively large population). The population may continue to fluctuate about the carrying capacity as limiting factors exert negative feedback on population growth when the population size is large. When the population size is small, limiting factors exert little negative feedback and population growth renews.

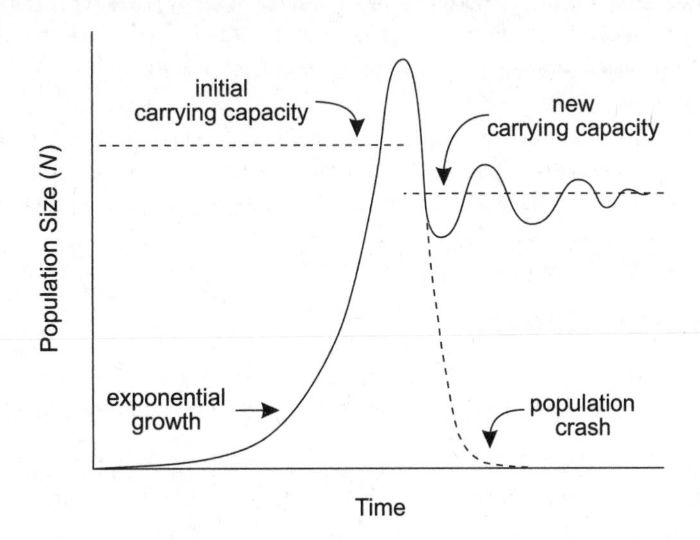

Effects of Carrying Capacity on Population Growth

Figure 11-5

Figure 11-6 shows population cycles in the snowshoe hare and its predator, the lynx. Since changes in the number of hares are regularly followed by similar changes in the number of lynx, it may appear that predation limits hare populations and that food supply limits lynx populations. Such fluctuation cycles are commonly observed between predator and prey. However, the data in Figure 11-6 indicate only an *association* between the two animals' populations, not that one population *causes* an effect in the other population. In fact, additional data suggest that the population size in hares is more closely related to the amount of available food (grass), which, in turn, is determined by seasonal rainfall levels.

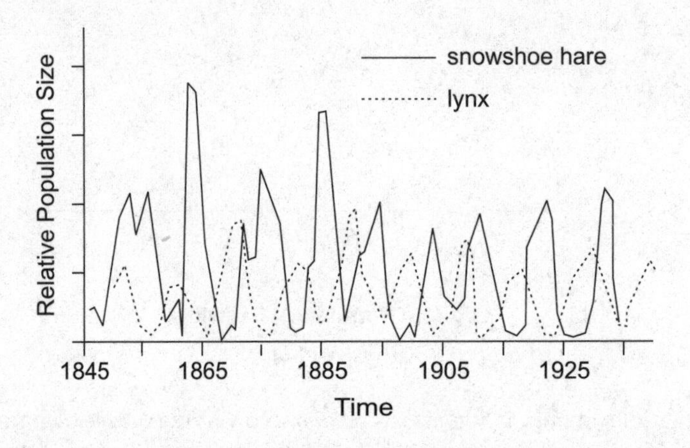

Population Cycles

Figure 11-6

The **life history** of an organism describes its strategy for maximum fitness. Reproductive success, a measure of fitness, depends upon four variables: the age of reproductive maturity, the frequency of reproduction, the number of offspring per reproductive event, and how long the organism lives. There are various ways in which these four variables can combine to maximize fitness given the resources available under different environmental conditions. In general, two major strategies emerge, as follows:

1. An ***r*-selected species** exhibits rapid growth (J-shaped curve). This type of reproductive strategy is characterized by **opportunistic species,** such as grasses and many insects, that quickly invade a habitat, quickly reproduce, and then die. They produce many offspring that are small, mature quickly, and require little, if any, parental care.
2. A ***K*-selected species** exhibits logistic growth (S-shaped curve), and the size of a mature population remains relatively constant (at the carrying capacity, *K*). Species of this type, such as humans, produce a small number of relatively large offspring that require extensive parental care until they mature. Reproduction occurs repeatedly during their lifetimes.

Human Population Growth

About a thousand years ago, the human population began exponential growth. By increasing the carrying capacity of the environment and by immigrating to previously unoccupied habitats, the following factors made exponential growth possible:

1. **Increases in food supply.** By domesticating animals and plants, humans were able to change from a hunter/gatherer lifestyle to one of agriculture. In the last half-century, food output from agriculture increased as a result of various technological advances, including the development and application of fertilizers and pesticides and the construction of irrigation systems.
2. **Reduction in disease.** Advances in medicine, such as the discoveries of antibiotics, vaccines, and proper hygiene, reduced the death rate and increased the rate of growth.

3. **Reduction in human wastes.** By developing water purification and sewage systems, health hazards from human wastes were reduced.
4. **Expansion of habitat.** Better housing, warmer clothing, and easy access to energy (for heating, cooling, and cooking, for example) allowed humans to occupy environments that were previously unsuitable.

Community Ecology

Community ecology is concerned with the interaction of populations. One form of interaction is **interspecific competition** (competition between different species). The following concepts describe the various ways in which competition is resolved:

1. The **competitive exclusion principle (Gause's principle).** When two species compete for exactly the same resources (or occupy the same niche), one is likely to be more successful. As a result, one species outcompetes the other, and eventually, the second species is eliminated. The competitive exclusion principle, formulated by biologist G. F. Gause, states that no two species can sustain coexistence if they occupy the same niche when resources are limiting.
   - Gause mixed two species of *Paramecium* that competed for the same food. One population grew more rapidly, apparently using resources more efficiently. Eventually, the second species was eliminated.
2. **Resource partitioning.** Some species coexist in spite of apparent competition for the same resources. Close study, however, reveals that they occupy slightly different niches. By pursuing slightly different resources or securing their resources in slightly different ways, individuals minimize competition and maximize success. Dividing up the resources in this manner is called resource partitioning.
   - Five species of warblers coexist in spruce trees by feeding on insects in different regions of the tree and by using different feeding behaviors to obtain the insects.
3. **Realized niche.** The niche that an organism occupies in the absence of competing species is its **fundamental niche.** When competitors are present, however, one or both species may be able to coexist by occupying their **realized niches,** that part of their existence where **niche overlap** is absent, that is, where they do not compete for the same resources.
   - Under experimental conditions, one species of barnacle can live on rocks that are exposed to the full range of tides. The full range, from the lowest to the highest tide levels, is its fundamental niche. In the natural environment, however, a second species of barnacle outcompetes the first species, but only at the lower tide levels, where desiccation is minimal. The first species, then, survives only in its realized niche, the higher tide levels.
4. **Character displacement (niche shift).** As a result of resource partitioning, certain characteristics may enable individuals to obtain resources in their partitions more successfully. Selection for these characteristics reduces competition with individuals in other partitions and leads to a divergence of features, or character displacement.
   - Two species of finches that live on two different Galápagos Islands have similar beaks, both suited for using the same food supply (seeds). On a third island, they coexist, but due to evolution, the beak of each bird species is different. This minimizes competition by enabling each finch to feed on seeds of a different size.

Predation is another form of community interaction. In a general sense, a predator is any animal that totally or partly consumes a plant or another animal. More specifically, predators can be categorized as follows:

1. A **true predator** kills and eats other animals.
2. A **parasite** spends most (or all) of its life living on another organism (the host), obtaining nourishment from the host by feeding on its tissues. Although the host may be weakened by the parasite, the host does not usually die until the parasite has completed at least one life cycle, though usually many more.
3. A **parasitoid** is an insect that lays its eggs on a host (usually another insect or a spider). After the eggs hatch, the larvae obtain nourishment by consuming the tissues of the host. The host eventually dies, but not until the larvae complete their development and begin pupation.

4. A **herbivore** is an animal that eats plants. Some herbivores, especially seed eaters, act like predators in that they totally consume the organism. Others animals, such as those that eat grasses **(grazers)** or leaves of other plants **(browsers),** may eat only part of the plant but may weaken it in the process.

Symbiosis is a term applied to two species that live together in close contact during a portion (or all) of their lives. A description of three forms of symbiosis follows. A shorthand notation for describing the relationship is provided, where a "+" indicates that one individual benefits, a "–" indicates one is harmed, and a "0" indicates no effect.

1. **Mutualism** is a relationship in which both species benefit (+, +).
   - Certain acacia trees provide food and housing for ants. In exchange, the resident ants kill any insects or fungi found on the tree. In addition, the ants crop any neighboring vegetation that makes contact with the tree, thereby providing growing space and sunlight for the acacia.
   - **Lichens,** symbiotic associations of fungi and algae, are often cited as examples of mutualism. The algae supply sugars produced from photosynthesis, and the fungi provide minerals, water, a place to attach, and protection from herbivores and from ultraviolet radiation. In some cases, however, fungal hyphae invade and kill some of their symbiotic algae cells. For this and other reasons, some researchers consider the lichen symbiosis closer to parasitism.
   - **Termites** harbor protists and bacteria for a mutualistic association, where termites supply wood (mostly cellulose) to the microbes in exchange for the breakdown of that wood. Protists are single-celled, eukaryotic organisms living inside the digestive tract of the termite, while the bacteria, living on or inside the protists, produce various **cellulases,** enzymes that break down wood.
   - The **digestive floras** are the bacteria and protists that live in the digestive tracts of animals. The flora receives food and a protected habitat. In exchange, the flora helps digest otherwise indigestible foods such as cellulose, produces vitamin K, prevents (by competitive exclusion) the growth of harmful bacteria, and carries out other beneficial functions. To digest cellulose and other polysaccharides, the flora carries out fermentation, the products of which provide the flora and the host with a source of energy and nutrients. Some mammals, the ruminants, which include cows, goats, and deer, have a specialized four-chambered stomach, the **rumen,** which serves as a fermentation chamber. Ruminants regurgitate their food for additional chewing to more completely break down food for more thorough fermentation. The digestive systems of other herbivorous mammals, such as rabbits and horses, lack a rumen and are less efficient. The digestive flora are part of a larger community that inhabits digestive tracts—the microbiome. In addition to mutualistic organisms, microbiota may include commensal, parasitic, and pathogenic organisms.
   - **Mycorrhizae** are mutualistic associations of certain fungi with the roots of plants. Plants provide carbohydrates to the fungus, and the filaments of the fungus increase the surface area of the roots, facilitating the absorption of water and minerals, especially phosphorus. Mycorrhizae are common; they form with most flowering plants and some conifers, ferns, and mosses.
2. In **commensalism,** one species benefits, while the second species is neither helped nor harmed (+, 0).
   - Many birds build their nests in trees. Generally, the tree is neither helped nor harmed by the presence of the nests.
   - Egrets gather around cattle. The birds benefit because they eat the insects aroused by the grazing cattle. The cattle, however, are neither helped nor harmed.
3. In **parasitism,** the parasite benefits from the living arrangement, while the host is harmed (+, –).
   - Tapeworms live in the digestive tract of animals, stealing nutrients from their hosts.

Coevolution

In the contest between predator and prey, some prey may have unique heritable characteristics that enable them to more successfully elude predators. Similarly, some predators may have characteristics that enable them to more successfully capture prey. The natural selection of characteristics that promote the most successful predators and the most elusive prey leads to coevolution of predator and prey. In other cases, two species may evolve so that mutual benefits increase. In general, coevolution is the evolution of one species in response to new adaptations that appear in another species. Some important examples of coevolution follow:

1. **Secondary compounds** are toxic chemicals produced in plants that discourage would-be herbivores.
   - Tannins, commonly found in oaks, and nicotine, found in tobacco, are secondary compounds that are toxic to herbivores. In many cases, metabolic adaptations have evolved in herbivores that allow them to tolerate these toxins. For example, monarch butterflies eat milkweed plants whose toxins accumulate in the bodies of the butterflies and serve to protect them from their predators.
2. **Camouflage** (or **cryptic coloration**) is any color, pattern, shape, or behavior that enables an animal to blend in with its surroundings. Both prey and predator benefit from camouflage.
   - The fur of the snowshoe hare is white in winter (a camouflage in snow) and brown in summer (a camouflage against the exposed soil).
   - The larvae of certain moths are colored so that they look like bird droppings.
   - The markings on tigers and many other cats provide camouflage in a forested background. In contrast, the yellow-brown coloring of lions provides camouflage in their savanna habitat.
   - Some plants escape predation because they have the shape and color of the surrounding rocks.
3. **Aposematic coloration** (or **warning coloration**) is a conspicuous pattern or coloration of animals that warns predators that they sting, bite, taste bad, or are otherwise to be avoided.
   - Predators learn to associate the yellow and black body of bees with danger.
4. **Mimicry** occurs when two or more species resemble one another in appearance. There are two kinds of mimicry:
   - **Müllerian mimicry** occurs when several animals, all with some special defense mechanism, share the same coloration. Müllerian mimicry is an effective strategy because a single pattern, shared among several animals, is more easily learned by a predator than would be a different pattern for every animal. Thus, bees, yellow jackets, and wasps all have yellow and black body markings.
   - **Batesian mimicry** occurs when an animal without any special defense mechanism mimics the coloration of an animal that does possess a defense. For example, some defenseless flies have yellow and black markings but are avoided by predators because they resemble the warning coloration of bees.
5. **Pollination** of many kinds of flowers occurs as a result of the coevolution of finely tuned traits between the flowers and their pollinators.
   - Pollen from flowers of the *Yucca* plant is collected by yucca moths. Pollination is accomplished when the moths roll the pollen into a ball, carry it to another *Yucca* plant, and deposit it on the stigma, the pollen receptor of a flower. The moth also deposits its eggs into some of the flower's ovules, but only about a third of the flower's seeds are eaten by the moth larvae after hatching from the eggs. There are no other pollinators for *Yucca* and no other hosts for yucca moth egg laying.
   - Red, tubular flowers with no odor have coevolved with hummingbirds, who are attracted to red and have long beaks and little sense of smell. The flowers provide a copious amount of nectar in exchange for the transfer of their pollen to other flowers.

Ecological Succession

Ecological succession is the change in the composition of species over time. The traditional view of succession describes how one community with certain species is gradually and predictably replaced by another community consisting of different species. As succession progresses, species diversity (the number of species in a community) and total biomass (the total mass of all living organisms) increase. Eventually, a final successional stage of constant species composition, called the **climax community,** is attained. The climax community persists relatively unchanged until destroyed by some catastrophic event, such as a fire.

Succession, however, is not as predictable as once thought. Successional stages may not always occur in the expected order, and the establishment of some species is apparently random, influenced by season, by climatic conditions, or by which species happens to arrive first. Furthermore, in some cases, a stable climax community is never attained because fires or other disturbances occur so frequently.

Succession occurs in some regions when climates change over thousands of years. Over shorter periods of time, succession occurs because species that make up communities alter the habitat by their presence. In both cases, the physical and biological conditions that made the habitat initially attractive to the resident species may no longer exist, and the habitat may be more favorable to new species. Some of the changes induced by resident species are listed below:

1. *Substrate texture* may change from solid rock to sand to fertile soil as rock erodes and the decomposition of plants and animals occurs.
2. *Soil pH* may decrease due to the decomposition of certain organic matter, such as acidic leaves.
3. *Soil water potential,* or the ability of the soil to retain water, changes as the soil texture changes.
4. *Light* availability may change from full sunlight to partial shade to near darkness as trees become established.
5. *Crowding,* which increases with population growth, may be unsuitable to certain species.

Succession is often described by the series of plant communities that inhabit a region over time. Animals, too, take up residence in these communities but usually in response to their attraction to the kinds of resident plants, not because of any way in which previous animals have changed the habitat. Animals do, however, affect the physical characteristics of the community by adding organic matter when they leave feces or decompose, and affect the biological characteristics of the community when they trample or consume plants or when they disperse seeds. But because animals are transient, their effects on succession are often difficult to determine.

The plants and animals that are first to colonize a newly exposed habitat are called **pioneer species.** They are typically opportunistic, ***r*-selected species** that have good dispersal capabilities, are fast growing, and produce many progeny rapidly. Many pioneer species can tolerate harsh conditions such as intense sunlight, shifting sand, rocky substrate, arid climates, or nutrient-deficient soil. For example, nutrient-deficient soils of some early successional stages harbor nitrogen-fixing bacteria or support the growth of plants whose roots support mutualistic relationships with these bacteria.

As soil, water, light, and other conditions change, *r*-selected species are gradually replaced by more stable ***K*-selected species.** These include perennial grasses, herbs, shrubs, and trees. Because *K*-selected species live longer, their environmental effects slow down the rate of succession. Once the climax community is established, it may remain essentially unchanged for hundreds of years.

There are two kinds of succession, as follows:

1. **Primary succession** occurs on substrates that never previously supported living things. For example, primary succession occurs on volcanic islands, on lava flows, and on rock left behind by retreating glaciers. Two examples follow:
   - *Succession on rock or lava* usually begins with the establishment of lichens. Hyphae of the fungal component of the lichen attach to rocks, the fungal mycelia hold moisture that would otherwise drain away, and the lichen secretes acids that help erode rock into soil. As soil accumulates, bacteria, protists, mosses, and fungi appear, followed by insects and other arthropods. Since the new soil is typically nutrient deficient, various nitrogen-fixing bacteria appear early. Grasses, herbs, weeds, and other *r*-selected species are established next. Depending upon local climatic conditions, *r*-selected species are eventually replaced by *K*-selected species such as perennial shrubs and trees.
   - *Succession on sand dunes* begins with the appearance of grasses adapted to taking root in shifting sands. These grasses stabilize the sand after about six years. The subsequent stages of this succession can be seen on the dunes of Lake Michigan. The stabilized sand allows the rooting of shrubs, followed by the establishment of cottonwoods. Pines and black oaks follow over the next 50 to 100 years. Finally, the beech-maple climax community becomes established. The entire process may require 1,000 years.

2. **Secondary succession** begins in habitats where communities were entirely or partially destroyed by some kind of damaging event. For example, secondary succession begins in habitats damaged by fire, floods, insect devastations, overgrazing, and forest clear-cutting and in disturbed areas such as abandoned agricultural fields, vacant lots, roadsides, and construction sites. Because these habitats previously supported life, secondary succession, unlike primary succession, begins on substrates that already bear soil. In addition, the soil contains a community of viable native seeds called the *soil seed bank.* Two examples of secondary succession follow:
   - *Succession on abandoned cropland* (called old-field succession) typically begins with the germination of *r*-selected species from seeds already in the soil (such as grasses and weeds). The trees that ultimately follow are region specific. In some regions of the eastern United States, pines take root next, followed by various hardwoods such as oak, hickory, and dogwood.
   - *Succession in lakes and ponds* begins with a body of water, progresses to a marsh-like state, then a meadow, and finally to a climax community of native vegetation. Sand and silt (carried in by a river) and decomposed vegetation contribute to the filling of the lake. Submerged vegetation is established first, followed by emergent vegetation whose leaves may cover the water surface. Grasses, sedges, rushes, and cattails take root at the perimeter of the lake. Eventually, the lake fills with sediment and vegetation and is subsequently replaced by a meadow of grasses and herbs. In many mountain regions, the meadow is replaced by shrubs and native trees, eventually becoming a part of the surrounding coniferous forest.

The Flow of Energy in Ecosystems

On average, only about 1% of the solar energy that reaches the surface of the earth is converted into organic matter. How that energy gets distributed to all the living things in an ecosystem helps us understand how ecosystems work.

Two types of flow charts demonstrate how energy flows through an ecosystem, showing who eats whom. The arrows used in these flow charts indicate the direction of energy flow.

1. A **food chain** is a linear flow chart of who eats whom. For example, a food chain depicting energy flow in a savanna may look like this:

   grass → zebra → lion → vulture

2. A **food web** is an expanded, more complete version of a food chain. It would show all of the major plants in the ecosystem, the various animals that eat the plants (such as insects, rodents, zebras, giraffes, and antelopes), and the animals that eat the animals (lions, hyenas, jackals, and vultures). Detritivores may also be included in the food web. Arrows connect all organisms that are eaten to the animals that eat them, pointing in the direction of energy flow.

Another way to illustrate energy flow and the production and utilization of energy is to organize plants and animals into groups called **trophic levels.** In general, organisms are either **autotrophs** that are able obtain energy from light or inorganic material or **heterotrophs** that must consume other organisms or organic material for their source of energy. Each of the following groups represents a trophic level that reflects its main energy source:

1. **Primary producers** are photoautotrophs that convert sun energy into chemical energy. They include plants, photosynthetic protists, and cyanobacteria. Primary producers can also be represented by chemoautotrophs when the sources of energy are inorganic substances.
2. **Primary consumers,** or herbivores, are heterotrophs that eat the primary producers.
3. **Secondary consumers,** or primary carnivores, are heterotrophs that eat the primary consumers.
4. **Tertiary consumers,** or secondary carnivores, are heterotrophs that eat the secondary consumers.
5. **Detritivores** are heterotrophs that obtain their energy by consuming dead plants and animals **(detritus).** The smallest detritivores, called **decomposers,** include fungi and bacteria. Other detritivores include nematodes, earthworms, insects, and scavengers such as crabs, vultures, and jackals.

Chemoautotrophs are the primary producers for hydrothermal vent communities where seawater comes in contact with super-hot rock, H_2S, and O_2. In these unique communities, chemoautotrophic prokaryotes live mutualistically in specialized organs of large (1.5 m long) tube worms. The tube worms absorb dissolved H_2S and O_2 and transport it to the prokaryotes, where they use the energy from H_2S to produce carbohydrates.

Ecological pyramids are used to show the relationship between trophic levels. Horizontal bars or tiers are used to represent the relative sizes of trophic levels, each represented in terms of energy (also called productivity), biomass, or numbers of organisms. The tiers are stacked upon one another in the order in which energy is transferred between levels. The result is usually a pyramid-shaped figure, although other shapes may also result. Several kinds of ecological pyramids are illustrated in Figure 11-7.

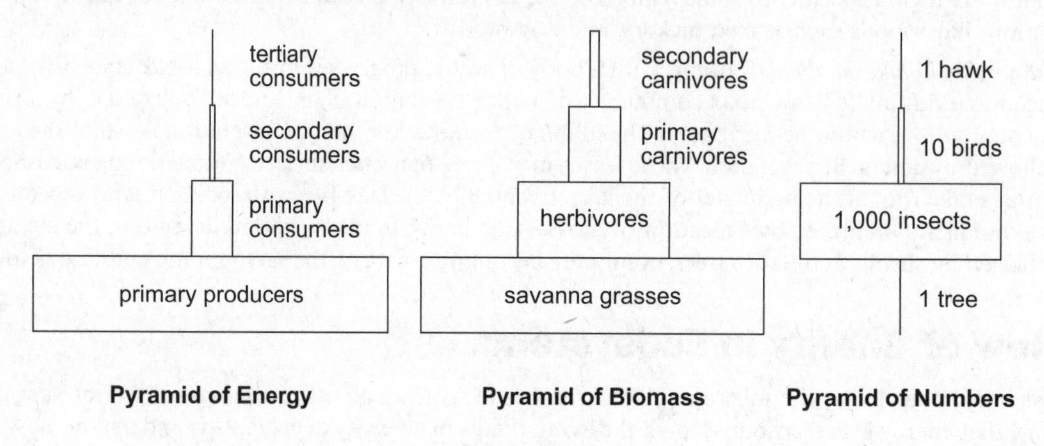

Figure 11-7

Primary productivity in an ecosystem is the amount of organic matter produced through photosynthetic (or chemosynthetic) activity per unit of time. Components of primary productivity include the following (note that the term *rate* means per unit time):

1. **Gross primary productivity (GPP)** is the rate at which producers acquire chemical energy before any of this energy is used for metabolism.
2. **Net primary productivity (NPP)** is the rate at which producers acquire chemical energy less the rate at which they consume energy through respiration. NPP represents the biomass available to herbivores.
3. **Respiratory rate (R)** is the rate at which energy is consumed through respiration (and other metabolic activities necessary to maintain life). This energy, much of it in the form of ATP, is ultimately lost as heat.

These components are related by the following equation.

$$NPP = GPP - R$$

NPP, generated at the bottom tier of the ecological pyramid of energy, supports all of the tiers above it. Energy stored in NPP is transferred to herbivores when they eat the primary producers. The herbivores, then, use that energy for respiration (and other activities necessary for life), and the remainder is used for growth, increasing biomass. Each successive tier above the herbivores, that is, the secondary and tertiary consumers, repeats the process, using the energy in the biomass of tiers below them to provide for respiration and growth.

Ecological efficiency describes the proportion of energy represented at one trophic level that is transferred to the next level. The relative sizes of tiers in an energy pyramid (or pyramid of productivity) indicate the ecological efficiency of the ecosystem. On average, the efficiency is only about 10%, that is, about 10% of the productivity of one trophic level is transferred to the next level. The remaining 90% is consumed by the individual metabolic activities of each plant or animal or is transferred to detritivores when they die.

It is important to note that much of the energy for respiration and other metabolic activities is ultimately lost as heat. Heat is energy that cannot be harnessed by organisms to do work and, thus, represents energy lost from the ecosystem. Ultimately, all energy originally gained through NPP is lost as heat. Thus, in contrast to nutrients and other forms of matter (discussed later in this chapter), heat cannot be recycled. Remember: Energy flows, matter recycles.

Because ecological efficiency is so low, nearly all domestic animals used for food or work are herbivores. If a carnivore were raised for food or work, the energy required to raise and sustain it would far exceed its value in food or work. The meat consumed by the carnivore would yield a greater return by merely using it directly for human food.

Species Diversity and Trophic Interactions

Certain species in a community can heavily influence the dynamics of that community:

1. The **dominant species** is the most abundant species or the species that contributes the greatest biomass to a community. A species becomes dominant in its particular habitat because, of all other species in the community, it is best able to compete for resources or escape predators or disease.
2. A **keystone species** is one that has a strong influence on the health of a community or ecosystem. Removal of a keystone species results in dramatic changes in the makeup of species that comprise other trophic levels.
   - Sea otters, a keystone species, eat sea urchins, and sea urchins eat kelp. When otter population size drops along the Pacific coast of North America due to excessive hunting (by humans) or predation (by killer whales), the number of urchins increases dramatically. In turn, kelp populations plummet, and the fish and marine invertebrates that lived in the kelp beds disappear.
   - The grey wolf, a keystone species, eats elk, deer, coyotes, and other large mammals. Because of threats to livestock, wolves were hunted to extinction in most U.S. states. In their absence, elk and deer populations exploded, vegetation was overgrazed, and erosion of the landscape ensued. Coyotes also flourished, and as a result, populations of smaller mammals diminished. When wolves were reintroduced to Yellowstone National Park, elk, deer, and coyote populations declined and much vegetation returned. Numbers of beavers, hares, and other small herbivores increased, camouflaged by the restored vegetation (especially near lakes and streams). These small herbivores, as well as the carcasses abandoned by wolves, attracted eagles, wolverines, and other meat-eating animals. Consequently, the return of wolves to Yellowstone restored the balance of living things that were historically native to the ecosystem.
3. An **invasive species** is an introduced species that proliferates and displaces native species because it is a better competitor or because its natural predators or pathogens are absent. Most invasive species have been accidently or deliberately introduced by humans.
   - **Kudzu** is a climbing vine that grows throughout the southeastern United States. It coils over trees and other vegetation, killing them by preventing sunlight from reaching leaves. Native to Asia, it was introduced to the United States at the first World's Fair in 1876.
   - **Dutch elm disease** is caused by a fungus that attacks and kills elm trees. The fungus is spread by a bark beetle. The native habitat of the fungus is uncertain, but it was introduced accidently in Europe and the United States in the early twentieth century.
   - **Potato blight** is caused by a fungus-like protist. Fungal spores overwinter on tubers (potatoes), germinate, and spread though plant tissues, eventually producing new spores on the leaves. Spores are dispersed by wind or water. Potatoes, native to South America, were introduced (deliberately) to Europe in the sixteenth century, where they remained disease-free until the middle of the nineteenth century. The fungus, possibly originating in Central or South America, first appeared in the United States, where spores were dispersed by wind, then delivered (accidently) to Europe on diseased potatoes.
   - **Smallpox** is an infectious human disease introduced to North America by Europeans in the sixteenth century. Because populations originating in the Western Hemisphere lacked any natural immunity, the disease caused widespread devastation among the Native American populations.

A number of factors influence the number and size of trophic levels in an ecosystem:

1. **Size of bottom trophic level.** Because primary producers provide the initial source of energy to the ecosystem, their numbers and the amount of biomass they generate govern the size and makeup of all other trophic levels. Thus, an ecosystem based on a small tier of primary producers cannot sustain many tiers above it.

2. **Efficiency of energy transfer between trophic levels.** On average, there is a 10% decline (that is, an exponential decline) in biomass and the energy it provides as it passes up through the tiers of the pyramid of energy. Thus, the number of trophic levels that can be supported declines rapidly. However, more efficient ecosystems, with higher photosynthetic efficiency, can generate longer food chains and more complex food webs. Such ecosystems occur in tropical rain forests.
3. **Stability of trophic levels.** In ecosystems with long food chains, top trophic levels are more susceptible to damage because there are more levels below them that can be weakened by environmental changes.
4. **Requirements of top predators.** The top predators that occupy the uppermost tier are usually large animals with proportionately large energy requirements. Thus, the size of top tiers is limited both because there is less biomass available at the top of the pyramid and because the individual energy requirements of the animals are large.

The size of trophic levels can also be regulated by interactions between the levels. Two interaction models are described:

1. A **bottom-up model** describes how changes in the structure of trophic levels are regulated by changes in the *bottom* trophic level. When primary productivity is low, few trophic levels above it can be supported. In the absence of predation by herbivores, primary producers expand without challenge. At some point, however, herbivores begin to respond to the increasing availability of primary production, and herbivore populations expand. As herbivore numbers increase, more primary producers are consumed, and their growth is checked. Then, predators respond to increasing numbers of herbivores, limiting herbivore growth, which, in turn, allows growth of primary producers to increase once again.
2. A **top-down model** describes how changes in the structure of trophic levels are regulated by changes in the *top* trophic level. The model is essentially the opposite of the bottom-up model. For example, sea otters occupying the top tier (as a keystone species, described above) limit the number of sea urchins. With sea urchins in check, the bottom trophic level of kelp forests increases in numbers. In contrast, if sea otters are removed, sea urchins increase and kelp forests dissipate. Thus, sea otters, comprising the top trophic tier, regulate the population size of the kelp, the bottom tier. Many examples of the top-down model are created by humans when they remove top predators (often keystone species) by overhunting or habitat destruction.

The **biodiversity** of an ecosystem is expressed as a function of the number of species, niches, and trophic levels in the ecosystem and the complexity of its food web. In addition to the interactions between species and trophic levels described above, a number of factors influence biodiversity:

1. **Climate** is a strong determinant of biodiversity. The amount, variability, and form of water (precipitation as rain or snow, presence of rivers or other bodies of water) strongly influence the abundance and type of primary producers and the number of species primary production can support. The range of temperatures between night and day, as well as throughout the year, is equally important.
2. **Latitude** is strongly correlated with climate, but it also determines solar energy exposure. Solar energy exposure controls the extent of photosynthesis and the biomass of the primary producers. Areas at the equator receive more solar energy than those at the poles. Also, because seasonal variations are minimized at lower latitudes (regions close to the equator), more constant environmental conditions are often able to support more species.
   - Tropical ecosystems are highly diverse, having many species but with smaller numbers of each species. In contrast, polar ecosystems have few species but with many individuals of those species.
3. **Habitat size and diversity** influence how many different kinds of organisms can be supported. The larger an ecosystem, the more likely that it contains greater kinds of soil texture, soil chemistry, and variations in the slope of the terrain. More diverse habitats support a greater variety of species.
4. **Elevation** also influences biodiversity. Temperature is strongly correlated with elevation, decreasing as elevation increases. Precipitation often increases with elevation, although the water content of snow deposited at higher elevations is only available to plants after it has melted.
   - Coniferous forests heavily laden with snow during the winter months become productive only after the snow has melted.

Simpson's Diversity Index can be used to calculate the biodiversity of a community:

Diversity Index $= 1 - \sum\left(\frac{n}{N}\right)^2$, where n = the number of individuals in a species population and N = the total number of all individuals among all species.

To calculate this index, calculate $\frac{n}{N}$ for each species and square it. Add together all of these values for each species and subtract the sum from 1. Values of D vary from 0 to 1, with higher numbers indicating greater diversity.

The **stability** of an ecosystem increases with increases in biodiversity. This occurs because, in a highly diverse system, disturbances may adversely affect only a few of the components (species) of the ecosystem, while one or more unaffected components can replace them. In systems with low biodiversity, disturbances may have a more permanent effect. Disturbances that threaten the stability of ecosystems include fires, floods, disease, and the various effects caused by humans (discussed in "Human Impact on Ecosystems," later in this chapter). Some environmental disturbances are extremely disruptive, however, and have major effects on ecosystems and the biosphere, as follows:

1. **El Niño** (El Niño–Southern Oscillation) is an atmospheric and oceanic phenomenon that precedes changes in weather patterns throughout the biosphere. During normal atmospheric conditions (La Niña), trade winds over the equatorial Pacific blow strongly from east to west. As the wind blows, it pushes surface water in the same direction, blowing it away from the west coast of South America. In its place, cold, nutrient-rich water from below rises to the surface (a process called **upwelling**) along the South American coastline. The nutrients promote a bottom-up ecological effect, stimulating algae growth first, then marine herbivores and fish. Sea birds eat the fish. But when an El Niño occurs, the trade winds and upwelling stop, algae populations decline, and the collapse of the local food web and local fish industry ensues. In contrast, many marine invertebrate populations skyrocket as they thrive on the carcasses of dead sea animals. The effects of El Niño stretch globally, increasing or decreasing rainfall from Peru to Australia, Indonesia, and the southwestern United States.
2. **Meteor impacts** and large **volcanic eruptions** have historically triggered global changes in atmospheric conditions. Both increase the amount of particulate matter in the atmosphere, reducing the amount of solar radiation reaching the surface of the earth. A decrease in primary production follows, and that initiates a bottom-up effect that devastates the ecosystem. The particulate matter released into the atmosphere by volcanic eruptions has a cooling effect on the global climate (by reducing solar radiation) much more so than any warming effect produced by the release of CO_2.
   - The last mass extinction occurred about 65 million years ago (mya), at the boundary of the Cretaceous and Tertiary geologic periods (the K-T boundary). It may have been caused by an asteroid impact or a volcanic eruption. About 75% of all species became extinct, including the (non-bird) dinosaurs.
   - A mass extinction occurred about 200 mya at the boundary of the Triassic and Jurassic geologic periods. It may have been caused by an asteroid impact or a volcanic eruption. About 50% of all species became extinct. The extinction opened up ecological niches that were filled by dinosaurs as they became the dominant terrestrial fauna.
3. **Plate tectonics** (continental drift) describes the movement of land masses, called plates, over the surface of the earth. Fault lines are plate boundaries. When plates collide, they generate earthquakes, create volcanos, and form mountains. Plates that are moving apart form ocean basins between them. The formation of continents arising from the separation of larger plates is a long-term isolating mechanism for speciation. As mountains and volcanos form and as plates move to new latitudes, environmental conditions change, creating new niches for speciation.
   - Australia, Antarctica, South America, Africa, Madagascar, and India were once joined as parts of a supercontinent called Gondwana (also called Gondwanaland). During their early evolution, mammals on Australia, South America, and Antarctica became isolated from other mammals when these continents separated. These mammals evolved into the marsupial mammals, while mammals on the remaining land masses evolved into the placental mammals.

Biogeochemical Cycles

Biogeochemical cycles describe the flow of essential elements from the environment to living things and back to the environment. The following list outlines the major storage locations (reservoirs) for essential elements, the processes through which each element incorporates into terrestrial plants and animals (assimilation), and the processes through which each element returns to the environment (release).

1. **Hydrologic cycle** (water cycle).
   - *Reservoirs:* Oceans, air (as water vapor), groundwater, glaciers. (Evaporation, wind, and precipitation move water from oceans to land.)
   - *Assimilation:* Plants absorb water from the soil; animals drink water or eat other organisms (which are mostly water).
   - *Release:* Plants lose water through their leaves (transpiration); animals and plants decompose.
2. **Carbon cycle.** Carbon is required for the building of all organic compounds.
   - *Reservoirs:* Atmosphere (as CO_2), bodies of water (as bicarbonate), fossil fuels (coal, oil, natural gas), peat, durable organic material (cellulose, for example).
   - *Assimilation:* Plants use CO_2 in photosynthesis; animals consume plants or other animals.
   - *Release:* Plants and animals release CO_2 through respiration and decomposition; CO_2 is released when organic material (such as wood and fossil fuels) is burned.
3. **Nitrogen cycle.** Nitrogen is required for the manufacture of all amino acids and nucleic acids.
   - *Reservoirs:* Atmosphere (N_2); soil (NH_4^+ or ammonium, NH_3 or ammonia, NO_2^- or nitrite, NO_3^- or nitrate).
   - *Assimilation:* Plants absorb nitrogen either as NO_3^- or as NH_4^+; animals obtain nitrogen by eating plants or other animals. The stages in the assimilation of nitrogen are as follows:
     - **Nitrogen fixation:** N_2 to NH_4^+ by nitrogen-fixing prokaryotes (in soil and root nodules); N_2 to NO_3^- by lightning and UV radiation.
     - **Nitrification:** NH_4^+ to NO_2^- and NO_2^- to NO_3^- by various nitrifying bacteria.
     - NH_4^+ or NO_3^- to organic compounds by plant metabolism.
   - *Release:* Bacteria and animals promote the release of nitrogen from organic and inorganic molecules.
     - **Denitrification:** NO_3^- converted to N_2 by denitrifying bacteria.
     - **Ammonification:** Organic compounds converted to NH_4^+ detritivorous bacteria.
     - Excretion of NH_4^+ (or NH_3), urea, and uric acid by animals.
4. **Phosphorus cycle.** Phosphorus is required for the manufacture of ATP and all nucleic acids. Biogeochemical cycles of other minerals, such as calcium and magnesium, are similar to the phosphorus cycle.
   - *Reservoirs:* Rocks and ocean sediments. (Erosion transfers phosphorus to water and soil; sediments and rocks that accumulate on ocean floors return to the surface as a result of uplifting by geological processes.)
   - *Assimilation:* Plants absorb inorganic PO_4^{3-} (phosphate) from soils; animals obtain organic phosphorus when they eat plants or other animals.
   - *Release:* Plants and animals release phosphorus when they decompose; animals excrete phosphorus in their waste products.

Human Impact on Ecosystems

Human activity damages the biosphere. Exponential population growth, destruction of habitats for agriculture and mining, pollution from industry and transportation, and many other activities all contribute to the damage of the environment. Some of the destructive consequences of human activity are summarized as follows:

1. **Global climate change.** The solar radiation that passes through the atmosphere consists mostly of visible light, some shorter wavelength ultraviolet (UV) radiation, and some longer wavelength infrared radiation (heat). Some of this solar radiation is reflected by the atmosphere back into space, while the remainder

passes through the atmosphere and is absorbed by the earth. The earth reemits some of the radiation back into the atmosphere, but as longer-wavelength, infrared radiation. As the infrared radiation passes through the atmosphere on its way out into space, it is absorbed by CO_2 and other gases. These gases, called **greenhouse gases,** are transparent to visible and UV radiation but absorb infrared radiation, reemitting it again as infrared, which, subsequently, is trapped and reemitted again by more greenhouse gases. As a result, the energy content of the atmosphere, and its temperature, increase. Because the glass enclosures of a greenhouse also trap heat, the trapping of heat in the atmosphere by energy-absorbing gases is called the **greenhouse effect.** As the food and energy needs of the human population rise, forests are burned for agricultural land and fossil fuels are burned for energy. Burning releases CO_2, and more CO_2 traps more heat. As a result, global temperatures are rising. Warmer temperatures raise sea levels (by melting more ice), decrease agriculture output (by affecting weather patterns), increase human disease (by broadening the range of tropical disease vectors), and threaten extinction to species (by disrupting the environmental conditions to which species are adapted).

2. **Ozone depletion.** The ozone layer forms in the upper atmosphere when UV radiation reacts with oxygen (O_2) to form ozone (O_3). The ozone absorbs UV radiation and, thus, prevents it from reaching the surface of the earth, where it would damage the DNA of plants and animals. Various air pollutants, such as chlorofluorocarbons (CFCs), enter the upper atmosphere and break down ozone molecules. Although their manufacture has been phased out, CFCs were historically used as refrigerants, as propellants in aerosol sprays, and in the manufacture of plastic foams. When ozone breaks down, the ozone layer thins, allowing UV radiation to penetrate and reach the surface of the earth. Areas of major ozone thinning, called **ozone holes,** have appeared over Antarctica, the Arctic, and northern Eurasia, but the ozone layer is predicted to return to normal by 2050 as a result of global cooperation to stop the use of CFCs.
3. **Acid rain.** The burning of fossil fuels (such as coal) and other industrial processes release into the air pollutants that contain sulfur dioxide and nitrogen dioxide. When these substances react with water vapor, they produce sulfuric acid and nitric acid. When these acids return to the surface of the earth (with rain or snow), they acidify soils and bodies of water, decreasing pH and adversely affecting plants and animals in lakes and rivers and on land.
4. **Desertification.** Overgrazing of grasslands that border deserts transform the grasslands into deserts. As a result, agricultural output decreases, or habitats available to native species are lost.
5. **Deforestation.** Clear-cutting of forests causes erosion, flooding, and changes in weather patterns. The slash-and-burn method of clearing tropical rain forests for agriculture increases atmospheric CO_2, which contributes to the greenhouse effect. Because most of the nutrients in a tropical rain forest are stored in the vegetation, burning the forest destroys the nutrients. As a result, the soil of some rain forests can support agriculture for only one or two years.
6. **Pollution.** Air pollution, water pollution, and land pollution contaminate the materials essential to life. Many pollutants do not readily degrade and remain in the environment for decades. Some toxins, such as the pesticide DDT, concentrate in plants and animals. As one organism eats another, the toxin becomes more and more concentrated, a process called **biological magnification.** Other pollution occurs in subtle ways. A lake, for example, can be polluted with runoff fertilizer or sewage. Abundant nutrients, especially phosphates, stimulate **algal blooms,** or massive growths of algae and other phytoplankton. The phytoplankton reduce oxygen supplies at night when they respire. In addition, when the algae eventually die, their bodies are consumed by detritivorous bacteria, whose growth further depletes the oxygen. The result is massive oxygen starvation for many animals, including fish and invertebrates. In the end, the lake fills with carcasses of dead animals and plants. The process of nutrient enrichment in lakes and the subsequent increase in biomass is called **eutrophication.** When the process occurs naturally, growth rates are slow and balanced. But with the influence of humans, the accelerated process often leads to the death of fish and the growth of anaerobic bacteria that produce foul-smelling gases.
7. **Reduction in species diversity.** As a result of human activities, especially the destruction of tropical rain forests and other habitats, plants and animals are becoming extinct at a faster rate than the planet has ever previously experienced. If they were to survive, many of the disappearing plants could become useful to humans as medicines, food, or industrial products.

Review Questions

Multiple-Choice Questions

The questions that follow provide a review of the material presented in this chapter. Use them to evaluate how well you understand the terms, concepts, and processes presented. Actual AP multiple-choice questions are often more general, covering a broad range of concepts, and often more lengthy. For multiple-choice questions typical of the exam, take the two practice exams in this book.

Directions: Each of the following questions or statements is followed by four possible answers or sentence completions. Choose the one best answer or sentence completion.

1. A group of interbreeding individuals occupying the same area is best called

A. a community
B. a population
C. an ecosystem
D. a symbiotic relationship

Questions 2–4 refer to the following age structure diagrams that represent four different populations.

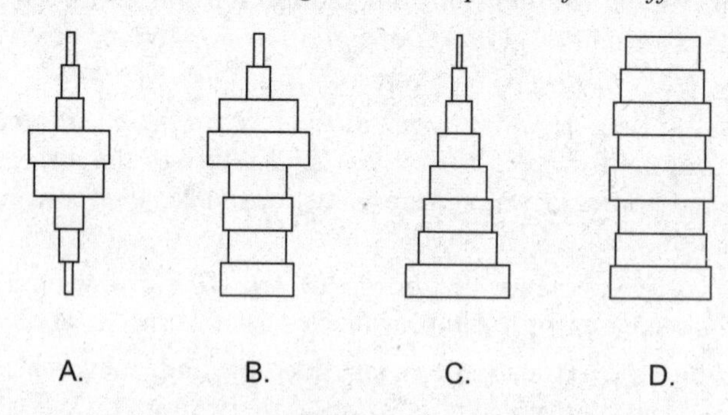

2. Which of the above populations is experiencing the fastest growth?

3. Which of the above populations is most nearly experiencing zero population growth over the time period represented by the diagram?

4. Which of the above populations is experiencing the effect of severe limiting factors?

5. All of the following populations would likely result in a uniform dispersion pattern EXCEPT:

A. nesting penguins on a small beach
B. perennial shrubs (of a given species) growing in a desert habitat
C. tropical trees (of a given species) in a tropical rain forest
D. lions on the savanna

Questions 6–7 refer to a population of 500 that experiences 55 births and 5 deaths during a one-year period.

6. What is the reproductive rate for the population during the one-year period?

A. 0.01/year
B. 0.05/year
C. 0.1/year
D. 50/year

7. If the population maintains the current growth pattern, a plot of its growth would resemble

A. exponential growth
B. fluctuating growth
C. *K*-selected growth
D. logistic growth

Use the following key for questions 8–12. Each answer in the key may be used once, more than once, or not at all.

A. commensalism
B. mutualism
C. Batesian mimicry
D. Müllerian mimicry

8. Burr-bearing seeds that are dispersed by clinging to the fur of certain mammals do not harm or help the mammals.

9. The monarch and viceroy butterflies both have orange wings with the same distinctive black markings. When the monarch caterpillar feeds on milkweed, a toxic plant, it stores the toxins, making both the monarch caterpillar and butterfly unpalatable and toxic. The viceroy caterpillar feeds on nontoxic plants, but because viceroy butterflies look like monarch butterflies, predators avoid eating them.

10. Oxpeckers are birds that ride rhinoceroses and other ungulates and eat various skin parasites, such as ticks.

11. Several species of poisonous snakes bear bright colors of red, black, and yellow.

12. Several species of brightly colored, harmless snakes look like poisonous coral snakes.

13. All of the following kinds of plants or animals characterize the initial stages of succession EXCEPT:

A. *r*-selected species
B. species with good dispersal ability
C. species that can tolerate poor growing conditions
D. species that invest large amounts of resources or time into development of progeny

14. Primary succession would occur

A. in a meadow destroyed by flood
B. in a meadow destroyed by overgrazing
C. on a newly created volcanic island
D. in a section of a forest destroyed by an avalanche

15. All of the following increase the concentration of CO_2 in the atmosphere EXCEPT:

A. photosynthesis
B. slash-and-burn clearing of tropical rain forests
C. burning of fossil fuels
D. burning of wood for cooking and heating

16. Nitrogen becomes available to plants by all of the following processes EXCEPT:

A. ammonification
B. denitrification by denitrifying bacteria
C. nitrification by nitrifying bacteria
D. nitrogen fixation in plant nodules

Question 17 refers to the following food chain.

diatoms → oysters → humans

17. In the above food chain, oysters represent

A. detritivores
B. herbivores
C. primary carnivores
D. secondary consumers

Question 18 refers to the following pyramid of biomass.

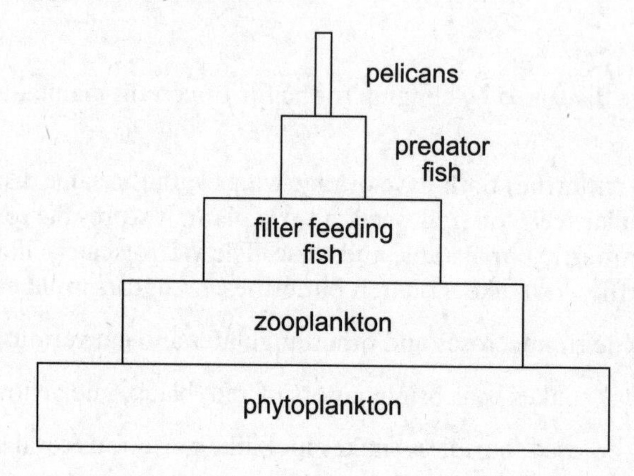

18. In which trophic level would the biological magnification of the pesticide DDT be most evident?

A. phytoplankton
B. filter feeding fish
C. predator fish
D. pelicans

Question 19 refers to the following table showing the population sizes for a community of three species.

Species	Number of Individuals
Species 1	500
Species 2	400
Species 3	100

19. Calculate Simpson's Diversity Index, D, for the data in the table above using $D = 1 - \Sigma\left(\frac{n}{N}\right)^2$, where n = the number of individuals in a species population and N = the total number of all individuals among all species.

A. 0.42
B. 0.49
C. 0.51
D. 0.58

Free-Response Questions

The AP exam has long and short free-response questions. The long questions have considerable descriptive information that may include tables, graphs, or figures. The short questions are brief but may also include figures. Both kinds of questions have four parts and generally require that you bring together concepts from multiple areas of biology.

The questions that follow are designed to further your understanding of the concepts presented in this chapter. Unlike the free-response questions on the exam, they are narrowly focused on the material in this chapter. For free-response questions typical of the exam, take the two practice exams in this book.

Directions: The best way to prepare for the AP exam is to write out your answers as if you were taking the exam. Use complete sentences for all your answers and do not use outline form or bullets. You may use diagrams to supplement your answers, but be sure to describe the importance or relevance of your diagrams.

1. A species of fly that does not bite or sting has alternating yellow and black stripes on its body. Explain the value these markings bring to these harmless flies.
2. A keystone species has a strong influence on the health of a community. Are you more likely to find a keystone species in a large, complex community or in a small, simple community? Explain your answer.
3. Describe how limiting factors regulate the growth of populations.
4. **a.** Describe the process of succession for a lake as it develops into a forest.
   b. Compare and contrast the succession of a lake and the eutrophication of a lake polluted by fertilizer or sewage.
5. Explain how two closely related species can occupy the same habitat, seemingly competing for the same kinds of resources.
6. Describe the cycling of nitrogen in an ecosystem.

Answers and Explanations

Multiple-Choice Questions

1. **B.** Since the individuals are interbreeding, they belong to a single species. A group of individuals of the same species is a population.
2. **C.** An age structure diagram that is pyramid-shaped with a broad base represents a population with fast growth.
3. **D.** An age structure diagram with all tiers of approximately the same width represents a population with a constant size.
4. **A.** This age structure diagram indicates high mortality rates among the older and younger generations, an indication that the weakest individuals are experiencing severe density-dependent limiting factors.
5. **D.** Lions are social animals and group in prides (or in bands of bachelor males). In contrast, the remaining answer choices are animals that are likely to be distributed in a uniform pattern, with each individual spaced equally from others. Whenever a population is subject to a limiting resource, there is likely to be competition. In order to minimize conflict, the result of competition is often to distribute the resource equally among all individuals in a population. By equally apportioning space, the penguins minimize conflict. The desert shrubs minimize competition for water. Tropical trees of a single species are also uniformly spaced through the forest, not because of competition, but probably because it minimizes the effect of herbivory and the transmission of parasites and disease. The greater the distance between trees, the more difficulty herbivores have in locating a tree.

6. **C.** The rate of growth, r, equals $\frac{55 \text{ births} - 5 \text{ deaths}}{500 \text{ per year}}$, or 0.1/year.

7. **A.** If the reproductive rate is greater than zero ($r > 0$), then the population is growing exponentially.

8. **A.** This is an example of commensalism because the seeds are helped (dispersal occurs) and the mammals are not harmed or helped by the seeds.

9. **C.** This is an example of Batesian mimicry because the defenseless viceroy butterfly benefits by mimicking the aposematic monarch butterfly.

10. **B.** Mutualism occurs when both organisms in the symbiosis benefit. In this example, the oxpeckers obtain food and the rhinos have parasites removed.

11. **D.** Snakes are Müllerian mimics when they both look alike and possess some kind of aposematic defense (poisonous or unpalatable, for example).

12. **C.** The brightly colored, harmless snakes are Batesian mimics because they benefit from the defense mechanism of the colorful poisonous snakes they resemble.

13. **D.** Species that make large parental investments in their offspring—such as oak trees, whose acorns require two years to mature, or mammals that nurse their young—are *K*-selected species and are characteristic of the later stages of succession.

14. **C.** Because a newly created volcanic island has a pristine substrate (fresh lava), never having been inhabited by living things, it is an example of primary succession. All of the other answer choices would initiate secondary succession.

15. **A.** Photosynthesis removes CO_2 from the atmosphere. On the other hand, respiration adds CO_2 to the atmosphere, as does the burning of any organic fuel such as wood, coal, oil, natural gas, and gasoline.

16. **B.** Depending on soil conditions and plant species, plants absorb nitrogen either in the form of NH_4^+ or NO_3^-. Denitrifying by soil bacteria converts NO_3^- to gaseous N_2, a form of nitrogen that plants cannot use. The other answer choices describe processes that generate either NH_4^+ or NO_3^-.

17. **B.** The beginning of a food chain must begin with producers, in this example, the diatoms. Since the oysters are eating the producers, the oysters are the herbivores.

18. **D.** DDT concentrates in the pelicans because they occupy the top of the food chain. Because the pesticide results in fragile eggshells that break before incubation is completed, many pelican populations had approached extinction before DDT was banned in the United States.

19. **D.** To determine the diversity index, calculate $\frac{n}{N}$ for each species, and then square each one. Add together the squares for species 1, 2, and 3. Finally, subtract the sum from 1. Values of D vary from 0 to 1, with higher numbers indicating greater diversity.

Species	n	$\frac{n}{N}$	$\left(\frac{n}{N}\right)^2$
Species 1	500	$\frac{500}{1,000} = 0.5$	$0.5^2 = 0.25$
Species 2	400	$\frac{400}{1,000} = 0.4$	$0.4^2 = 0.16$
Species 3	100	$\frac{100}{1,000} = 0.1$	$0.1^2 = 0.01$
Totals	**1,000**		**0.42**

$$\text{Diversity Index} = 1 - \sum\left(\frac{n}{N}\right)^2 = 1 - 0.42 = 0.58$$

Free-Response Questions

1. These flies are Batesian mimics. By looking like similarly colored bees and wasps, they fool predators into thinking that they are armed and dangerous.

2. Keystone species are more likely to be found in small, simple communities. Large, complex communities have many interacting populations, and the removal of one species can more likely be replaced with another species whose niche may be comparable enough to maintain the stability of the community. If a species can be removed without major negative consequences to the community, then it isn't a keystone species.

3. *For this question, you should begin by describing intrinsic, exponential population growth. Describe how populations experiencing rapid growth can deplete their resources and subsequently crash to extinction. Then, describe the different kinds of limiting factors (density-dependent and density-independent) that restrict populations to logistic growth and that limit the size of populations to the carrying capacity of the habitat. Describe the interaction of predator and prey and why population sizes fluctuate around the carrying capacity. For this question, it is important that you supplement your discussion with graphs of each kind of population growth pattern because they demonstrate your ability to express and interpret data in analytical form.*

4. a. Succession describes the series of communities that occupy an area over time. If the process begins on a newly exposed surface, it is called primary succession. If it occurs on a substrate previously supporting life, such as is the case with most lakes, then it is called secondary succession. Succession occurs because each community changes the habitat in such a way that it becomes more suitable to new species. Soil, light, and other growing conditions change as each community occupies a region. The final, climax community is a stable community that remains unchanged until destroyed by some catastrophic event, such as fire. Then the process begins again.

   The first community to occupy the lake consists of pioneer species with *r*-selected characteristics. These characteristics include good dispersal ability, rapid growth, and rapid reproduction of many offspring. The lake is first populated by algae and protists, followed by rotifers, mollusks, insects, and other arthropods. Various vegetation, such as grasses, sedges, rushes, and cattails, grows at the perimeter of the lake. Submerged vegetation (growing on the lake bottom) is replaced by vegetation that emerges from the surface, perhaps covering the surface with leaves. As the plants and animals die, they add to the organic matter that fills the lake. In addition, sediment is deposited by water from streams that enter the lake. Eventually, the lake becomes marshy as it is overrun by vegetation. When the lake is completely filled, it becomes a meadow, occupied by plants and animals that are adapted to a dry, rather than marshy, habitat. Subsequently, the meadow is invaded by shrubs and trees from the surrounding area. In a temperate mountain habitat, the climax community may be a deciduous forest consisting of oaks or maples. In colder regions, the climax community is often a coniferous forest, consisting of pines, firs, and hemlocks.

   b. Eutrophication is the increase in inorganic nutrients and biomass of a lake. Eutrophication occurs in both unpolluted and polluted lakes. In the natural succession of the lake described above, eutrophication occurred slowly over a period of dozens of years (perhaps over a hundred years). As a result, changes in the chemical and physical nature of the lake allowed for the orderly change of communities, each new community suitable for the new conditions.

   In polluted lakes, eutrophication is accelerated by effluent from sewage or fertilizer. As a result, algae growth occurs rapidly over a period of months. At high densities, algae reduce oxygen levels when they respire at night. When the algae die as shorter winter days approach, rapidly growing aerobic bacteria that feed on the dead algae further deplete the oxygen content of the water. In the absence of oxygen, many of the plants and animals die. The bottom of the lake fills with dead organisms that, in turn, stimulate growth of anaerobic bacteria (some of which produce foul-smelling sulfur gases). In addition, the surface of the lake may become littered with dead fish.

Essay questions about succession are common on the AP exam. You may be asked to describe succession for a particular kind of habitat, as in this question, or you may be given a choice of different kinds of successions. Note that the first paragraph and the beginning of the second paragraph in part a contain very general information that applies to all successional processes. Once you have completed the generalities, describe the successional events for the specific habitat requested. In part b of the question, the first paragraph makes a comparison (states that unpolluted and polluted lakes are both the result of the same eutrophication processes), and the second paragraph makes a contrast (how the rate of eutrophication is faster in a polluted lake).

5. *For this question, you should describe the competitive exclusion principle and how it results in resource partitioning, character displacement, and realized niches.*

6. Although the atmosphere consists of 80% nitrogen gas (N_2), plants can utilize nitrogen only in the form of ammonium (NH_4^+) or nitrate (NO_3^-). Nitrogen fixation, the conversion of N_2 to NH_4^+, occurs by nitrogen-fixing bacteria that live in the soil, or in the root nodules of certain plants such as legumes. In turn, nitrification occurs by certain nitrifying bacteria that convert NH_4^+ to NO_2^- (nitrite) and by other nitrifying bacteria that convert NO_2^- to NO_3^-. In addition, some N_2 is converted to NO_3^- by the action of lightning. On the other hand, denitrification occurs when denitrifying bacteria convert NO_3^- back to N_2.

From the NH_4^+ or NO_3^- absorbed, plants make amino acids and nucleic acids. When animals eat the plants (or other animals), they, in turn, obtain a form of nitrogen that they can metabolize. When animals break down proteins, they produce ammonia (NH_3). Since NH_3 is toxic, many animals, such as aquatic animals, excrete it directly. Other animals convert NH_3 to less toxic forms, such as urea (mammals) or uric acid (insects and birds). When plants and animals die, they decompose through the process of ammonification, in which bacteria convert the amino acids and other nitrogen-containing compounds to NH_4^+, which then becomes available again for plants.

The AP exam may ask you to describe any of the biogeochemical cycles. The nitrogen cycle is the most common request. You may want to supplement your discussion with a drawing using arrows to show the various conversions of nitrogen. If you do make a drawing, however, you must still provide a complete discussion.

PART 2

LABORATORY REVIEW

Chapter 12

Review of Laboratory Investigations

Review

The College Board requires that 25% of the AP Biology instructional time be devoted to laboratory activity. The laboratory experience is important, as it provides the laboratory experience typical of a first-year college course in biology. The College Board provides 13 laboratory investigations for use in the AP Biology course. In addition, your AP Biology instructor may substitute or supplement these investigations with other labs. Of the 13 investigative labs and instructor supplementary labs, 8 labs are *required*, with 2 labs from each of the Big Ideas: evolution, cellular processes, genetics, and interactions (see Table 12-1).

Questions about lab investigations on the AP exam will either be of a general nature or provide a clear description of all aspects of the lab with enough information to answer the question. However, you will be at a distinct advantage if you have done the lab or are familiar with it. Unfortunately, there is not enough time in the school year to do all 13 labs and another possible half-dozen or so of instructor supplementary labs. To help you gain an advantage for AP lab investigations you have not done, this chapter provides full descriptions of all the AP lab investigations.

The AP free-response part of the exam consists of two long questions and six short questions. At least one of the long questions will address laboratory analysis or design. Other questions, long and short, may also involve laboratory activities or the application of math skills to biological questions. A four-function calculator (with a square root function) is required for the exam. The question types fall into one of the following five categories:

1. **Experimental analysis.** In this type of essay question, you are given some experimental data and asked to interpret or analyze the data. The question usually includes several parts, each requesting specific interpretations of the data. In addition, you are usually asked to prepare a graphic representation of the data. Graph paper is provided. Guidelines for preparing a graph are given in the next section.
2. **Experimental design.** This type of essay question asks you to design an experiment to answer specific questions about given data or an experimental situation. Guidelines for designing an experiment are given in the next section.
3. **Modeling.** This type of question requires you to develop a mathematical expression that explains or can be used to predict a biological process.
4. **Math applications.** In this kind of question, you will explain how a mathematical expression describes a biological process.
5. **Predicting and justifying.** In this type of question, you will be given data, often in tabular or graphical form, that you will use to make predictions or explanations.

If these types of questions all sound the same to you, don't fret. The laboratory investigations that the College Board provides and the ones your teacher uses as supplements will provide you with exposure and familiarity with these kinds of questions. Although the data or situations on the AP exam may differ from your classroom activities, you will be able to draw from lab experiences to answer the questions on the exam.

The material in this chapter summarizes each of the 13 laboratory exercises with a brief description of its experimental design and conclusions. This material is not intended to substitute for an actual laboratory experience, but to provide you with a review that will help you answer the AP Biology questions.

Summary of AP Biology Laboratory Investigations

Table 12-1

Big Idea	Investigation	Description
1. Evolution	1. Artificial Selection	Investigating the effect of selection for trichome number in plants
	2. Modeling Evolution	Investigating evolution in a model population
	3. Comparing DNA Sequences	Using BLAST to find evolutionary relationships
2. Cellular Processes	4. Diffusion and Osmosis	Observing diffusion and osmosis in artificial and living cells
	5. Photosynthesis	Identifying factors that influence the rate of photosynthesis
	6. Cellular Respiration	Identifying factors that influence the rate of cellular respiration
3. Genetics	7. Mitosis and Meiosis	Investigating the cell division processes of mitosis and meiosis
	8. Bacterial Transformation	Transferring foreign DNA into a bacterium
	9. Restriction Enzyme Analysis of DNA	Mapping DNA
4. Interactions	10. Energy Dynamics	Investigating energy flow in an ecosystem
	11. Transpiration	Identifying factors that influence the rate of transpiration
	12. Animal Behavior	Observing behavior of fruit flies in response to stimuli
	13. Enzyme Catalysis	Investigating factors that influence enzyme activity

Graphing and Interpreting Data

The laboratory question in the free-response part of the AP exam will often ask you to create a graph using data provided in the question or ask you to predict and graph data as part of an experiment that you design. Include the following in your graph (Figure 12-1):

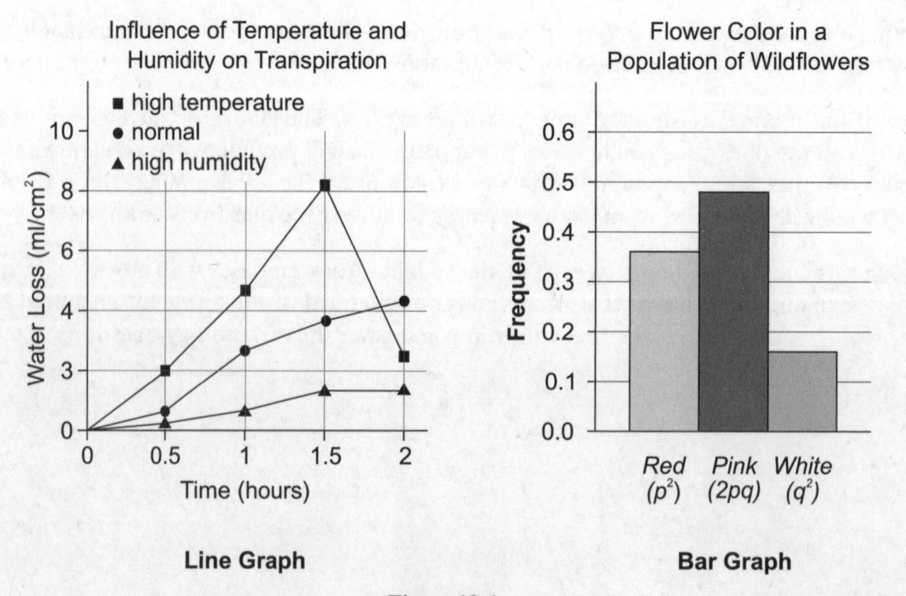

Figure 12-1

1. **Label each axis.** Indicate on each axis what is being measured and in what unit of measurement. For example, "Time (minutes)," "Distance (meters)," or "Water Loss (ml/m^2)" are appropriate labels.
2. **Provide values along each axis at regular intervals.** Select values and spacing that will allow your graph to fill as much of the graphing grid as possible. Be sure that the units on each axis are linear—that is, one unit on the axis must have the same length anywhere on the axis.
3. **Use the *x*-axis for the independent variable and the *y*-axis for the dependent variable.**
   - The **independent variable** is the variable that you choose to manipulate—that is, the variable to which *you* assign values. For example, suppose you wanted to investigate fruit fly longevity as a function of diet. The independent variable is the type of diet. *You* choose to investigate diet and *you* assign different "values" of diet (for example, high protein, high calorie, etc.).
   - The **dependent variable** is how the experiment responds to the independent variable. Its values are the data that you collect. (Think "*d*ata is for the *d*ependent variable.") The dependent variable measures how the experiment is responding to the independent variable that you have chosen to investigate.

   If you are plotting the progress of an event, then you have chosen time as the independent variable and the data you collect that measure changes in the event (such as weight change, distance traveled, or carbon dioxide released) are represented by the dependent variable.
4. **Connect the plotted points.** Either connect the plotted points with straight lines or draw a straight line through the plotted points estimating the line of best fit. Avoid smooth curves, as they usually imply knowledge about intermediate points not plotted or a mathematical equation that fits the experimental results. If the question asks you to make predictions beyond the data actually graphed, extrapolate (extend) the plotted line with a different line form (for example, dotted or dashed).
5. **In graphs with more than one plot, identify each plot.** If you plot more than one set of data on the same graph, identify each plot with a short phrase. Alternately, you can draw the points of each plot with different symbols (for example, circles, squares, or triangles) or connect the plotted points using different kinds of lines (solid line, dashed line, or dash-dot line) and then identify each kind of symbol or line in a legend.
6. **Provide a title for the graph.** Your title should be brief but descriptive.
7. **Use bar graphs for discrete data.** Line graphs, as described above, are appropriate for displaying data when the independent variable is continuous; that is, the independent variable has a countless number of values in an interval. Time, temperature, concentration, and light intensity are examples of continuous variables. For discrete variables, variables that have a limited number of values, a bar graph is more appropriate. A discrete variable may represent drug treatments, diets, genotypes (such as blood type alleles), or phenotypes (such as eye color). To construct a bar graph, draw a vertical column for each value of the independent variable from the *x*-axis to a height on the *y*-axis that reflects the value of its dependent variable. Other aspects of the bar graph, as described above in items 1–3 and 6, are the same.

Designing an Experiment

A laboratory free-response question may ask you to design an experiment to test a given hypothesis or to solve a given problem. In many cases, the question will ask you not only to design an experiment, but also to discuss expected results. Since the form of these questions can vary dramatically, it is not possible to provide a standard formula for preparing your answer. However, the following list provides important elements that you should include in your answer if they are appropriate to the question:

1. **Identify the independent and dependent variables.** The independent variable is the variable you are manipulating to see how the dependent variable changes.
   - You are investigating how the crustacean *Daphnia* responds to changes in temperature. You expose *Daphnia* to temperatures of 5°C, 10°C, 15°C, 20°C, 25°C, and 30°C. You count the number of heartbeats/sec in each case. Temperature is the **independent variable** (you are manipulating it), and the number of heartbeats/sec is the **dependent variable** (you observe how it changes in response to different temperatures).

- You design an experiment to investigate the effect of exercise on pulse rate and blood pressure. The physiological conditions (independent variable, or variable you manipulate) include sitting, exercising, and recovering at various intervals following exercise. You make two kinds of measurements (two dependent variables) to evaluate the effect of the physiological conditions—pulse rate and blood pressure.

2. **Describe the experimental treatment.** The experimental treatment (or treatments) is the various values that you assign to the independent variable. The experimental treatments describe how you are manipulating the independent variable.
   - In the *Daphnia* experiment, the different temperature values (5°C, 10°C, 15°C, 20°C, 25°C, and 30°C) represent six experimental treatments.
   - In the experiment on physiological conditions, the experimental treatments are exercise and recovery at various intervals following exercise.
3. **Identify a control treatment.** The control treatment, or control, is the independent variable at some normal or standard value. The results of the control are used for comparison with the results of the experimental treatments.
   - In the *Daphnia* experiment, you choose the temperature of 20°C as the control because that is the average temperature of the pond where you obtained the culture.
   - In the experiment on physiological conditions, the control is sitting, when the subject is not influenced by exercising.
4. **Use only one independent variable.** Only one independent variable can be tested at a time. If you manipulate two independent variables at the same time, you cannot determine which is responsible for the effect you measure in the dependent variable.
   - In the physiological experiment, if the subject also drinks coffee in addition to exercising, you cannot determine which treatment, coffee or exercise, causes a change in blood pressure.
5. **Random sample of subjects**. You must choose the subjects for your experiments randomly. Since you cannot evaluate every *Daphnia,* you must choose a sample population to study. If you choose only the largest *Daphnia* to study, it is not a random sample and you introduce another variable (size) for which you cannot account.
6. **Describe the procedure.** Describe how you will set up the experiment. Identify equipment and chemicals to be used and why you are choosing to use them. If appropriate, provide a labeled drawing of the setup.
7. **Describe expected results.** Use graphs to illustrate the expected results, if appropriate.
8. **Provide an explanation of the expected results in relation to relevant biological principles.** The results you give are your expected results. Describe the biological principles that led you to make your predictions.
   - In the experiment on physiological conditions, you expect blood pressure and pulse rate to increase during exercise in order to deliver more O_2 to muscles. Muscles use the O_2 for respiration, which generates the ATP necessary for muscle contraction.

Statistical Analyses

When you take the AP exam, you will receive two pages of equations and formulas to use while you answer questions for both sections of the exam. These pages provide you with all of the equations and formulas you will need to solve any questions that evaluate your ability to perform data analysis and solve quantitative problems. You can see these pages at the beginning of Part 3 of this book. Included on these pages are statistical equations and tables that provide you with the tools to perform statistical analysis.

There are two kinds of statistics that you will need to know for the AP exam:

1. **Descriptive statistics** are used to summarize data. In particular, you may need to calculate averages (mean, median, and mode) or measures of variation (range, standard error, and standard deviation). Definitions and formulas for these statistics are provided on the Equations and Formulas pages.

2. The **chi-square (χ^2) test for goodness-of-fit** is used to compare observed results with expected results. The statistic tells you whether the two are significantly different. In particular, the statistic determines whether the difference is due to chance (at an accepted probability) or whether some other factor is responsible. To understand this statistic, consider the following two values of χ^2:
   - $\chi^2 = 0$. In this case, the observed results and the expected results are exactly the same.
   - $\chi^2 = 100$. In most cases, this value of χ^2 is a relatively large value and will indicate that the difference between the observed and expected results did not occur by chance and that they should be considered significantly different. Such a result often indicates that your expectations were wrong or your data collection was faulty.

   More typically, the value you calculate for χ^2 will be between these two extremes. To determine whether the χ^2 value obtained indicates that the difference between the observed data and expected data occurred by chance or because the expectations or data are wrong, a χ^2 statistical table is consulted (provided on the Equations and Formulas pages). If chance explains 95% or more of the deviation of χ^2 from zero, then the experimental results are acceptable. If not, there are errors in the data collection or a problem with the expectations.

   As an example, consider the trait of eye color in fruit flies, where red eyes (*R*) are dominant to brown (sepia) eyes (*r*). A cross between a heterozygous red-eyed fly (*Rr*) and a brown-eyed fly (*rr*) is expected to produce ½ heterozygous red-eyed flies (*Rr*) and ½ brown-eyed flies (*rr*). What can you say about this investigation if you counted 55 red-eyed flies and 45 brown-eyed flies from this cross when you expected 50 of each (½ and ½)? To calculate the χ^2 statistic, create a contingency table of all the values that are needed to calculate the χ^2 statistic, where *O* and *E* represent observed values and expected values, respectively:

	O	E	$O-E$	$(O-E)^2$	$\frac{(O-E)^2}{E}$
Red-eyed flies	55	50	5	25	0.5
Brown-eyed flies	45	50	−5	25	0.5

$$\chi^2 = \Sigma \frac{(O-E)^2}{E} = 0.5 + 0.5 = 1.0$$

   Next, you need to determine the degrees of freedom (*df*) of the data using the formula provided on the Equations and Formulas pages:

$$df = \text{number of } E \text{ values} - 1 = 2 - 1 = 1$$

   To determine if the observed data values obtained in the fruit fly cross differ from the expected values by chance or for another reason, you compare your calculated value of χ^2 to the table of critical values in the chi-square table provided on the Equations and Formulas pages:

Chi-Square Table

	Degrees of freedom (*df*)							
p	1	2	3	4	5	6	7	8
0.05	3.84	5.99	7.81	9.49	11.07	12.59	14.07	15.51
0.01	6.63	9.21	11.34	13.28	15.09	16.81	18.48	20.09

   The table provides critical values for χ^2 for two values of probability, *p*, and at eight different degrees of freedom. A critical value of χ^2 at $p = 0.05$ indicates that 95% of the difference between observed and expected values can be explained by chance. Thus, for the fruit fly calculations, the calculated value of $\chi^2 = 1.0$, at $df = 1$ and $p = 0.05$, is *less* than the critical value. What do you do with this value?

- *Reject the null hypothesis*. The **null hypothesis** is that the observed values are not significantly different from the expected values; that is, the differences are the result of chance. If the calculated value for χ^2 is *more* than the critical value, you reject the null hypothesis.
- *Accept the null hypothesis*. If the calculated value for χ^2 is *less* than the critical value, you do not reject the null hypothesis. "Do not reject the null hypothesis" is not quite the same as saying you accept the null hypothesis because accepting the null hypothesis implies that you proved it to be true. However, proof doesn't happen in statistics (or science). Instead, you were not able to disprove it.

If your calculated value for χ^2 is greater than the critical value, and you rejected the null hypothesis, then something other than chance is influencing your results. Crossing over, deleterious mutations, or nonrandom mating (perhaps some traits produce poor breeders) are factors to consider. High values of χ^2 may also result if the expected values used do not account for sex linkage or are otherwise in error.

Investigation 1: Artificial Selection

In this investigation, you explore the effect of artificial selection. Artificial selection is selection by humans. We have used this kind of selection for thousands of years to breed animals for special purposes and to raise plants to yield larger or more flavorful fruits and vegetables. The investigation suggests using fast-growing plants to examine trichomes (plant hairs) on leaves, but it also encourages using other observable, heritable traits on these plants. Your teacher may substitute other organisms.

In this investigation, you gain experience observing biological phenomena and collecting and analyzing data. You learn firsthand how selection influences the genetic makeup of a population.

Part I: Artificial Selection of a Trait

1. Choose a trait upon which to carry out artificial selection. In order for selection to work—that is, to change the genetic makeup of populations of subsequent generations—the trait must be heritable. If available, examine Wisconsin Fast Plants. Consider trichome density, the color intensity of pigments in leaves, or plant height.
2. Select the plants with the trait of interest. If you choose trichome density, decide whether you want higher or lower density. Select 10% of the class population of plants (12–15 plants) that possess the trait.
3. Among the selected plants, measure the trait. For example, count and record a sample number of trichomes in a specific position on a specific leaf (same leaf from every plant).
4. Allow the selected plants to flower. Cross-pollinate the flowers and collect the seeds when the fruits mature.
5. Plant the seeds from the first generation to start the second generation.
6. When the second-generation plants are mature, score the trait a second time.
7. Compile the data and present them in a graph, such as a histogram.

Part II: Further Investigation of a Trait

Design an experiment to investigate a different trait or further investigate some aspect of the trait you selected for in part I. For example, if you investigated trichomes in part I, consider investigating what, if any, advantage trichomes contribute to the survival or reproductive success of the plant.

Investigation 2: Modeling Evolution

The goal of this investigation is to create a mathematical model of evolution. You enter data into a spreadsheet, make "what if" changes in the data, and observe the results. By doing this, you discover the variables that influence evolution and what kind of influence those variables have. Ultimately, you confirm that Hardy-Weinberg equilibrium describes a population that is *not* evolving; deviations from equilibrium indicate that the population *is* evolving.

Evolution is the change in allele frequencies in a population over time. If a population can be described by Hardy-Weinberg equilibrium, then evolution is not occurring. So that evolution does not occur, there must be:

1. no selection
2. no mutation
3. no gene flow (no migration)
4. large populations (so that genetic drift does not occur)
5. random mating

Part I: A Mathematical Model for Evolution

Depending upon whether you use the AP-provided investigation or another similar activity, the steps you follow will generally look like this:

Step 1: Choose frequencies for p and q. You begin the activity by choosing values for the frequencies, p and q, of two alleles, *A* and *B*. Since there are only two alleles for this gene, the values must sum to 1:

$$p + q = 1$$

Step 2: Mix gametes. An individual can contribute one of these two alleles to a gamete (egg or sperm), either the *A* allele or the *B* allele. A zygote, or fertilized egg, will have one of these alleles from each parent and will become *AA*, *AB*, *BA*, or *BB*. To document Hardy-Weinberg equilibrium, the mixing of the gametes must be random (selection is absent). If you are creating a spreadsheet in this part of the investigation, you can use the random number generator of the spreadsheet (RAND). If you choose $p = 0.6$ and $q = 0.4$, you essentially tell the spreadsheet to randomly assign to each gamete an *A* or a *B* allele, except that 60% of the time it needs to be an *A* and 40% of the time it needs to be a *B*.

Step 3: Determine frequencies of zygotes. If the union of gametes is a random event, then the chance that any two gametes come together depends upon how many of each gamete is present. The chance that two *A* alleles come together to form an *AA* zygote is $p \times p$, or p^2. For the *AB* and *BA* zygotes, the chance is $p \times q + q \times p$, or $2pq$. For the *BB* zygote, the chance is $q \times q$, or q^2. This reasoning results in the Hardy-Weinberg equation:

$$p^2 + 2pq + q^2 = 1$$

The equation is equal to 1 because the terms represent all (100%) of the possible combinations of alleles.

Step 4: Generate multiple generations. Using the allele frequencies of the new generation (the zygotes), repeat steps 2 and 3 for several new generations. The allele frequencies, p and q, for each generation may be slightly different from the previous generation because the mixing of alleles to form zygotes is random. You can reexamine your population with a new set of random combinations of alleles by telling your spreadsheet to recalculate the data. For this, it will select new random numbers.

Part II: Generate "What If" Scenarios

Use your mathematical model of Hardy-Weinberg to generate "what if" scenarios. When you first create your population in your spreadsheet, you will probably begin with a small population (perhaps $N = 16$) spread over several generations. You can change various attributes of your model to see how it affects your population and how it relates to Hardy-Weinberg equilibrium. For example:

1. **Add additional generations.** If you stretch your population to 10 or more generations, you might find that one of the alleles disappears from that generation (its frequency becomes zero). If one allele disappears, the remaining allele is said to be "fixed." Once an allele is fixed in a population, it remains so until some evolutionary process (mutation or gene flow) reintroduces it.
2. **Increase the population size.** The fluctuations in gene frequencies that you see from generation to generation are examples of **genetic drift,** changes in gene frequencies as a result of chance. The effect of genetic drift decreases as population size increases, so increasing the size of the population should reduce its effect.
3. **Investigate the effect of natural selection.** You can simulate the effect of selection for or against one of the alleles. Selection can influence any of the stages during the life cycle of an individual but ultimately will determine whether that individual will contribute gametes to the next generation. One way to simulate selection is to manipulate the formula in your spreadsheet that determines whether an individual passes its alleles to the next generation. To simulate selection against *BB* individuals, for example, adjust the formula that counts the number of *BB* individuals that contribute gametes to the next generation. In the spreadsheet in Figure 12-2, if no selection occurs, then the number of *BB* individuals is calculated in cell G261 with the following formula:

   2*SUM(G5:G254) + SUM(F5:F254)

   If only 50% of the *BB* individuals survive, the formula becomes the following:

   <u>0.5</u>*2*SUM(G5:G254) + SUM(F5:F254)

   The formulas for all the cells are shown in Figure 12-3.
4. **Investigate the effect of mutation and gene flow.** You can simulate a mutation of *A* to *B* or *B* to *A* in ways similar to an investigation of natural selection, above.

By manipulating your population model, you should make the following conclusions:

1. The frequencies of alleles should remain unchanged over time unless one of the five mechanisms of evolution is active.
2. Small populations are more subject to genetic drift than larger populations.
3. The dominant or recessive characteristic of an allele does not influence its selective value. A dominant or recessive allele has selective value only if the phenotype it generates has selective value. Thus, a recessive trait may have a greater selective value than a dominant trait. Without selection, the allele frequencies remain unchanged, regardless of the dominant or recessive characteristic of the allele.
4. Selection can affect an individual at any stage during its life cycle: gamete, zygote, egg, larva, juvenile, or adult.

Ultimately, you should conclude that any of the five processes listed at the beginning of this investigation can influence the evolution of populations. But which of these has the greatest influence? Although each population is different and although evolution may progress differently in the same population at different times, studies of evolution in populations in the real word (in the "wild") generally demonstrate that natural selection is the strongest driving force for evolution.

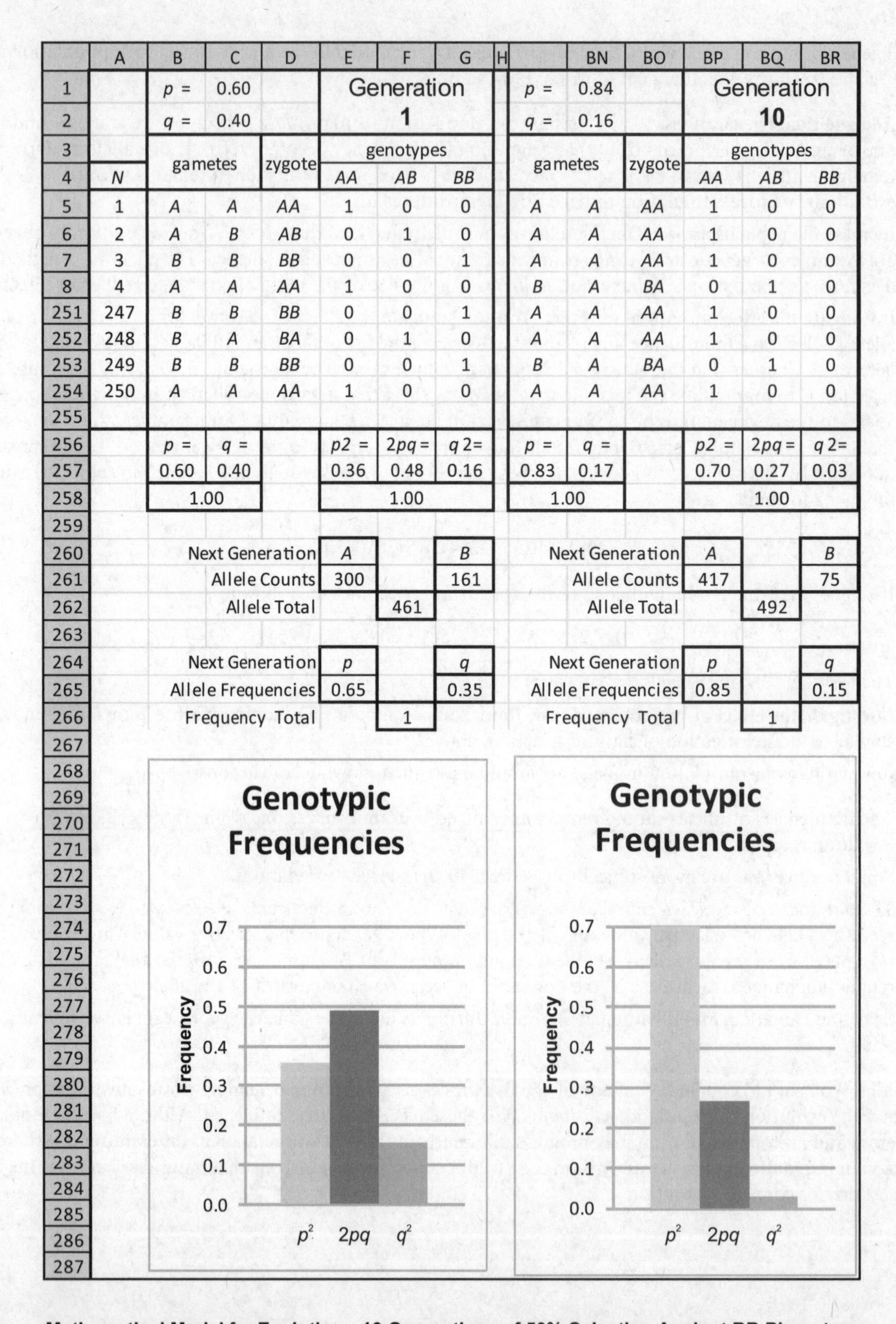

	A	B	C	D	E	F	G	H		BN	BO	BP	BQ	BR
1		p =	0.60		Generation				p =	0.84		Generation		
2		q =	0.40		1				q =	0.16		10		
3		gametes		zygote	genotypes				gametes		zygote	genotypes		
4	N				AA	AB	BB					AA	AB	BB
5	1	A	A	AA	1	0	0		A	A	AA	1	0	0
6	2	A	B	AB	0	1	0		A	A	AA	1	0	0
7	3	B	B	BB	0	0	1		A	A	AA	1	0	0
8	4	A	A	AA	1	0	0		B	A	BA	0	1	0
251	247	B	B	BB	0	0	1		A	A	AA	1	0	0
252	248	B	A	BA	0	1	0		A	A	AA	1	0	0
253	249	B	B	BB	0	0	1		B	A	BA	0	1	0
254	250	A	A	AA	1	0	0		A	A	AA	1	0	0
255														
256		p =	q =		p2 =	2pq =	q 2=		p =	q =		p2 =	2pq =	q 2=
257		0.60	0.40		0.36	0.48	0.16		0.83	0.17		0.70	0.27	0.03
258		1.00				1.00			1.00				1.00	
259														
260				Next Generation	A		B				Next Generation	A		B
261				Allele Counts	300		161				Allele Counts	417		75
262				Allele Total		461					Allele Total		492	
263														
264				Next Generation	p		q				Next Generation	p		q
265				Allele Frequencies	0.65		0.35				Allele Frequencies	0.85		0.15
266				Frequency Total		1					Frequency Total		1	
267														
268														
269														
270														
271														
272														
273														
274														
275														
276														
277														
278														
279														
280														
281														
282														
283														
284														
285														
286														
287														

Mathematical Model for Evolution—10 Generations of 50% Selection Against BB Phenotype

Figure 12-2

	A	B	C	D	E	F	G
1		p =	0.60 (1)		Generation		
2		q =	=1-C1 (2)		1		
3		gametes		zygote	genotypes		
4	N				AA	AB	BB
5	1	=IF(RAND()<=C$1, "A","B")	=IF(RAND()<=C$1, "A","B")	=CONCATENATE (B5,C5)	=IF (D5="AA",1,0)	=IF(D5="AA",0, IF(D5="BB",0,1))	=IF (D5="BB",1,0)
6	2	=IF(RAND()<=C$1, "A","B") (3)	=IF(RAND()<=C$1, "A","B")	=CONCATENATE (B6,C6) (4)	=IF (D6="AA",1,0) (5)	=IF(D6="AA",0, IF(D6="BB",0,1)) (6)	=IF (D6="BB",1,0) (7)
7	3	=IF(RAND()<=C$1, "A","B")	=IF(RAND()<=C$1, "A","B")	=CONCATENATE (B7,C7)	=IF (D7="AA",1,0)	=IF(D7="AA",0, IF(D7="BB",0,1))	=IF (D7="BB",1,0)
252	248	=IF(RAND()<=C$1, "A","B")	=IF(RAND()<=C$1, "A","B")	=CONCATENATE (B252,C252)	=IF (D252="AA",1,0)	=IF(D252="AA",0, IF(D252="BB",0,1))	=IF(D252="BB",1,0)
253	249	=IF(RAND()<=C$1, "A","B")	=IF(RAND()<=C$1, "A","B")	=CONCATENATE (B253,C253)	=IF (D253="AA",1,0)	=IF(D253="AA",0, IF(D253="BB",0,1))	=IF (D253="BB",1,0)
254	250	=IF(RAND()<=C$1, "A","B")	=IF(RAND()<=C$1, "A","B")	=CONCATENATE (B254,C254)	=IF (D254="AA",1,0)	=IF(D254="AA",0, IF(D254="BB",0,1))	=IF (D254="BB",1,0)
255							
256		p =	q =		p2 =	2pq =	q 2=
257		=COUNTIF (B5:C254,"A") /COUNTA (8) (B5:C254)	(9) =COUNTIF (B5:C254,"B") /COUNTA (B5:C254)		=COUNTIF (E5:E254,1) /COUNTA (E5:E254)	=COUNTIF (F5:F254,1) /COUNTA (F5:F254)	=COUNTIF (G5:G254,1) /COUNTA (G5:G254)
258		=B257+C257				=E257+F257+G257	
259							
260				Next Generation	(10) A		B
261				Allele Counts	=2*SUM(E5:E254) +SUM(F5:F254)	selection →	= 0.5* (11) 2*SUM(G5:G254) +SUM(F5:F254)
262				Allele Total		(12) =E261+G261	
263							
264				Next Generation	p		q
265				Allele Frequencies	=E261/F262 (13)		(14) =G261/F262
266				Frequency Total		=E265+G265	

	I	J	K	L	M	N
1	p =	=E265 (15)		Generation		
2	q =	=1-K2		2		

(1) Cell C1 - Enter a value for *p* here.

(2) Cell C2 - Calculates the value for *q*.

(3) Cells B5 & C5 - Enters "A" if a random number (between 0 and 1) is less than or equal to 0.6 (*p*); if greater than 0.6, "B" is entered. ("A" is entered 60% of the time and "B" is entered 40% of the time.)

(4) Cell D5 - Combines two alleles from gametes columns to make zygote genotype.

(5) Cell E5 - Enters 1 if zygote is "AA"; if not, enters 0.

(6) Cell F5 - Enters 1 if zygote is "AB" or "BA"; if not, enters 0.

(7) Cell G5 - Enters 1 if zygote is "BB"; if not, enters 0.

(8) Cell B257 - Counts number of A gametes in columns B & C and divides by total number of gametes.

(9) Cell C257 - Counts number of B gametes in columns B & C and divides by total number of gametes.

(10) Cell E261 - Enters total number of A alleles from AA and AB individuals.

(11) Cell G261- Enters total number of B alleles from BB and AB individuals. *Also, formula allocates 50% selection against BB individuals.*

(12) Cell F262 - Enters sum of A and B alleles.

(13) Cell E265 - Enters allele frequency for *p*.

(14) Cell G265 - Enters allele frequency for *q*.

(15) Cell J1 - Enters *p* from E265 into next generation columns.

Formulas for a Mathematical Model for Evolution—50% Selection Against BB Phenotype

Figure 12-3

Investigation 3: Comparing DNA Sequences

In this investigation, you use bioinformatics to establish evolutionary relationships. **Bioinformatics** is the computer analysis of biological information, such as DNA or protein sequences, for understanding biological processes.

Part I: Using BLAST to Find Evolutionary Relationships

You begin this investigation by observing an image of a fossil obtained from an excavation in China. Using features that you've detected in the image, you decide where on a provided cladogram the fossil organism should be positioned. A **cladogram** (or phylogenetic tree) is a "tree-like" graphical representation of the relatedness of species. Each branch of the phylogenetic tree represents the divergence of a species (or group of species) from a common ancestor. Each branch may also show the **shared derived character** or the trait that species along that branch share (Figure 12-4).

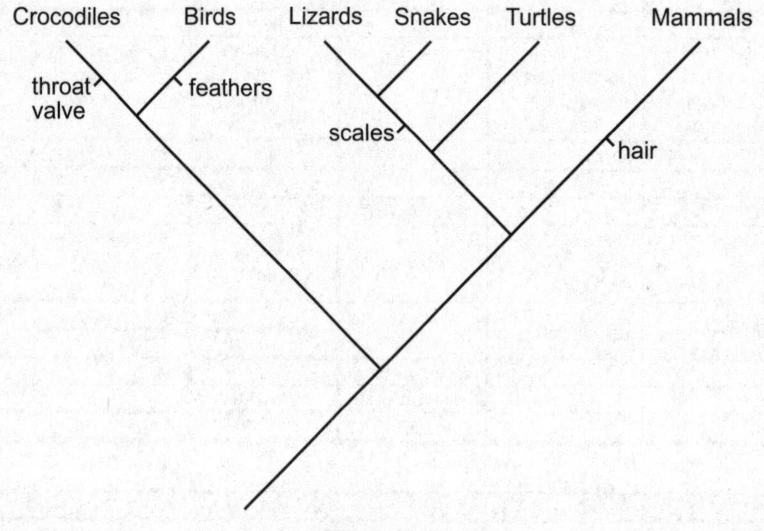

Figure 12-4

To confirm your placement of the fossil organism on the tree, you access BLAST, an online bioinformatics program, to analyze DNA obtained from tissue surviving in the fossil. You are provided with several gene sequences from the fossil (which you download from an AP Biology website) that you submit to BLAST. BLAST then compares your gene sequences with others in the database and provides a list of organisms with gene sequences that are most similar to the fossil sequence. A typical BLAST output appears in Figure 12-5.

The top section of a BLAST report provides a graphical summary of the results. Note the following information (the numbered items below correspond to the numeric callouts in Figure 12-5):

1. **Color Key.** The color key (shown in the figure as shades of gray) is a legend defining the color of each of the horizontal bars below. The bars represent alignment sequences of other organisms. The number on each color segment in the key provides an alignment score range; the higher the alignment score, the greater the similarity between the submitted sequence (from the fossil) and the sequence of another organism.
2. **Query Bar.** The bar below the color key represents the DNA sequence submitted for the fossil.
3. **Nucleotide Number.** The numbers below the query bar enumerate the nucleotides in the submitted sequence.
4. **Aligning Sequences.** Each bar in the group shown here represents a sequence in another organism that matches the submitted sequence. Match the color of the sequence to the color key to get an alignment score range for the sequence. Bars are arranged so that the best matches are at the top. The span of each bar can be compared to the submitted sequence in the query bar to determine the parts of the sequences that match and the parts that are absent (gaps).

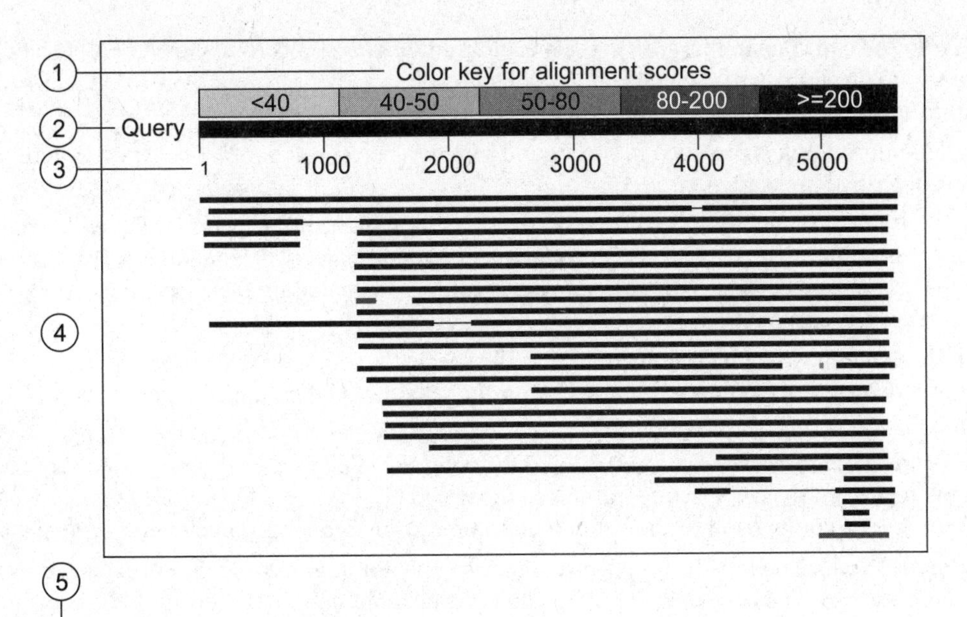

5

Distance tree of results

Sequences producing significant alignments

```
                                                                        Score      E
                                                                        (Bits)     Value
ref | NM 204790.1    | Gallus gallus collagen, type  V, (COL...          1.029e+04  0.0     UEGM
ref | XR 118465.1    | PREDICTED: Mmeleagris gallopavo coll...          3917       0.0     UGM
ref | XM 003757475.1 | PREDICTED: Sarcophilus harrisii c...             3487       0.0
ref | XM 001506246.2 | PREDICTED: Ornithorhynchus anatin...             3476       0.0     GM
ref | XM 001372383.2 | PREDICTED: Monodelphis dommetica ...             3465       0.0     GM
ref | XM 003794863.1 | PREDICTED: Otolemur garnettii col...             3227       0.0
ref | XR 131578.1    | PREDICTED: Equs caballus collagen al...          3221       0.0     GM
ref | NM 000093.3    | Homo sapiens collagen, type W, alpha...          3177       0.0     UEGM
ref | XM 003470915.1 | PREDICTED: Cavia porcellus collag...             3064       0.0     GM
ref | XM 003222940.1 | PREDICTED: Anolis carolinensis collagen, ...     3022       0.0     UGM
ref | XM 537804.3    | PREDICTED: Canis lupus familiaris collage...     3020       0.0     GM
ref | XR 054662.1    | PREDICTED: Taeniopygia guttata misc_RNA (...     3001       0.0     GM
ref | NM 001014971.1 | Sus scrofa collagen, type V, alpha 1 (COL...     2992       0.0     UGM
ref | XM 002920704.1 | PREDICTED: Ailuropoda melanoleuca collage...     2976       0.0     GM
ref | XM 0003312016  | PREDICTED: Meganin dadlonglegs nycusa...         2420       0.0
ref | XM 001118214.2 | PREDICTED: Macaca mulatta hypothetical LO...     2314       0.0     GM
ref | XM 003449634.1 | PREDICTED: Oreochrois niloticus collagen ...     2279       0.0     GM
Ref | XM 002935230.1 | PREDICTED: Xenopus (Silurana) tropicalis ...     2132       0.0     GM
                     .
          (6)        .              (7)                                  (8)       (9)     (10)
                     .
ref | XM 002833191.2 | PREDICTED: Pongo abelii collagen ...              717        0.0
ref | XR 123063.1    | PREDICTED: Nomascus leucogenys colla ...         481        4e-132  GM
ref | XM 002724221.1 | PREDICTED: Oryctolagus cuniculus ...             431        4e-117  UGM
ref | XM 01118209.2  | PREDICTED: Macaca mulatta collagen ...           403        8e-109  UGM
ref | XM 002833217.1 | PREDICTED: Pongo abelii collagen ...              398        4e-107  GM
ref | NM 134452.1    | Rattus norvegicus collagen, type V, ...          398        4e-107  UEGM
ref | XR 109855.2    | PREDICTED: Homo sapiens hypothetical ...         390        6e-105  M
```

```
□ref|NM 204790.1| UEGM Gallus gallus collagen, type V,alpha 1 (COL5A1), mRNA
gb|AF137273.1| Af137273 UGM Gallus gallus alpha 1(V)collagen mRNA,complete cds
Length=5575

GENE ID: 395568 COL5A1 | collagen,typeV,alpha[Galllus gallus]
(10 or fewer PubMed links)

Score = 1.029e+04 bits (5571), Expect = 0.0            (11)
Identities = 5575/5575 (100%), Gaps = 0/5575 (0%)
Sttrand=Plus/Plus

 Query 1    ATGGATACGCACACGCGATGGAAAAGGCGCAGTTGGATACGGAACTGGCAGCTGCACGTC 60
            ||||||||||||||||||||||||||||||||||||||||||||||||||||||||||||
 Sbjct 1    ATGCATACGCACACGCGATGGAAAAGGCGCAGTTGGATACGGAACTGGCAGCTGCACGTC 60

 Query 61   GCGCTGGTGCTGCTGGGCGCCGCGGCCCTGGGCCGGGCAGCTGAACCTGCAGACCTCTTG 120
            ||||||||||||||||||||||||||||||||||||||||||||||||||||||||||||
 Sbjct 61   GCGCTGGTGCTGCTGGGCGCCGCGGCCCTGGGCCGGGCAGCTGAACCTGCAGACCTCTTG 120
```

Figure 12-5

5. **Distance Tree.** You can click in this area to see a cladogram of the results (not shown here). The displayed cladogram shows the submitted sequence highlighted in yellow. Other sequences provide species names. A legend (at the far right) provides a common description for the species.

   More detailed information is provided in the second section. Here, each bar of the list of matching sequences is described in detail.

6. **References.** You can use a link in this column to get a common name for a species (e.g., *Gallus gallus* is a chicken), a description of the sequence (e.g., the sequence is associated with the mRNA that codes for the connective-tissue protein collagen), the species classification, the matching DNA sequence, the translated amino acid sequence, and published references.
7. **Species.** In this column, you are given the name of the species in which the sequence was found and the beginning of a description of the sequence. Click on the reference (1st column) to get the complete description.
8. **Alignment Score.** *The bigger the alignment score,* the greater the similarity to the submitted sequence for the fossil and the higher on the list. Clicking on this number will take you to a location (lower on this same page) that provides a comparison of the submitted and matching sequences, nucleotide by nucleotide.
9. **Expect Value (E Value).** The E value provides a numeric value for alignments that are expected by chance. If an alignment occurs by chance, then it is unlikely to indicate a biological relationship. *The smaller the E value,* the more likely the alignment is the result of common ancestry. Small E values use a notation where e represents the logarithmic base of 10. Thus, the expression "3e–08" represents 3×10^{-8}.
10. **Related Databases.** Clicking on the icons (U, E, G, M) in this column takes you to relevant information in other databases.

    In the bottom section of the BLAST report, the matching sequence from each of the other organisms is paired with the submitted sequence of the fossil, nucleotide by nucleotide.

11. **DNA Sequences.** The matching sequence and the submitted sequence are paired, nucleotide by nucleotide, for each species. A fraction and percentage are given for the nucleotides that are identical (e.g., 5575/5575, 100%) and for the gaps that occur where parts of the sequence are missing (e.g., 0/5575, 0%). The alignment score takes into account both of these values.

Once you've obtained a BLAST result for your sequence, use the alignment scores to support your placement of the fossil on your cladogram. Only consider sequences with E values smaller than 0.01% (0.0001 or 1e–04), as values larger than this suggest that the match occurs by chance and not from a biological relationship.

Part II: Additional Investigations Using BLAST

For the remainder of the lab, you explore a gene or protein of your choice using the BLAST resource.

Investigation 4: Diffusion and Osmosis

This investigation provides exercises that examine the movement of water across selectively permeable membranes. Refer to Chapter 3, "Cell Structure and Function," for a complete review of diffusion, osmosis, and plasmolysis.

In animal cells, the direction of osmosis, in or out of a cell, depends on the concentration of solutes inside and outside the plasma membrane. In plant cells, however, osmosis is also influenced by turgor pressure, the internal pressure of water exerted on the cell wall. To account for differences in both concentration and pressure, a more general term, **water potential,** is used to describe the tendency of water to move across a selectively permeable membrane. Water potential is the sum of the **pressure potential** (from any externally applied force) and the **solute potential** (osmotic potential):

$$\Psi = \Psi p + \Psi s$$

water potential = pressure potential + solute potential

Water potential has the following properties:

1. Water moves across a selectively permeable membrane from an area of *higher* water potential to an area of *lower* water potential.
2. Water potential can be positive or negative. Negative water potential is called **tension.**
3. In a living cell, water potential is always zero or negative.
4. Solute potential results from the presence of solutes and is always negative. A higher concentration of solutes generates a smaller (or more negative) solute potential.
5. Pressure potential is zero unless some force is applied, such as that applied by a cell wall.
6. Pure water at atmospheric pressure has a water potential of zero (pressure potential = 0 and solute potential = 0).
7. Water potential is measured in bars (1 bar is approximately equal to 1 atmosphere pressure) or megapascals (1MPa = 10 bars).
8. The solute potential can be calculated using the formula $\Psi_s = -iCRT$, where i is the ionization constant, C is the molar concentration, $R = 0.0831$ (pressure constant, in liter-bar/mole °K), and T is the temperature in Kelvin (273 + °C). The ionization constant is equal to the number of ions that a substance will produce in water: 2 for NaCl (two ions: $Na+$ and Cl^-) and 1 for a substance that does not ionize (like sucrose).

   You do not need to memorize the solute potential formula or the value of R. They will be provided to you on the Equations and Formulas pages of the AP exam.

Think of water potential as potential energy, the ability to do work. The water at the top of a dam has a high water potential, a high potential energy, and a large capacity for doing work, such as the ability to generate electricity as it runs downhill.

Part I: Cell Surface Area

This part of the investigation examines the effect of surface area and cell size on the rate of diffusion. You cut up agar or gelatin that has been soaked in phenolphthalein into pieces of different sizes to model (represent) cells. Then you immerse the "cells" into a solution of NaOH, which turns phenolphthalein red. Cutting the "cells" in half reveals how far the NaOH has diffused.

For each size or shape "cell," the depth to which the NaOH (or red color) diffuses will be about the same for the same amount of time. But the smaller the "cell," the sooner the NaOH will reach its center. A calculation of the surface area-to-volume ratio (S/V) of the cells should reveal that "cells" with the larger S/V will have a larger portion of their volume penetrated by diffusion (turned red) of the NaOH in a fixed amount of time. Since the plasma membrane is the surface through which substances must pass (going in or out), a smaller cell, with a larger S/V, is more able to accommodate its metabolic need compared to a larger cell.

If a question on the AP exam asks you to design an experiment to investigate the effect of temperature or concentration on diffusion, you can apply the procedures of this investigation but at various temperatures or by using different concentrations of NaOH.

Part II: Diffusion and Osmosis

This part of the investigation examines the influence of solute type and concentration on diffusion and osmosis. Model cells are represented by bags made with dialysis tubing. The tubing limits the passage of solutes and models a selectively permeable membrane. It will allow small molecules to pass through, such as water and monosaccharides (glucose), but not disaccharides (sucrose), polysaccharides (starch), or proteins.

In this investigation, diffusion and osmosis are manipulated by two variables:

1. **Solute type.** Smaller molecules (water, NaCl, glucose) can diffuse across the model "plasma membrane" (dialysis tubing), while larger molecules (sucrose, proteins) cannot.
2. **Solute concentration.** The relative concentration of solutes inside and outside the bag determines the net flow of solutes or water across the membrane; the greater the solute concentration gradient, the faster the initial flow of solutes.

To investigate the influence of solute type and concentration, nine bags are prepared. Four bags are each filled with a different solute: sucrose, NaCl, glucose, and ovalbumin (a protein). These four bags are immersed in water. Another four bags are filled with water, and each is immersed in one of the four different solute solutions. A ninth bag, a control, is filled with water and immersed in water. The direction of water movement can be determined by weighing the "cells" before and after immersion.

There are several things to remember for this investigation:

1. Diffusion is *net* movement of substances from an area of higher concentration to an area of lower concentration.
2. Diffusion results from *random* movement of molecules.
3. Osmosis is the diffusion of *water* molecules across a *selectively permeable membrane.*

Also, keep in mind that the rate of diffusion (and osmosis) is influenced by these factors:

1. The initial rate of diffusion is faster when the concentration gradient is greater. But equilibrium is reached sooner when the concentration gradient is smaller. This is because the movement of a fewer number of solute molecules is necessary to establish equilibrium.
2. The rate of diffusion is faster when the temperatures of the solutions are greater. This is because molecules at higher temperatures have more kinetic energy and are moving faster.
3. The rate of diffusion is faster when solute weight is smaller. All molecules at the same temperature have the same average kinetic energy, but because kinetic energy equals $\frac{1}{2} \times \text{mass} \times \text{velocity}^2$, lighter molecules move faster.

By measuring the weights of the dialysis bags before and after immersion into the solutions, you are able to determine the direction of osmosis, the movement of *water* across a selectively permeable membrane. But how do you know if the *solutes* were able to pass across the membrane? Several qualitative tests can be used to determine the presence of the solutes inside and outside of the dialysis bags:

1. A commercial glucose testing tape can be used to test for the presence of glucose.
2. A Benedict's test can be used to test for the presence of glucose or fructose.
3. A biuret test can be used to test for the presence of proteins.
4. A flame test can be used to test for the presence of sodium.
5. Lugol's solution can be used to test for the presence of starch. Lugol's solution (iodine and potassium iodide, or IKI) is yellow-brown but turns dark blue in the presence of starch.

Part III: Observing Osmosis in a Living Cell

Cells contain water and various solutes. As a result, the cell has a negative solute potential, and pure water (water potential = 0) will enter the cell because water moves from higher to lower pressure potentials. In many plant cells, the water and solutes are stored in the central vacuole. As water enters the cell, the vacuole expands until the pressure, called **turgor pressure,** exerts a force on the cell wall. In a mature cell, the cell wall cannot stretch and expand. Thus, in a healthy cell, when the cell is fully expanded, the cell wall presses back with a pressure equal to and opposite of the turgor pressure.

In part II of this investigation, you determined the direction of osmosis across a dialysis bag, a cell *model.* This part of the investigation repeats the investigation for *real* cells.

First, you observe a plant cell, such as the aquatic plant *Elodea* or the moss *Mnium.* The most prominent features of a plant cell are the cell wall and the chloroplasts. You may also see the nucleus. The central vacuole will be apparent because of what you don't see: Much of the cell appears empty because it is occupied by the central vacuole, which contains colorless water and solutes.

Second, you use a microscope to observe the effect of osmosis on a plant cell as it responds to one of the four solutions used in part II (sucrose, NaCl, glucose, or ovalbumin). The solution you use creates one of three environmental treatments:

1. When the solution surrounding the cell is **hypertonic** (*higher* solute concentration) relative to the contents of the cell, the net movement of water is *out of the cell,* cell turgor decreases, and the plasma membrane collapses. This is called **plasmolysis** and can be identified when the cell, viewed under a microscope, appears to shrivel, leaving a gap between the cell contents and the cell wall.
2. When the solution surrounding the cell is **hypotonic** (*lower* solute concentration) relative to the contents of the cell, the net movement of water is *into the cell.* A healthy cell may not appear to change because the cell is already exerting its maximum turgor pressure. To observe a change, replace a hypertonic solution (when the cell is exhibiting plasmolysis) with a hypotonic solution to observe the expansion of the cell as turgor pressure is restored. Note, however, that the plasmolyzed cell is not equivalent to the healthy cell, as the plasmolyzed cell has a lower (more negative) solute potential as a result of having lost water (solutes are more concentrated).
3. When the solution surrounding the cell is **isotonic** (*equal* solute concentration) relative to the contents of the cell, there is *no net movement* of water into or out of the cell. Similar to the hypotonic solution above, no observable changes will occur in a healthy cell. But unlike the hypotonic solution, no change will also be observed if an isotonic solution is applied to a plasmolyzed cell. But again, be aware that the solution is isotonic with the plasmolyzed cell (not a healthy cell) and the plasmolyzed cell has a lower (more negative) solute potential (more concentrated solution in the cell) than a healthy cell.

Animal cells do not have a cell wall. As a result, immersing an animal cell in a hypotonic solution will cause the cell to expand. Unless the cell has a mechanism for removing excess water, the cell will continue to expand until it breaks apart, a process called **lysis.**

Third, you design an investigation to determine the concentration of a solution and the water potential of a plant cell. Although there are various alternative setups for this investigation, the most common one is to provide you with five sucrose solutions (0.2M, 0.4M, 0.6M, 0.8M, and 1.0M) that are labeled A, B, C, D, and E. You do not know which of the labeled solutions is which concentration. You are also provided with living cells (potatoes, sweet potatoes, or yams). One method you can use to determine the concentration of each unknown solution follows:

1. Cut, dry, and weigh five samples of living material.
2. Immerse each sample into a beaker containing one of the unknown solutions.
3. After 45 minutes, remove each sample, dry it, and weigh it.
4. Calculate the percentage change in the weight of each sample.
5. Plot your results (percent weight change vs. unknown solutions). Since a greater solute concentration in the solution will generate a greater change in potato weight, plotting the solutions on the *x*-axis in order of their increasing effect on the potato weight change will display the solutions in order of concentration.

From earlier parts of this investigation, you know that osmosis will occur if the water potentials on each side of the plasma membrane are unequal; the greater the difference in water potentials, the greater the net movement of water across the membrane. A cell, then, will gain (in a hypotonic solution) or lose (in a hypertonic solution) the greatest amount of water (and weight) when the difference in water potentials across the membrane is greatest. When the water potentials are equal (in an isotonic solution), there is no net movement of water, and the weight of the cell remains unchanged. Thus, you can identify the sucrose concentrations of the various solutions by the degree to which they result in a weight change. A graph of your data, with the solutions plotted such that they are arranged in increasing effect on the potato weight, will reveal the solutions in order of solute concentration.

When there is no change in the weight of a potato, the surrounding sucrose solution is isotonic relative to the potato, and the water potentials of the water and surrounding solutions are equal. The water potential (Ψ) of the sucrose solution is the sum of its solute potential (Ψ_s) and pressure potential (Ψ_p), but for a solution in an open beaker, the pressure potential = 0.

$$\Psi(\text{sucrose solution}) = \Psi_s + \Psi_p = \Psi_s(\text{isotonic solution}) + 0 = \Psi_s(\text{isotonic solution})$$

Now, the water potential of the sucrose solution and the cell can be calculated using $\Psi_s = -iCRT$.

$$\Psi(\text{potato cell}) = \Psi_s(\text{isotonic solution}) = -iCRT$$

Note that the pressure potential (Ψ_p) of the potato cell is *not* equal to zero due to the turgor pressure exerted by the contents of the water on the cell wall. Thus, the individual values of Ψ_s and Ψ_p of the potato cell remain unknown. Only the sum, Ψ, can be determined in this investigation.

In some alternative forms of this investigation, you may not be given any information about the sucrose concentrations of the four solutions. One way of dealing with this is to mix your own solutions, observe their effects, and compare them to the unknown solutions. Plotting the weight changes (vs. concentrations) for both your solutions and the unknown solutions together on the same graph will reveal approximations for the unknown concentrations.

Investigation 5: Photosynthesis

In this exercise, you investigate factors that influence the rate of photosynthesis. Photosynthesis can be summarized by the following reaction.

$$CO_2 + H_2O + \text{light} \rightarrow CH_2O + O_2$$

To determine the rate of this reaction, you can measure the disappearance of CO_2 or the accumulation of O_2. Both of these gases fill the air spaces among the spongy mesophyll of leaves. You must consider, however, that while photosynthesis is occurring in the leaf, some parts of the leaf may be generating energy through aerobic respiration. Thus, the gases that fill the air spaces are being altered by a concurrent process of respiration:

$$CH_2O + O_2 \rightarrow CO_2 + H_2O + \text{ATP}$$

Measuring O_2 accumulation, then, is a measurement of *net* photosynthetic activity:

$$O_2 \text{ production} = \text{gross photosynthesis} - \text{respiration} = \text{net photosynthesis}$$

Part I: Determining the Rate of Photosynthesis

To measure the rate of net photosynthesis, you record the rate that submerged pieces of leaves rise in water as O_2 accumulates in their air spaces. For the treatment group, you submerge leaves in a bicarbonate solution (to supply CO_2). For the control group, you submerge leaves in a solution without bicarbonate. Here is a brief summary of the procedure:

1. Cut disks of leaf tissue from leaves using a hole punch.
2. Fill one beaker with a bicarbonate solution. Fill a second beaker with a solution without bicarbonate. Add a drop of soap into each solution, as it will reduce the hydrophobic characteristic of the leaf's cuticle and will facilitate submerging the leaf disks.
3. Use a syringe to remove the gases that fill the air spaces in the spongy mesophyll. For each treatment, withdraw the plunger of a syringe, insert the cut leaf disks, and replace the plunger. For the treatment group, add bicarbonate solution (taken from the beaker) by dipping the syringe into the beaker and pulling on the syringe. Repeat for the control group, except use the solution without bicarbonate.
4. Submerge each group of disks into its respective beaker—treatment disks into a solution of bicarbonate and control disks into a solution without bicarbonate.
5. Expose both beakers to a light source and begin timing.
6. As photosynthesis progresses, oxygen will accumulate in the spongy mesophyll, increasing the buoyancy of the leaf disks. Record the time it takes each disk to reach the surface.
7. Summarize the data by calculating the median, the value at which half of the data is above and below. The median is a better statistic than the mean for this data because it reduces the impact of outliers. To obtain a *rate* for photosynthetic activity, calculate the time it takes for 50% of the disks to reach the surface $\left(\frac{\text{\# of disks}}{\text{time for 50\%}}\right)$. This value, called the effective time (ET_{50}), has been shown to be a reliable number for photosynthetic rate for comparison across treatments.

Part II: Investigating Factors That Affect the Rate of Photosynthesis

Light intensity, light wavelength, temperature, and concentration of bicarbonate are environmental factors that affect photosynthetic rates. Biotic factors can also influence photosynthetic rates. Leaf characteristics vary among individual plants and even vary among leaves of a single plant. A leaf taken from a shady part of a plant may differ from a sun-exposed leaf by its density of stomata, density and length of leaf hairs, and thickness of cuticle. All these and other factors influence photosynthetic rates.

In this part of the investigation, you design an experiment to investigate a factor that may influence photosynthetic rates. Follow the procedure for part I, manipulating (giving different values to) the independent variable you investigate. For example, if you investigate light intensity, you manipulate the variable by assigning different values of light intensity.

After collecting the data, graph ET_{50} vs. the independent variable (the variable you are investigating). For some independent variables, such as light intensity, it makes sense to graph $1/ET_{50}$ vs. light intensity because such a graph rises as photosynthetic rate increases (Figure 12-6).

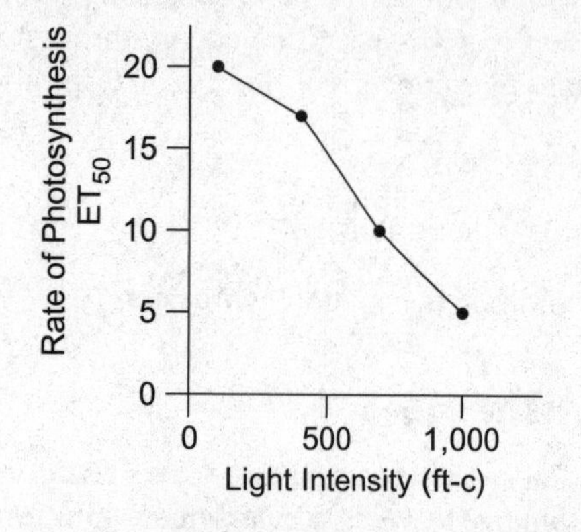

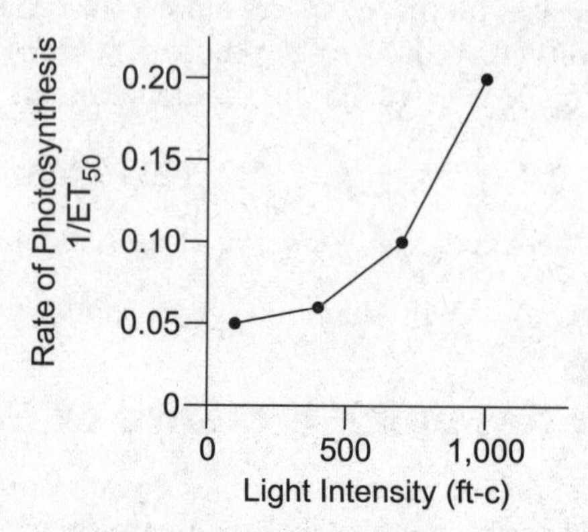

Photosynthetic Rate Expressed as ET_{50} and $1/ET_{50}$ as a Function of Light Intensity

Figure 12-6

Investigation 6: Cellular Respiration

This investigation provides a method for measuring the rate of cellular respiration. Cellular respiration is the breakdown of glucose with oxygen to produce carbon dioxide, water, and energy, as follows:

$$C_6H_{12}O_6 + O_2 \rightarrow 6\ CO_2 + 6\ H_2O + \text{energy}$$

In this lab, respiratory rate is measured in seeds or other organisms by observing the changes in the volume of gas surrounding the organisms at various times. The rate of respiration can be determined by measuring the rate at which O_2 is consumed or the rate at which CO_2 is generated. In this investigation, O_2 consumption is measured, but since the volume of gas can be affected by both the consumption of O_2 and the production of CO_2, the CO_2 is removed by using potassium hydroxide (KOH). KOH reacts with CO_2 gas to produce solid K_2CO_3, as follows:

$$2\ KOH + CO_2 \rightarrow K_2CO_3 + H_2O$$

Since CO_2 gas is removed by its reaction with KOH, volume changes can be attributed only to the following:

1. Consumption of oxygen (due to respiration)
2. Changes in temperature
3. Changes in atmospheric pressure

Unless you are specifically investigating the effect of temperature or pressure on respiration, you must ensure that these two variables remain constant when comparing the results of experimental treatments.

Part I: Determining the Rate of Respiration

Either microrespirometers or gas pressure sensors can be used to measure O_2 generation. Here is a brief summary for using microrespirometers:

1. To assemble a microrespirometer, use a glue gun to attach a capillary tube to the needle end of a syringe. Remove the syringe plunger and insert cotton slightly moistened with a drop of KOH. Insert germinating seeds (or other living organisms) and reinsert the plunger. Attach a calibrated gauge (ruler or other graduated scale) to the capillary tube.
2. Assemble a second microrespirometer, but substitute baked seeds or glass beads for the germinating seeds. This microrespirometer is the control.
3. Place both microrespirometers (weighted down with washers) in a water bath to maintain a constant temperature.
4. Place a drop of a weak solution of soap and red food coloring to the open end of the capillary tubes. The soap reduces the adhesion of water to the glass tube, allowing it to flow freely, and the red food dye makes the water more visible. If the drop does not flow into the tube on its own, pull lightly on the plunger.
5. At regular time intervals, record changes in volume in the capillary tubes. As O_2 is consumed, the red drop should move downward toward the syringe, indicating a decrease in volume.
6. Changes in volume observed in the control microrespirometer represent responses to changes in temperature and pressure. Use these values to correct values observed in the treatment microrespirometer.
7. Graph the accumulated change in volume vs. elapsed time. The rate of respiration is the slope of the line of best fit through the plotted points.

Part II: Investigating Factors That Affect the Rate of Respiration

In this part of the investigation, you design an experiment to investigate a factor that may influence the rate of respiration. For environmental factors, consider temperature, light intensity, and light wavelength. For biotic factors, consider age or size of seed, or seeds of species that occupy different habitats. Also consider measuring respiratory rates for small insects, comparing, for example, the respiratory rates of solitary insects and pairs of same-sex and opposite-sex individuals.

Investigation 7: Mitosis and Meiosis

In this lab you explore the events that occur in cell division during mitosis and meiosis. You also investigate the influence of environmental effects on mitosis and examine the effects of mutations and cancer on cell division. As part of the review for this lab, you should read Chapter 7, "Cell Cycle." In particular, review the activities that occur during mitosis, meiosis, and cytokinesis, the cell structures involved in cell division, and the similarities of and differences between plant and animal cell division.

Part I: Modeling Mitosis

In this part of the investigation, you use clay, pipe cleaners, socks, or beads to represent chromosomes. You display the various stages of the cell cycle using these model chromosomes.

Part II: Evaluating an Environmental Effect on Mitosis

Various chemicals in the environment have the potential to influence mitotic activity. **Lectins,** a class of proteins that bind specifically to carbohydrate molecules, occur naturally throughout nature and serve various functions. In animals, they serve as cell-membrane receptors or facilitate cell adhesion or immune functions. In plants, their functions include defensive proteins that disrupt the digestion of predators that eat them. In addition, certain soil fungi secrete lectins that act as **mitogens** (substances that induce mitosis), influencing mitotic activity in the growing root tips of plants.

In this part of the investigation, you explore the effect of lectins on the roots of onions. The lectins are extracted from kidney beans and are used to simulate the effect of fungal lectins. You are provided with two kinds of onion roots—roots that have been treated with lectins and roots that have not been treated. You follow the following steps:

1. Prepare microscope slides of treated root tips (treatment group) and untreated root tips (control group).
2. Under the microscope, count the number of cells undergoing mitosis (all stages) and the number of cells not actively dividing (cells in interphase) for both treated and untreated root tips.
3. Calculate the percentages of cells undergoing mitosis and cells in interphase for both treated and untreated root tips. Use a chi-square (χ^2) statistical analysis to evaluate whether the difference between the treated root tips (the observed data) and the untreated root tips (the expected data) is statistically significant.

Part III: Cancer and the Cell Cycle

The cell cycle is tightly controlled by checkpoints and other mechanisms that evaluate cell status to ensure that each stage of the cell cycle is complete before advancing to the next stages. When DNA is damaged from replication errors or from radiation, chemicals, or other environmental effects, the mechanisms that monitor cell cycle status may fail. Cancer, characterized by uncontrolled cell division, may result.

This part of the investigation involves a discussion of two kinds of cancer:

1. **HeLa cells.** HeLa cells are cancer cells taken from Henrietta Lacks, an African-American woman, who died of cervical cancer in 1951. Although normal cells will divide only up to 50 times in culture, HeLa cells have been dividing continuously in culture in labs all over the world since they were removed from Lacks. Even for cancer cells, the tenacity of HeLa cells is exceptional. As a result, these "immortal" cells have been used for research and have contributed greatly to our understanding and treatment of disease. The cells were commercialized, and the original cells are freely distributed to researchers.
2. **Philadelphia chromosome.** Philadelphia chromosome is a chromosome translocation in which a segment of chromosome 9 has exchanged with a segment of chromosome 22. A karyotype of the condition displays a chromosome 9 that is longer and a chromosome 22 that is shorter than their respective normal chromosomes. The result of the translocation is the formulation of a gene on the chromosome 22 that codes for a protein that accelerates cell division. Various kinds of leukemia (cancers of the blood or bone marrow) result. However, a drug that limits the effect of this protein has been identified.

The AP exam may ask you to discuss the controversial issues associated with scientific research and its application to society. For example, the HeLa cells were taken from Lacks without her permission and used for research without her knowledge or compensation. In addition, medical research leading to the discovery of drugs to treat genetic disorders or illnesses requires clinical studies using human patients. In some cases, these patients may suffer from the side effects of the experimental treatment, while others, in the control group, may suffer because they do not receive an effective new treatment. Although there may not be a right or wrong answer for these issues, you should be prepared to discuss (by giving examples) the privacy, ethical, social, and legal issues that result from the application of scientific and medical research to society.

Part IV: Modeling Meiosis

Like part I of this investigation, you use clay, pipe cleaners, socks, or beads to represent chromosomes. In this part, however, you display the various stages of meiosis.

Part V: Crossing Over in Meiosis

Crossing over is the exchange of DNA between homologous chromosomes during meiosis. The process contributes to genetic recombination and genetic variation among offspring.

The results of crossing over during meiosis can be readily visualized under the microscope in the **asci** of the fungus *Sordaria*. Fungi grow as filaments called hyphae. Specialized hyphae, called asci, are reproductive filaments that contain eight haploid ascospores. The asci are embedded in other hyphae that form a fruiting body. One kind of fruiting body, a **perithecium,** surrounds the asci, except for a passageway that allows for the escape of the ascospores.

Sexual reproduction begins when hyphae from two strains fuse. Nuclei from one strain pair with nuclei from the second strain. Subsequently, these pairs of unlike nuclei fuse to produce diploid nuclei, which then undergo *meiosis*. During meiosis I, homologous chromosomes pair and separate. During meiosis II, each chromosome separates into two chromatids. At the end of meiosis, there are four daughter cells, each possessing a single chromatid originating from one of the four chromatids that made up each pair of homologous chromosomes. Each of the four daughter cells then divides by *mitosis* to produce two ascospores. The resulting order of the eight ascospores in the ascus corresponds to the alignment of chromatids during meiosis. If no crossing over occurs, then each set of four adjacent ascospores represents a single parent strain and will possess the same traits. This is illustrated in the center of Figure 12-7 for a trait that determines spore color.

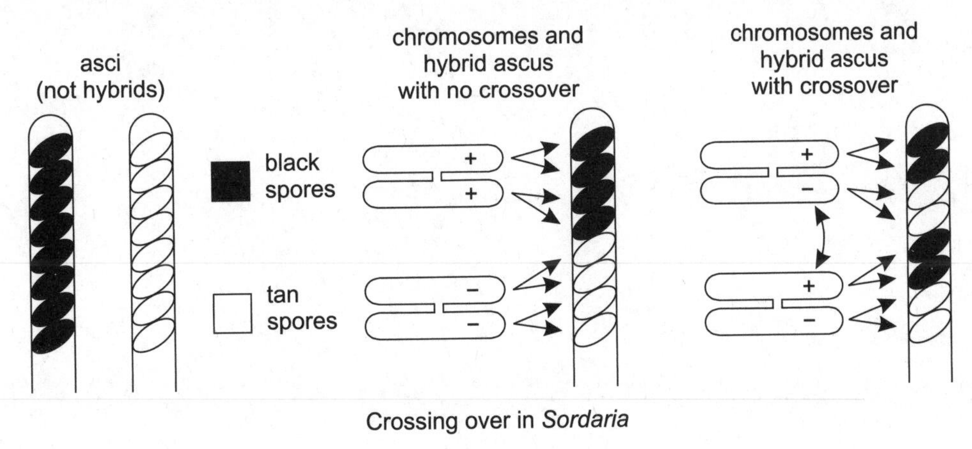

Crossing over in *Sordaria*

Figure 12-7

If crossing over occurs (Figure 12-7, right), then traits on two nonsister chromatids exchange, producing alternating patterns of ascospore pairs with and without the traits included in the crossover.

In this part of the investigation, you observe asci under the microscope to determine crossover frequency. Here is a summary of the steps:

1. Grow two fungal strains of the same species, each bearing ascospores of a different color, on a plate of nutrient-enriched agar. Perithecia form at the interface between the two strains and exhibit hybrid asci (spores from both strains).
2. Under the microscope, count the number of hybrid asci that contain ascospores with crossovers and the number of asci that contain ascospores without crossovers.
3. To determine crossover frequency, first divide the number of asci with crossovers by the total hybrid asci observed (asci with and without crossovers); then divide this number by 2 and convert to a percent by multiplying by 100.

Investigation 8: Bacterial Transformation

Transformation is the uptake of external DNA by cells. In this investigation, you will direct transformation by facilitating the transfer of a bacterial plasmid into *E. coli* bacteria.

Plasmids are common vectors (or carriers) for transferring DNA into bacteria. This investigation uses the plasmid pAMP, a plasmid that contains a gene that provides resistance to the antibiotic ampicillin. As a result, bacteria that have successfully absorbed the plasmid possess resistance to the antibiotic and can be separated from other bacteria by ampicillin treatment. Only those bacteria that have the pAMP plasmid survive the ampicillin treatment. In addition, you may use plasmids that also have a colored-marker gene, such as green fluorescent protein (GFP), that allows you to observe the growth of transformed cells.

Many bacteria take up extracellular DNA readily and, thus, transform naturally. Others can be induced to absorb DNA only under specific laboratory conditions. In either case, cells that are able to absorb DNA are designated **competent** cells.

Part I: Transferring a Plasmid into a Bacterium

In the first part of this lab, the bacterial plasmid, pAMP, is transferred to *E. coli* bacteria. The following steps summarize the procedure:

1. Add $CaCl_2$ to two tubes. $CaCl_2$ induces competence in the bacteria.
2. Add *E. coli* bacteria to both tubes. One tube will be transformed with pAMP plasmids. The second will be a control.
3. Transfer pAMP plasmids to one of the test tubes.
4. Facilitate the absorption of DNA (transformation) by giving the bacteria a heat shock (short pulse of heat).
5. Incubate both tubes overnight to allow for the growth of bacterial colonies. Then, transfer bacteria from one of the tubes to an agar plate with ampicillin and to an agar plate without ampicillin. Repeat for the second tube.
6. Record your results in a table (Table 12-2).
7. Interpret you results: Only bacteria transformed with pAMP plasmids can grow in the presence of ampicillin and form colonies. The control tube of bacteria confirms that without pAMP, growth cannot occur in the presence of ampicillin. It also confirms that the untransformed *E. coli* are not ampicillin resistant.

Table 12-2

	Agar Plate without Ampicillin	Agar Plate with Ampicillin
Tube 1: Bacteria with pAMP plasmids	Growth	Growth
Tube 2: Bacteria without pAMP plasmids (control)	Growth	No growth

Part II: Determine Transformation Efficiency

The transformation efficiency is a quantitative measurement of the effectiveness of the DNA transfer procedure. It is defined as follows:

$$\text{Transformation efficiency} = \frac{\text{Number of colonies}}{\text{Amount of DNA transferred } (\mu g)}$$

Each colony of bacteria on a plate is a population that began with the reproduction of a single bacterium. So the number of colonies is equivalent to the number of cells successfully transferred (transformants).

To calculate the amount of DNA transferred, you need to know each of the values listed below. Values for common protocols are provided as an example.

From protocol:	Concentration of DNA plasmid solution used = 0.005 μg/μl
From protocol:	Amount of DNA plasmid solution transferred to each tube = 10 μl
Calculate:	Amount of DNA transferred to each tube = 10 μl × 0.005 μg/μl = 0.05 μg
From protocol:	Volume of transformation suspension spread from tube to agar plate = 100 μl
From protocol:	Volume of transformation suspension in tube (before transfer to plate) = $CaCl_2$ solution + plasmid solution + nutrient broth = 250 μl + 10 μl + 250 μl = 510 μl
Calculate:	Fraction of suspension transferred from tube to plate $= \frac{100\ \mu g}{510\ \mu g} = 0.196$
Calculate:	Amount of DNA transferred from tube to plate = 0.196 × 0.05 μg = 0.01 μg

If there were 22 bacteria colonies on the agar plate, then

Calculate:	$\text{Transformation efficiency} = \frac{\text{\# of colonies}}{\text{amount of DNA transferred}} = \frac{22\ \text{transformants}}{0.01\ \mu g} = 2.200\ \text{transformants}/\mu g = 2.2 \times 10^3\ \text{transformants}/\mu g$

Part III: Design Your Own Investigation

Design an experiment to investigate any questions that may have occurred to you while doing parts I and II. Consider the effects of mutations from UV light or chemicals (caffeine, theobromine, food preservatives, etc.) or changes in the transformation protocol. If available, consider the use of other plasmids, such as those that produce fluorescent markers (e.g., green fluorescent protein, GFP).

Investigation 9: Restriction Enzyme Analysis of DNA

Part I: Review Restriction Enzymes

Restriction endonucleases (restriction enzymes) are used to cut up DNA. A restriction enzyme cleaves a DNA molecule at a specific sequence of nucleotides, producing cuts that are usually jagged, with one strand of the DNA molecule extending beyond the second strand. These jagged ends, or "sticky ends," allow DNA fragments with complementary jagged ends to fit back together. This technique is important for inserting genes (DNA fragments) into bacterial plasmids. When a plasmid and a foreign source of DNA (with a gene of interest) are treated with the same restriction enzyme and mixed together, the sticky ends of the foreign DNA will match the sticky ends of the plasmid DNA. The foreign DNA fragment can then be bonded to the plasmid DNA by treatment with DNA ligase. The recombinant plasmid can then be used to introduce the foreign DNA into a cell.

Part II: DNA Mapping

When DNA is collected at a crime scene, its analysis requires that it be cut up into pieces and then separated. The pieces are arranged by size to form a DNA chart or map. Because the DNA of every individual (except identical twins) is different, the map of DNA pieces from the crime scene can be compared to other DNA in an effort to determine its origin. DNA maps are "DNA fingerprints" that identify the individual from whom they came. They can help recognize mutations, identify species, produce family trees, or establish evolutionary relationships.

Part III: Separating DNA Fragments

In this part of the investigation, DNA fragments (restriction fragment length polymorphisms, or **RFLPs**) are separated using **gel electrophoresis.** The steps are summarized here:

1. To prepare a gel, a liquefied porous material (usually agarose) is poured into a tray fitted with combs, a device that allows the formation of sample wells (holes). After the gel cools and solidifies, the combs are removed to expose the wells.
2. The DNA to be analyzed is digested (cleaved) with a restriction enzyme. (This already may have been done for you.) A tracking dye is added to the sample. This allows the leading edge of DNA migration to be observed.
3. Load the DNA sample into a well in the gel in the tray of the electrophoresis apparatus. Other samples can be loaded in additional wells, including a sample of DNA of known fragment sizes to be used as a comparison standard.
4. Begin electrophoresis. The electrophoresis apparatus applies a voltage to opposite ends of the gel. DNA carries an overall negative charge due to its phosphate groups, and, as a result, fragments migrate from the negative to the positive electrode. Turn off the apparatus when the tracking dye nears the end of the gel.
5. Immerse the gel in methylene blue, a dye that allows the fragments to be observed.
6. Record the distance each fragment has migrated from the well. The migration distance for a fragment is inversely proportional to the $\log_{10}$ of its molecular weight—longer, heavier fragments move more slowly and travel shorter distances than smaller fragments. The number of base pairs (bp length) in a fragment can be substituted for its molecular weight.
7. Using semi-log graph paper, prepare a standard curve using the observed migration distances and known fragment sizes for the standard sample. The standard curve is a plot of fragment size (base pairs) against migration distance. Since migration distance is inversely proportional to fragment size, plotting migration distance against the log of the number of base pairs produces a straight line.
8. Use the standard curve to determine the size of each fragment produced by the sample under investigation.

Investigation 10: Energy Dynamics

The primary productivity of a community is a measure of the amount of biomass produced by autotrophs through photosynthesis (or chemosynthesis) per unit of time. Primary productivity can be determined by measuring the rate at which CO_2 is consumed, O_2 is produced, or biomass is stored. In this investigation, primary productivity is determined by the amount of biomass produced by plants.

Primary productivity can be examined with respect to the following factors. (Note that the term **rate** means *per unit time.*)

1. **Gross primary productivity** is the rate at which producers acquire chemical energy through photosynthesis before any of this energy is used for metabolism.
2. **Net primary productivity (NPP)** is the rate at which producers acquire chemical energy through photosynthesis less the rate at which they consume energy through respiration.
3. **Respiratory rate** is the rate at which energy is consumed through respiration.

Part I: Estimating Net Primary Productivity

Net primary productivity for Wisconsin Fast Plants is evaluated at various intervals by weighing them. This provides the mass of a live plant, which contains a considerable amount of water. Dry mass can be determined by weighing a group of live plants, then drying them in an oven, and weighing the dry matter to determine the percent of dry matter in the live plants. Chemical energy can be determined by multiplying biomass by 4.35 kcal/g. The net *rate* of chemical energy gain can be determined by calculating the change in energy per day.

Part II: Estimating Energy Flow between Producers and Consumers

Energy flow through a system is evaluated by allowing larvae of cabbage white butterflies to feed on Brussels sprouts, a close relative to Wisconsin Fast Plants (both are in the genus *Brassica*). To follow energy flow, make regular measurements of Brussels sprouts and larval masses. Dry larval masses can be approximated by multiplying live mass by 40%, but a more accurate conversion can be determined by sacrificing a group of larvae, drying them, and weighing them. Dry larval mass can be converted to kcal by multiplying the mass by 5.5 kcal/g.

To follow the energy flow, account for mass changes in the Brussels sprouts and butterfly larvae and in the mass of fecal matter (frass) produced by the larvae.

Part III: Investigating Additional Energy Flow Questions

The final exercise in this investigation is to design your own experiment. Consider the relationship between the survival and reproductive strategies of a plant and how it allocates energy to plant defenses or to various organs (root, shoot, and leaf growth and reproductive organs). Also consider how biomass varies as a function of wet mass and how that might influence energy flow to herbivores.

Investigation 11: Transpiration

Transpiration is the evaporation of water from plants. Differences in water potential move water from the soil to the leaves. From leaves (and, less so, other plant parts), water evaporates mostly through the stomata.

Review these properties of water potential:

1. Water moves across a selectively permeable membrane from an area of *higher* water potential to an area of *lower* water potential.
2. Water potential can be positive or negative. Negative water potential is called **tension.**
3. In a living cell, water potential is always zero or negative.
4. Solute potential (osmotic potential) results from the presence of solutes and is always negative. A higher concentration of solutes generates a more negative solute potential.
5. Pressure potential is zero unless some force is applied, such as that applied by a cell wall.
6. Pure water at atmospheric pressure has a water potential of zero (pressure potential = 0 and solute potential = 0).

The dominant mechanisms for the movement of water through a plant are transpiration, adhesion, cohesion, and tension **(TACT)**. These and other contributing factors are described below.

1. **Osmosis.** Water enters root cells by osmosis because the water potential is higher outside the root in the surrounding soil than inside the root. Dissolved minerals contribute to a lower water potential inside the root by decreasing the solute potential.
2. **Root pressure.** As water enters the xylem cells (plants cells that transport water), the increase in solute potential produces root pressure. Root pressure, however, causes water to move only a short distance up the stem.
3. **Transpiration.** In leaves, water moves from mesophyll cells (the principal plant cells in a leaf) to intercellular air spaces and then out the stomata. This occurs because the water potential is highest in the mesophyll cells and lowest in the relatively dry air outside the leaf. Evaporation of water from plant surfaces is called transpiration.
4. **Adhesion-cohesion-tension.** Because water is a polar molecule, it forms weak hydrogen bonds with other water molecules and other polar molecules. As water moves up through a plant to the leaves, the adhesion (attraction of unlike molecules) of water molecules to cell walls opposes the downward pull of gravity. Also, cohesion (attraction of like molecules) between water molecules makes water act as a continuous polymer from root to leaf. As transpiration removes molecules of water from the leaves, water molecules are pulled up from the roots. The transpirational pull of water through the xylem decreases the pressure potential in the xylem, resulting in negative water potential, or tension. The cohesion-tension condition produced by transpiration is the dominant mechanism for the movement of water up a stem.

On very hot or dry days, the loss of water by transpiration may exceed the rate by which water enters the roots. Under these conditions, the stomata may close to prevent wilting.

Part I: Density of Stomata

In the first part of this lab, you investigate the relationship between the density of stomata and habitat. To calculate the density, you divide the average number of stomata by the average surface area of a leaf. To count the stomata, you prepare a stomatal peel and examine it under a microscope.

Although stomatal density varies among species, it also varies among plants of the same species and among leaves of a single plant. You should consider a variety of factors that may be responsible, many of which are described here:

1. **Temperature.** When the temperature of liquid water rises, the kinetic energy of the water molecules increases. As a result, the rate at which liquid water is converted to water vapor increases; thus, the warmer

the environment, the greater the potential for transpiration. Plants that live in hot environments must have mechanisms to conserve water, and a lower density of stomata may be a solution.

2. **Humidity.** A decrease in humidity decreases the water potential in the surrounding air. In response, the rate of transpiration increases. Dry habitats often accompany hot environments. Plants that live in dry habitats must have mechanisms that conserve water loss, and a lower density of stomata may be a solution. However, not all dry habitats experience hot or even warm temperatures. Alpine habitats and other habitats with prolonged snow coverage are dry habitats except during the short period of the year when the snow is melting. (Plants can't absorb water in the form of snow or ice.) In contrast, plants that live in moist habitats, where water is readily available, can easily remediate water loss from transpiration.
3. **Air movement.** Moving air removes recently evaporated water away from the leaf. As a result, the humidity and the water potential in the air around the leaf drop, and the rate of transpiration increases.
4. **Light intensity.** When light is absorbed by the leaf, some of the light energy is converted to heat. Transpiration rate increases with temperature. Leaves of plants growing in bright sunlight sustain higher temperatures than plants growing in shade, and such exposure may influence stomatal density. However, somewhat countering the increase in temperature is the cooling effect caused by heat removal through transpiration.
5. **Leaf size.** Because larger leaves have a greater surface area than smaller leaves, there is a greater potential for heating and transpiration. Plants with larger leaves often grow in shade where they balance surface area and the need for light against the need to minimize transpiration. Stomatal density may also be affected.
6. **Leaf orientation.** A horizontally oriented leaf maximizes surface area for the capture of sunlight but can also introduce heat stress. Some plants adapted to hot environments remediate this problem by orienting leaves vertically to minimize surface area exposure to the sun. In these leaves, you may find stomata abundant on both sides of the leaf.
7. **Leaf hairs.** By removing moisture from the leaf, air movement increases the moisture gradient across leaf cells and transpiration increases. Hairs on the surface of leaves slow air movement, reducing transpiration.
8. **Stomatal structure.** Some plant species have leaves with stomata located below the leaf surface in depressions. This reduces air movement across the stomatal opening and slows transpiration.
9. **Leaf color.** White or light-colored leaves reflect sunlight and can reduce leaf heating.
10. **Cuticle thickness.** The cuticle is a waxy coating on the surface of leaves that reduces water loss. A thick cuticle in many plants is an adaptation to hot or dry habitats. You will need to examine a cross section of the leaf to evaluate cuticle thickness, as the stomatal peel you prepare only allows you to examine the surface.

Part II: Factors That Influence Transpiration Rate

In this part of the investigation, you design an experiment to investigate the effect of an environmental factor on transpiration rate. In brief, the procedure is as follows:

1. **Choose an environmental factor to investigate.** Many of the factors above, such as temperature, humidity, light intensity, and air movement, are appropriate environmental factors for this investigation because they can easily be manipulated (varied over a range of different values).
2. **Set up apparatus.**
   - To measure transpiration rate using a single cutting of a plant, assemble a potometer by inserting a calibrated pipette into one end of a flexible plastic tube. Bend the flexible tube into a U shape so that the open ends of the pipette and tube are pointing up, and mount the tube and pipette on a ring stand. If available, you can substitute a gas pressure sensor (connected to a computer) for the pipette. Fill the tube and pipette completely with water. Cut the stem of a plant seedling under water, and quickly insert the seedling into the open end of the tube. Cutting the stem under water reduces the chance of air entering the xylem. An air bubble in the xylem (or in the plastic tube) will expand under the low pressure generated by transpiration (a process called **cavitation**) and break the continuity of the water column. Apply petroleum jelly to seal the space between the stem of the seedling and the tube.

- If equipment for a potometer is not available, you can measure the transpiration of an entire plant. Enclose the root ball in a plastic bag and remove all the flowers (as they are likely to fall off). Most weight change over a period of several days is the result of transpiration.
- Repeat the above process for each environmental condition to be investigated. Alternatively, separate groups of students can each investigate a different environmental variable.
- Prepare one additional potometer for normal conditions, a control to which the transpiration rate for the applied environmental condition can be compared.

3. **Collect data.** Record the change in water level in the pipette over several intervals of time. If you are using gas pressure sensors, the associated computer software will generate a transpiration rate. If you are using an entire plant, weigh the plant at various intervals over a period of several days while maintaining constant environmental conditions. After recording this data, you will need to remove the leaves to determine their total surface area because the number of leaves and the sizes of leaves will influence the amount of transpiration.
4. **Calculate transpiration rate.** Calculate water loss per surface area per time (e.g., ml/cm^2/min) by dividing water loss by the total surface area and by the time recorded for the intervals. If using gas pressure sensors, divide the computer-generated transpiration rate (kPa/min) by the leaf surface area to obtain kPa/cm^2/min.
5. **Analyze data.** Graph your data together with the data of other students (if available) who have investigated the effect of other environmental variables. One way to prepare the graph is to plot accumulated water loss as a function of time. Once you have a visual display of the effects of different environmental variables, explain the results, applying what you know about how plant habitat, stomatal density, and other plant adaptations influence transpiration rate.

Investigation 12: Animal Behavior

In this lab, you investigate the behavior of fruit flies (*Drosophila melanogaster*) in response to various environmental stimuli. Alternative subjects for observation include pill bugs (*Armadillidium,* terrestrial crustaceans) and brine shrimp (*Artemia,* aquatic crustaceans).

Before working with fruit flies, you should familiarize yourself with some important fruit fly characteristics:

1. **Life cycle.** Depending upon temperature, *D. melanogaster* requires 10 to 14 days to complete the stages from egg to adult.
   - **Eggs** hatch into maggot-like larvae after about 1 day.
   - **Larvae** undergo three growing stages, or instars, over a 4- to 7-day period. They molt (shed their skins) after the first two stages.
   - **Pupae** form after the third larval instar. A hardened outer case (cocoon) forms around the larva. Inside, a larva undergoes metamorphosis and emerges as an adult fly after 5 to 6 days.
   - **Adults** may live for several weeks. Females may begin mating 10 hours after they emerge from the pupa.
2. **Sex of fruit flies.** To properly identify behavior that may be sex-dependent, you must be able to distinguish the males from the females. The following *typical* (but variable) characteristics are used:
   - A female is larger than a male and has four to six solid dark stripes across the dorsal side (top) of her abdomen. The posterior end of the abdomen is somewhat pointed.
   - A male is smaller than a female and has fewer (two to three) stripes on his abdomen. The posterior end of the abdomen is rounded and heavily pigmented (as if two or three stripes have fused). A male also has a small bundle of black hairs, or sex combs, on the uppermost joint of his front legs.
3. **Virgin females.** If your class conducts genetic experiments with the fruit flies as a follow-up to this investigation, you should be aware that after mating, female flies store male sperm to fertilize their eggs. To ensure that the female does not use sperm from a mating that occurred before the investigation begins, only virgin, or unmated, females can be used. Since a female does not mate until 10 hours after emerging from the pupa, isolating the female soon after emergence will ensure a virgin fly.
4. **Fly mutations.** Your investigation of fly behavior will probably involve wild-type flies. However, you may want to investigate behavior of flies with genetic mutations. Common mutations include eye color variations, wing deformities, and antennae irregularities.

Part I: Fruit Fly Response to Light and Gravity

You begin this investigation by observing the behavior of fruit flies that you have put in a vial. Consider the following behaviors:

1. **Kinesis** is an *undirected* movement in response to a stimulus. The response is a change in the speed of an animal's movement, often movement from a stationary posture. Insects' scurrying about after lifting a rock is an example of kinesis.
2. **Taxis** is a *directed* movement toward (*positive* taxis) or away (*negative* taxis) from a stimulus.
   - **Phototaxis** is a response to light. Positive phototaxis is movement toward light, and negative phototaxis is movement away from light.
   - **Geotaxis** is a response to gravity and may be positive or negative.

Part II: Fruit Fly Response to Chemicals

In this part of the investigation, you assemble a "choice chamber" in which about 20 to 30 flies respond by moving toward or away from a chemical you put in one end of the chamber. You introduce the chemical by adding it to cotton balls that you place at opposite ends of the chamber. You can assess the strength of the response by the number of flies that move toward the chemical after a given amount of time (e.g., 30 seconds). For each chemical test, you should rule out the influence of environmental variables such as light, gravity, and background colors or motion. You can do this by evaluating fruit fly behavior with the choice chamber positioned in different orientations. Chemicals to consider testing include the following:

1. **Water.** Use water as your first chemical to establish a control to which other chemicals can be compared. If there is a significant movement toward or away from the water, evaluate the integrity of your choice chamber or surrounding environment to determine to what, if anything, the flies are responding.
2. **Alcohol.** In alcohol fermentation, glucose is broken down anaerobically (in the absence of oxygen) by yeasts to produce ethanol (ethyl alcohol) and CO_2. ATP is produced through substrate-level phosphorylation.
3. **Vinegar.** In acetic acid fermentation, ethanol is broken down by acetic acid bacteria to produce vinegar (acetic acid).
4. **Alka-Seltzer.** Alka-Seltzer is a medication that relieves stomach pain by neutralizing excess stomach acid. When dissolved in water, it releases CO_2. Add a few drops of water to some Alka-Seltzer before adding it to the choice chamber.
5. **Various laboratory chemicals.** Diluted solutions of NaOH and HCl are usually available from your teacher.
6. **Various condiments.** Consider testing mustard, capers, or salad dressing. Check their ingredients to see if they have anything in common with other chemicals you are testing.
7. **Fruit in various stages of ripeness or decay.** Fruit flies, after all, are attracted to fruit. But why are they attracted to fruit? Try to determine if they have preferences to a fruit's ripeness and, together with your other data, determine what it is about fruit that they pursue.

Part III: Comparing Fruit Fly Preferences

In this part of the investigation, you design an experiment to determine the *degree* of preference that fruit flies have for chemicals. In other words, you determine which chemical is their favorite, which is their least favorite, and preferences of the remaining chemicals that fall in between. Pair each chemical with every other chemical in a two-pole (two-ended) choice chamber, or test multiple chemicals at once with a multi-pole chamber. Collect your data into a table (Figure 12-8).

	Substance in Corner B													
	Substance 1		Substance 2		Substance 3		Substance 4		Substance 5		Substance 6		Substance 7	
Substance in Corner A	Prefer A	Prefer B	Prefer A	Prefer B	Prefer A	Prefer B	Prefer A	Prefer B	Prefer A	Prefer B	Prefer A	Prefer B	Prefer A	Prefer B
Substance 1	15	15												
Substance 2	25	5	14	16										
Substance 3	25	5	10	20	15	15								
Substance 4	21	9	8	22	7	23	16	14						
Substance 5	6	24	0	30	3	27	6	24	15	15				
Substance 6	8	22	3	27	8	22	13	17	30	0	15	15		
Substance 7	28	2	30	0	30	0	28	2	30	0	30	0	16	14

Number of Fruit Flies Preferring Substances in Corner A or Corner B of Choice Chamber

Figure 12-8

Use a χ^2 statistical analysis to see if the combined data of your class are significant.

The AP exam may ask you to design an experiment similar to the one you completed in this lab. Begin by selecting a type of organism and observing them in the absence of applied stimuli. Then, change the environment by introducing a stimulus. Only one stimulus should be applied at a time. In the investigation that you completed above, that one stimulus was a chemical. You manipulated the variable (gave the variable different values) by changing the kind of chemical stimulus. Be sure to include the following in your experimental design:

1. One or more experimental treatments (different values for your independent variable); some independent variables you can consider include the following:
   - Physical stimuli such as humidity, temperature, light, sound, gravity, pH, and chemicals (salt, drugs, nicotine, alcohol, caffeine, aspirin, and pesticides).
   - Biotic stimuli such as the introduction of members of the same species (males or females) or other species (predators or prey). If members of the same species are introduced, the sex of the introduced individual may influence behavior (mating or agonistic behaviors). Multiple members of the same species may elicit social behaviors.
2. A control treatment to which the experimental treatments can be compared.
3. A graph, histogram, or other graphic representation of the data.
4. An interpretation or discussion of the data.
5. A statistical analysis. If an analysis is not requested in the question, you should still suggest an appropriate analysis (e.g., χ^2).

Investigation 13: Enzyme Catalysis

In this investigation, the effect of a catalyst on the rate of a reaction is measured. The reaction investigated is the breakdown of hydrogen peroxide (H_2O_2) into H_2O and O_2. Two kinds of enzymes catalyze the decomposition of H_2O_2—catalase and peroxidase. Catalase decomposes H_2O_2 directly, whereas peroxidase decomposes H_2O_2 using an organic reducing agent. These reactions are compared in the first two equations below. The letter A represents the organic substance.

$$2\,H_2O_2\ (\text{gas}) \xrightarrow{\text{catalase}} 2\,H_2O + O_2\ (\text{gas})$$

$$4\,H_2O_2 + 4\,AH_2 \xrightarrow{\text{peroxidase}} 8\,H_2O + 2\,A_2$$

$$4\,H_2O_2 + 4\,\underset{(\text{colorless})}{\text{guaiacol}} \xrightarrow{\text{peroxidase}} H_2O + \underset{(\text{orange-brown})}{\text{tetraguaiacol}}$$

One option in this investigation is to use peroxidase with the organic reducing agent guaiacol. In the presence of peroxidase, H_2O_2 is broken down to H_2O as four molecules of guaiacol are converted to tetraguaiacol (third equation above). Because guaiacol is colorless and tetraguaiacol is orange-brown, the amount of product produced can be measured by the color intensity of tetraguaiacol.

At the conclusion of this lab, you should know the following:

1. The rate of a reaction is determined by measuring the accumulation of one of the products or by measuring the disappearance of the substrate (reactant).
2. The rate of a reaction is the slope of the linear (straight) part of the graph that describes the accumulation of product (or decrease in substrate) as time progresses.
3. Reaction rate may be affected by temperature, pH, substrate concentration, and enzyme concentration.

Part I: Determine a Baseline for Peroxidase Activity

The goal for this part of the investigation is to establish a **baseline** for measuring the amount of H_2O_2 that is broken down. Once a baseline is established, it can serve as a comparison to subsequent reactions that you investigate in parts II and III. You can measure the progress of the H_2O_2 reaction by measuring the disappearance of H_2O_2 or the accumulation of H_2O or O_2. If a reducing agent (like guaiacol) is used, its disappearance or the accumulation of its product (tetraguaiacol) can be measured.

In this investigation of H_2O_2 decomposition, you document the progress of the reaction by measuring the accumulation of tetraguaiacol or the accumulation of O_2. Here are three ways to do to that:

1. **Color intensity by comparing to a color palette.** A color palette can be created in which a series of test tubes with increasing color intensity are correlated with increasing concentration of tetraguaiacol product. To prepare the palette, run the H_2O_2 decomposition reaction with peroxidase and guaiacol to completion and then make a series of product dilutions (10%, 20%, etc.). You can use this palette to evaluate the progress of the reaction because each dilution bears the color of the product (tetraguaiacol) at a concentration proportional to the reactant (guaiacol). Note that *four* guaiacol molecules are converted to *one* tetraguaiacol.
2. **Color intensity by measuring absorbance.** A spectrophotometer can be used to determine the amount of tetraguaiacol product produced by measuring the absorbance of the solution. The darker the solution, the greater its absorbance, and the more product produced.
3. **O_2 generation.** An oxygen probe can be used to measure the oxygen product directly as it is being produced. In this case, an organic reducing agent (like guaiacol) is not used.

To determine the rate of a reaction in the presence of an enzyme using a color palette, follow these steps:

1. Mix the reactants (H_2O_2, guaiacol) with the enzyme (peroxidase).
2. Take pictures at 1-minute intervals.
3. Compare pictures with the color palette to determine the amount of product formed at the end of each time interval.
4. Plot product accumulation vs. time.
5. Draw a line of best fit through the plotted points. The slope of the line is the rate of the reaction.

Part II: The Effect of pH on Enzyme Activity

Now that you have a baseline for comparison, you can investigate the effect of different treatments on the rate of H_2O_2 decomposition. In this part of the investigation, you investigate the effect of pH. You prepare a series of test tubes, substituting various pH solutions for the H_2O component in the baseline solution. For each pH investigated, record the amount of accumulated product (measured by color intensity, absorbance, or O_2 generated), at various time intervals and graph that data to determine the rate of each reaction. Then plot the rates together on a graph as a function of pH.

Part III: Investigate Other Effects on Enzyme Activity

In this part of the investigation, you explore how the rate of peroxidase activity is influenced by a variable of your choice. Consider enzyme concentration, substrate concentration, or temperature. Be sure that you test only one variable at a time. For example, if you investigate the effect of enzyme or substrate concentration by adding additional enzyme or substrate, be sure to reduce the amount of H_2O by the same amount to maintain a constant volume.

Review Questions

Multiple-Choice Questions

The questions that follow provide a review of the material presented in this chapter. Use them to evaluate how well you understand the terms, concepts, and processes presented. Actual AP multiple-choice questions are often more general, covering a broad range of concepts, and often more lengthy. For multiple-choice questions typical of the exam, take the two practice exams in this book.

Directions: Each of the following questions or statements is followed by four possible answers or sentence completions. Choose the one best answer or sentence completion.

1. A dialysis bag is filled with a 3% starch solution. The bag is immersed in a beaker of water containing a 1% IKI (iodine and potassium iodide) solution. IKI is yellow-brown but turns blue in the presence of starch. The dialysis bag is permeable to IKI but impermeable to starch. All of the following observations are correct EXCEPT:

- **A.** When the bag is first placed in the beaker, the water potential inside the bag is negative.
- **B.** When the bag is first placed in the beaker, the solution in the beaker is yellow-brown.
- **C.** After 15 minutes, the solution in the bag turns blue.
- **D.** After 15 minutes, the mass of the dialysis bag has decreased.

Questions 2–3 refer to five potato cores that are placed in five beakers containing different concentrations of sucrose. The following graph shows the change in mass of each of the potato cores after 24 hours in the beakers.

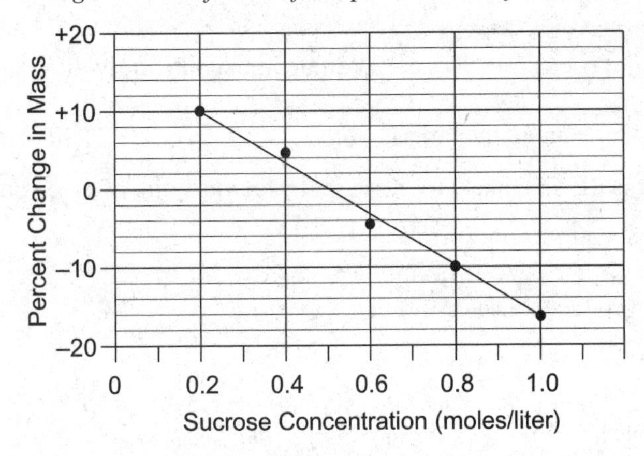

2. The water potential of the potato core can be calculated using which of the following sucrose concentrations?

- **A.** 0.2 M
- **B.** 0.4 M
- **C.** 0.5 M
- **D.** 0.10 M

3. All of the following statements are true EXCEPT:

- **A.** When first immersed in the beaker with 0.2 M sucrose, the water potential of cells in the potato core is more negative than that of the sucrose solution.
- **B.** After 24 hours in the beaker with 0.2 M sucrose, the pressure potential of cells in the potato core has increased.
- **C.** All of the sucrose solutions have a negative water potential.
- **D.** When the net movement of water into a potato core is zero, the water potential of the potato core is zero.

For questions 4–6, use the following graph of an enzyme-mediated reaction.

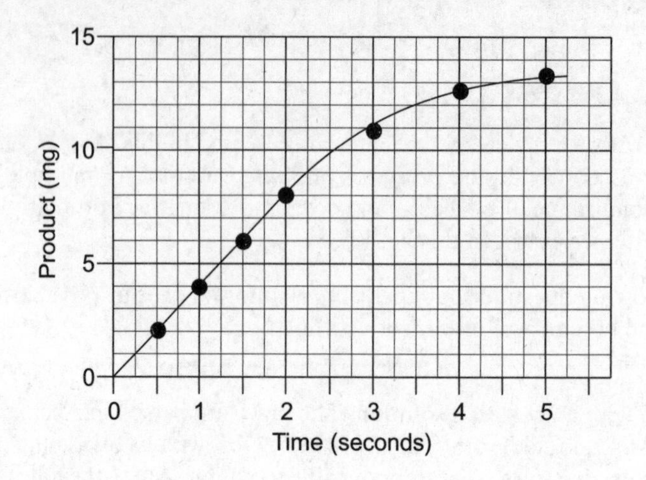

4. What is the initial rate of the reaction?

- **A.** 0.25 mg/sec
- **B.** 4 mg/sec
- **C.** 4.5 mg/sec
- **D.** 5 mg/sec

5. What will be the effect on the reaction if the enzyme is heated to 100°C before being mixed with the substrate?

- **A.** The reaction rate will increase.
- **B.** The reaction will occur at a slower rate.
- **C.** The reaction will not occur or will occur at a rate not significantly different from a reaction rate with no enzyme at all.
- **D.** The reaction rate will remain unchanged.

6. Which of the following is LEAST likely to increase the forward rate of an enzyme-mediated reaction?

- **A.** an increase in the substrate concentration
- **B.** an increase in the enzyme concentration
- **C.** an increase in the product concentration
- **D.** an increase in pH

Question 7 refers to the following figure that illustrates four different arrangements of ascospores, all resulting from a cross between a strain homozygous for the wild type of spores (black spores) and a strain homozygous for the mutant color (tan).

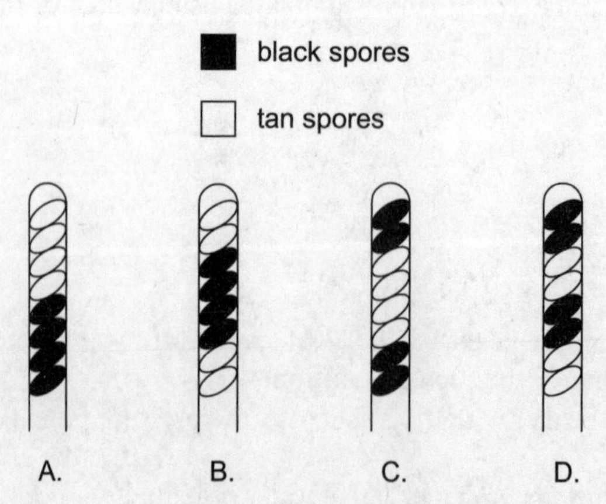

7. Which of the asci in the figure contains ascospores produced during meiosis *without* crossing over?

Question 8 refers to an experiment that measures the rate of photosynthesis by using DPIP as a substitute for NADP⁺. Oxidized DPIP (the lower energy state of DPIP) is blue, and reduced DPIP (the higher energy state of DPIP) is clear. Various wavelengths of light are used to illuminate samples that contain chloroplasts, DPIP, and an appropriate buffer. The results of the experiment are shown here.

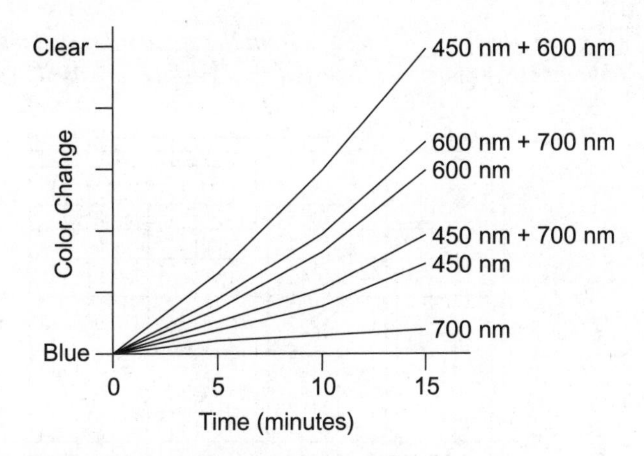

8. Which of the following is a reasonable interpretation of the results?

A. All wavelengths of light are equally absorbed and utilized during photosynthesis.
B. Blue (at 450 nm) is the most efficient wavelength for photosynthesis.
C. There are at least two wavelengths of light utilized during photosynthesis.
D. A blue solution indicates maximum photosynthetic activity.

Questions 9–10 refer to an experiment designed to measure the respiratory rate of crickets. Three respirometers are prepared as follows.

Respirometer	Contents
1	KOH, 1 cricket weighing 2 grams
2	KOH, 1 cricket weighing 3 grams
3	KOH

All respirometers (consisting of a syringe connected to a capillary tube) are immersed in the same water bath. One end of the pipette is connected to the jar; the other end of the pipette is open to the surrounding water.

9. A pressure change registered by respirometer 1 or 2 may indicate any of the following EXCEPT:

A. the cricket is alive
B. a change in temperature of air inside the jar
C. a change in the amount of oxygen inside the jar
D. a change in the amount of CO_2 inside the jar

10. The purpose of respirometer 3 is to act as a control for all of the following EXCEPT:

A. changes in the amount of O_2 inside the jar
B. changes in water bath temperature
C. changes in atmospheric pressure
D. changes produced by water pressure variations in the water surrounding the respirometers

11. Competent bacteria

A. are resistant to antibiotics
B. can be induced to accept foreign DNA
C. cannot reproduce
D. cause disease

Question 12 refers to the following semi-log graph of results from the standard restriction enzyme used in a gel electrophoresis procedure. Phage lambda DNA molecules digested with HindIII *are used as the standard.*

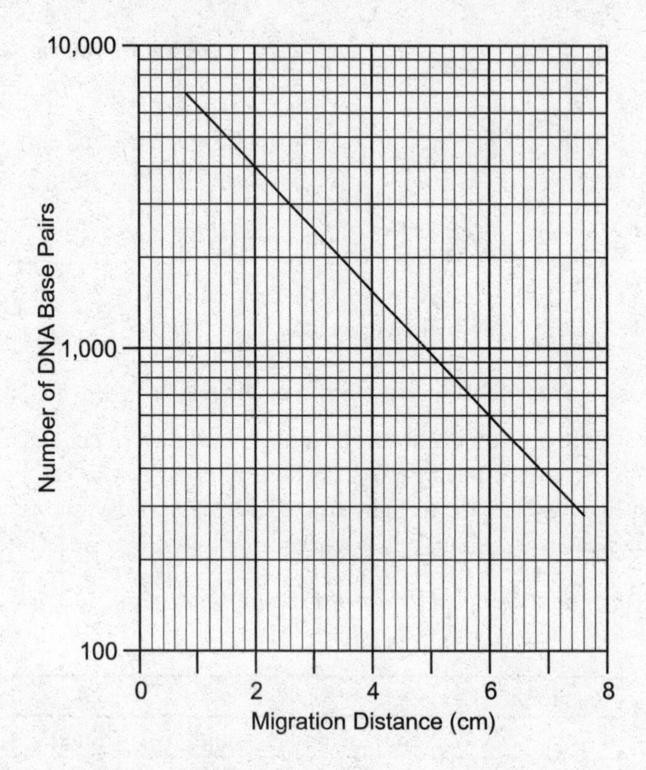

12. A fragment of phage lambda DNA produced by *Eco*RI endonuclease migrates 6 cm. If this fragment is produced during the same electrophoresis procedure as the standard shown in the graph, how large is the fragment?

A. 150 base pairs
B. 600 base pairs
C. 800 base pairs
D. 6,000 base pairs

Question 13 refers to the following diagram, representing the bands produced by an electrophoresis procedure using DNA from four human individuals. Each DNA sample is treated with the same restriction enzyme.

Individual 1 Individual 2 Individual 3 Individual 4

13. Which of the following is a correct interpretation of the gel electrophoresis data?

A. Individual 1 could be an offspring of individuals 3 and 4.
B. Individual 1 could be an offspring of individuals 2 and 3.
C. Individual 2 could be an offspring of individuals 1 and 3.
D. Individual 3 could be an offspring of individuals 2 and 4.

14. A population consists of 20 individuals, of which 64% are homozygous dominant for a particular trait and the remaining individuals are all heterozygous. All of the following could explain this situation EXCEPT:

A. Genetic drift is occurring.
B. The recessive allele is deleterious.
C. All homozygous recessive individuals emigrate.
D. Only heterozygous individuals mate.

15. Which of the following series of terms correctly indicates the gradient of water potential from lowest water potential to highest water potential?

A. air, leaf, stem, root, soil
B. soil, root, stem, leaf, air
C. root, leaf, stem, air, soil
D. air, soil, root, leaf, stem

Question 16 refers to the following graph that shows the rate of water loss for three plants. One plant is exposed to normal conditions, a second plant is exposed to high temperature, and a third plant is exposed to high humidity.

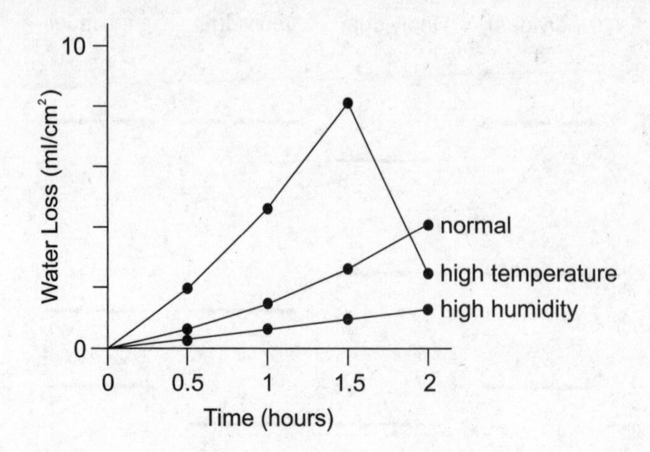

16. The sudden decrease in water loss after 1.5 hours for the plant exposed to high temperatures is probably caused by

A. the burning of the leaves
B. a lack of CO_2 to maintain photosynthesis
C. a lack of O_2 to maintain photosynthesis
D. stomatal closure

Questions 17–18 refer to a laboratory experiment that allows Artemia *brine shrimp to move into water with different concentrations of salt. The results of the experiment are shown in the histogram below.*

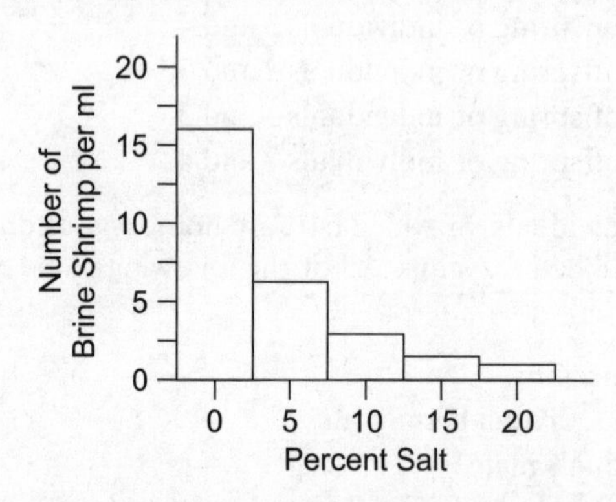

17. According to the preceding histogram, most brine shrimp are found in a habitat in which the salt concentration is

A. 0%
B. 5%
C. 15%
D. 20%

18. Under natural conditions, *Artemia* brine shrimp are rarely found in bodies of water with salt concentrations below 5%. This is probably because

A. the brine shrimp prefer low levels of salt concentration
B. the brine shrimp prefer high levels of salt concentration
C. predators of the brine shrimp are common in fresh water and water with low salinity
D. the brine shrimp cannot survive in fresh water

19. The net primary productivity for a temperate forest was measured at 2,000 mg carbon fixed/L/day. The respiratory rate of the community was determined to be 1,000 mg carbon fixed/L/day. The gross primary productivity for this community is

A. 1,000 mg carbon fixed/L/day
B. 2,000 mg carbon fixed/L/day
C. 3,000 mg carbon fixed/L/day
D. 4,000 mg carbon fixed/L/day

Free-Response Questions

The AP exam has long and short free-response questions. The long questions have considerable descriptive information that may include tables, graphs, or figures. The short questions are brief but may also include figures. Both kinds of questions have four parts and generally require that you bring together concepts from multiple areas of biology.

The questions that follow are designed to further your understanding of the concepts presented in this chapter. Unlike the free-response questions on the exam, they are narrowly focused on the material in this chapter. For free-response questions typical of the exam, take the two practice exams in this book.

Directions: The best way to prepare for the AP exam is to write out your answers as if you were taking the exam. Use complete sentences for all your answers and do *not* use outline form or bullets. You may use diagrams to supplement your answers, but be sure to describe the importance or relevance of your diagrams.

1. Many factors influence transpiration rate. Name three of these factors and explain how they operate.

2. In an exercise to model evolution, it was found that after 10 generations of reproduction, the frequency of one of the two alleles in the model became zero. Explain why an allele would "disappear" after such a short number of reproductive cycles.

3. In a gel electrophoresis procedure, DNA samples are placed on a gel and an electric voltage is applied. In two or three sentences, explain the purpose of applying an electric voltage.

4. An experiment was conducted to measure the effect of light on photosynthetic rate. The following three treatments were evaluated:

- Treatment I: Healthy chloroplasts exposed to light
- Treatment II: Boiled chloroplasts exposed to light
- Treatment III: Healthy chloroplasts incubated in darkness

Oxidized DPIP was added to each treatment to simulate $NADP^+$. When oxidized DPIP is reduced (energized) by photosynthesis, it turns from blue to clear. The degree to which the DPIP was reduced in each treatment was determined by using a spectrophotometer. A spectrophotometer measures the amount of light that is transmitted through a sample. The spectrophotometer in this experiment was set to measure light at a wavelength of 605 nm.

The following data were collected for the healthy chloroplasts exposed to light.

	Treatment I					
Time (minutes)	0	5	10	15	20	25
Average % transmittance (5-minute intervals)	30	45	60	70	75	78

a. Construct and label a graph for the healthy chloroplasts exposed to light. On the same set of axes, draw and label two additional lines representing your prediction of the data obtained for treatments II and III.
b. Justify your predicted data for treatments II and III.
c. Describe the process that causes the reduction (energizing) of DPIP.

Answers and Explanations

Multiple-Choice Questions

1. **D.** After 15 minutes, water will move from the beaker (higher water potential) into the bag (lower water potential), and the mass of the bag will increase. After 15 minutes, the solution inside the *bag* turns blue because the IKI that diffuses into the bag mixes with the starch. When the bag is first placed in the beaker, the water potential in the bag is negative—the sum of the negative solute potential and a zero pressure potential (flaccid bag).

2. **C.** When the net movement into and out of the potato core is zero, the water potentials inside and outside the potato core are the same, and there is no change in the mass of the potato core. A sucrose solution of 0.5 M shows a 0% change in mass on the graph.

3. **D.** When the net movement of water into a potato core is zero, the water potentials inside and outside the potato core are the same, but not zero. Because the potato core immersed in the 0.2 M sucrose solution gained weight, the water potential of its cells must have been smaller than the water potential of the sucrose solution. After 24 hours, water that enters the potato core in the 0.2 M sucrose solution causes the potato cells to expand and gain weight. Since the rigid cell walls cannot expand, pressure potential increases as the cell walls exert a restraining pressure on the cell contents.

4. **B.** The initial rate of reaction is the slope of the plotted curve at the beginning of the reaction. Since the straight-line portion of the curve from 0 to 2 seconds indicates a constant rate of reaction, the slope at any point along this portion of the line will provide the initial rate. For the entire interval from 0 to 2 seconds, the slope, determined by the change in product formed divided by the change in time, is $\frac{8\text{ mg} - 0\text{ mg}}{2\text{ sec} - 0\text{ sec}} = 4\text{ mg/sec}$.

5. **C.** An enzyme that is heated to 100°C will be structurally damaged. As a result, the reaction, no longer under the influence of the enzyme, is not likely to occur. If any activity persists, the reaction rate would be equivalent to the same reaction in the absence of the enzyme (extremely slow).

6. **C.** Since enzyme-mediated reactions are reversible (they convert product back to substrate), increasing the concentration of the product will slow the forward direction of the reaction and accelerate the reverse reaction. Conversely, an increase in the substrate concentration will increase the forward rate of the reaction. Increasing the enzyme concentration will not slow the reaction rate but may increase it if the substrate concentration is high enough to utilize additional enzymes. An increase in pH may change the rate of reaction, but the nature of the enzyme must be known in order to determine whether the rate is increased or decreased.

7. **A.** The ascus containing ascospores with two groups of four adjacent ascospores of the same color results when no crossovers occur. If no crossing over takes place, the order of ascospores corresponds to each of the four chromatids of a homologous pair of chromosomes (two ascospores from each chromatid). Thus, the first four ascospores possess traits from one parent and the second four ascospores possess traits from the second parent. If crossing over occurs, traits on two nonsister chromatids will exchange, resulting in a swap of traits between one pair of ascospores and another pair and producing an ascus that may look like any one of the images in the other answer choices.

8. **C.** Because the graph shows that 450 nm and 600 nm each produce a high photosynthetic rate, and together produce the highest rate, the graph indicates that the photosynthetic process depends on these two wavelengths. Although light of 450 nm does induce significant photosynthetic activity, light of 600 nm induces more. A blue solution indicates that little or no photosynthesis has occurred because DPIP has not been reduced.

9. **D.** An increase in CO_2 gas (from respiration) cannot be detected because it immediately reacts with KOH to produce solid K_2CO_3. A live cricket will decrease the amount of O_2 gas detected by the respirometer. Changes in temperature and atmospheric pressure also cause the respirometer to register a change in volume because temperature and atmospheric pressure affect the water pressure on the pipette inlets.

10. **A.** Since there is no insect and, thus, no O_2 consumption in respirometer 3, the purpose of this respirometer is to control all the variables that might influence the volume changes in respirometers 1 and 2, other than O_2 consumption by insects.

11. **B.** Competence refers to a stage of rapid population growth during which bacteria are most receptive to absorbing foreign DNA.

12. **B.** The vertical line at 6 cm and the horizontal line at 600 base pairs intersect on the standard curve. On the log scale for the *y*-axis, each horizontal line between 100 and 1,000 represents an increase of 100 base pairs.

13. **D.** This is an example of DNA fingerprinting using gel electrophoresis. The horizontal bands represent fragments of DNA, produced by a restriction enzyme, that have migrated across a gel, with lighter fragments moving farthest. Each lane (column) represents the DNA fragments from a different individual. Fragments in different lanes that have migrated the same distance are fragments consisting of the same DNA. Since an offspring's DNA originates from a combination of DNA from both of its parents, the DNA fragments of the offspring must match the fragments of one or the other parent. The DNA fragments from individual 3 can be found in *either* individuals 2 or 4. As a result, it is possible (but not certain) that individual 3 inherited his or her DNA from individuals 2 and 4. In an actual DNA fingerprinting analysis, many different restriction enzymes are used so that many different DNA fragments can be compared.

14. **D.** If only heterozygotes mate, 25% of the offspring, on average, should be homozygous recessive. Thus, this answer cannot explain the absence of homozygous recessive individuals in the population. Because the population is so small, genetic drift may be responsible. Alternately, natural selection against individuals with a deleterious homozygous recessive genotype may also explain why this genotype is absent from the population. Note, however, that the recessive allele remains in the population because it is masked by the dominant allele in heterozygous individuals. The absence of homozygous recessive individuals can also be explained if these individuals leave the population (emigrate) for another location.

15. **A.** Water potential is highest in the soil, decreases from root to leaf, and is lowest in the air. Water moves from the soil into the roots and through the plant and transpires from the leaf because water moves from the area of highest water potential to the area of lowest water potential.

16. **D.** When water entering roots cannot adequately replace the water loss by transpiration, the stomata close to prevent wilting.

17. **A.** The tallest vertical bar (above 0% salt concentration) indicates the preference of the greatest number of brine shrimp.

18. **C.** This question asks why brine shrimp in the wild are rarely found in water with low salt concentrations when laboratory experiments demonstrate that is where they prefer to be. The answer is that under natural conditions, shrimp predators are found in these waters and either the shrimp are eaten or the shrimp avoid these waters because of the predators. Although brine shrimp prefer low-salt water, they can still survive in water with higher salt concentrations where predators are absent.

19. **C.** The gross primary productivity is the sum of the net primary productivity (2,000 mg carbon fixed/L/day) and the respiratory rate (1,000 mg carbon fixed/L/day), for a total of 3,000 mg carbon fixed/L/day.

Free-Response Questions

1. A major factor influencing transpiration rate is temperature. The hotter the environment, the greater the kinetic energy of water molecules, and the faster water molecules change from a liquid state to a gas state. Wind is also a factor because it removes moisture around the leaf and increases the water gradient, which increases the rate of movement of water molecules from inside to the outside of the leaf. Leaf size is also a factor because it increases the surface area from which water transpires.

2. The population size entered into the model was probably small, less than 50 individuals. When populations are small, the effect of genetic drift (changes in allele frequencies by chance) has a strong influence on allele frequencies.

3. DNA is negatively charged due to the negatively charged phosphate groups in nucleotides. When an electric voltage is applied, DNA fragments are attracted to the positively charged end of the voltage field. As a result, the fragments move across the gel, but because larger, heavier fragments move more slowly than smaller, lighter fragments, the various fragments in the sample DNA are separated.

4. a.

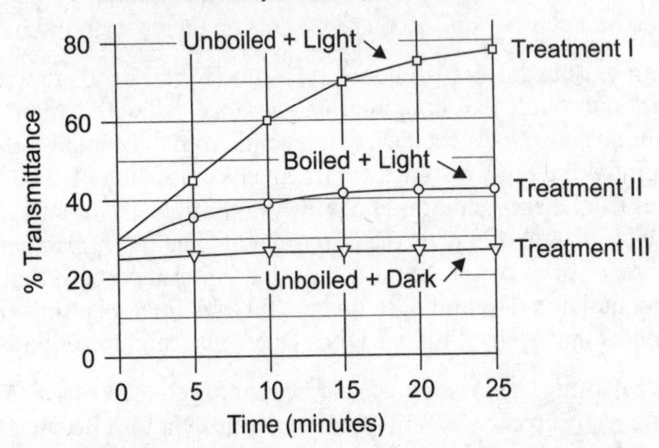

b. In treatment II, the chloroplasts were boiled. Boiling damages the chloroplasts, disrupting the thylakoid membranes. Also, the enzymes and photosynthetic pigments embedded in the membranes are denatured. When these molecules lose their secondary and tertiary structures, they can no longer function properly. As a result, photosynthesis in treatment II is greatly reduced, occurring only in those few membranes that may still be intact with functioning enzymes and pigments. Only a small amount of DPIP is reduced, and the transmittance of a treatment II sample remains low (blue).
In treatment III, no DPIP is reduced because photosynthesis cannot occur in the absence of light. If photosynthesis cannot occur, electrons in the pigment systems are not reduced (energized); thus, there are no energized electrons to reduce DPIP. The DPIP remains oxidized (blue), and transmittance remains low.

c. When healthy chloroplasts are exposed to light, electrons of pigment molecules in photosystem II are energized (reduced). These electrons are eventually passed to the special chlorophyll *a* (P_{680}). Two electrons are then passed to a primary electron acceptor, the first molecule of an electron transport chain. (An electronic transport chain is a series of electron carriers, such as cytochromes.) As the two electrons are passed from one carrier to the next, energy from the electrons is used to generate 1.5 ATP molecules (on average). At the end of the chain, the two electrons are accepted by pigment molecules in photosystem I, where they are again energized by light energy, passed to a special chlorophyll *a* (P_{700}), and then passed to a primary electron acceptor. The two electrons are then used to reduce $NADP^+$. With H^+ (from the splitting of H_2O), NADPH is formed. NADPH then supplies energy for the fixation of CO_2 during the Calvin cycle. In this experiment, DPIP is added so that the process can be visualized. As oxidized DPIP is energized by electrons from photosystem I to form reduced DPIP, DPIP turns from blue to clear.

PART 3

AP BIOLOGY PRACTICE EXAMS

Equations and Formulas

A list of equations and formulas, below and on the next page, is provided with the AP exam.

Standard Error

$$SE_{\bar{x}} = \frac{s}{\sqrt{n}}$$

Mean

$$\bar{x} = \frac{1}{n}\sum_{i=1}^{n} x_i$$

Standard Deviation

$$s = \sqrt{\frac{\sum (x_i - \bar{x})^2}{n-1}}$$

Chi-Square

$$\chi^2 = \sum \frac{(O-E)^2}{E}$$

s = standard deviation of a sample population
$\bar{x}$ = mean
n = sample size

Σ = sum of all
O = observed individuals
E = expected individuals
df = number of E values minus one

Chi-Square Table

p	Degrees of freedom (df)							
	1	2	3	4	5	6	7	8
0.05	3.84	5.99	7.81	9.49	11.07	12.59	14.07	15.51
0.01	6.63	9.21	11.34	13.28	15.09	16.81	18.48	20.09

Metric Prefixes

10^9	giga	(G)
10^6	mega	(M)
10^3	kilo	(k)
10^{-2}	centi	(c)
10^{-3}	milli	(m)
10^{-6}	micro	(μ)

Laws of Probability

Addition rule for mutually exclusive events: $P(A \text{ or } B) = P(A) + P(B)$

Multiplication rule for independent events: $P(A \text{ and } B) = P(A) \times P(B)$

Hardy-Weinberg Equations

$p^2 + 2pq + q^2 = 1$ $\quad p$ = frequency of dominant allele

$p + q = 1$ $\quad q$ = frequency of recessive allele

mode = value that occurs most frequently
median = central value with equal number of values above and below
mean = sum of all entries divided by the number of entries
range = largest value minus smallest value

Rate

$\frac{dY}{dt}$ = change in a variable (Y) per unit of time

dY = change in a variable

dt = change in time

Population Growth

$\frac{dN}{dt} = B - D$

B = birth rate

D = death rate

Exponential Growth

$\frac{dN}{dt} = r_{max}N$

Logistic Growth

$\frac{dN}{dt} = r_{max}N\left(\frac{K-N}{K}\right)$

N = population size

K = carrying capacity

r_{max} = maximum rate of growth

Water Potential (Ψ)

$\Psi = \Psi_p + \Psi_s$

Ψ_p = pressure potential

Ψ_s = solute potential

Solution in Open Container

$\Psi_p = 0$

$\Psi = 0 + \Psi_s = \Psi_s$

Solute Potential

$\Psi_s = -iCRT$

i = ionization constant (sucrose = 1 because it does not ionize in water)

C = molar concentration

R = 0.0831 liter-bar/mole °K

T = temperature in Kelvin = °C + 273

$pH = -\log[H^+]$

Simpson's Diversity Index

Diversity Index $= 1 - \Sigma\left(\frac{n}{N}\right)^2$

n = number of individuals in a species population

N = total number of all individuals among all species

Not Provided on AP Exam

(memorize if not available on your calculator)

$\pi = 3.14$

Surface Area of a Sphere

$SA = 4\pi r^2$

Volume of a Sphere

$V = \frac{4}{3}\pi r^3$

Surface Area of a Cube

$SA = 6s^2$

Volume of a Cube

$V = s^3$

Surface Area of a Cylinder

$SA = 2\pi r^2 + 2\pi rh$

Volume of a Cylinder

$V = \pi r^2 h$

Surface Area of a Rectangular Solid

$SA = \Sigma$ (area of each surface)

$= 2lh + 2lw + 2wh$

Volume of a Rectangular Solid

$V = lwh$

r = radius

l = length

h = height

w = width

Σ = sum of

s = area of one side of a cube

Chapter 13

Practice Exam 1

Answer Sheet for Practice Exam 1

(Remove This Sheet and Use It to Mark Your Answers)

Multiple-Choice Questions

1 Ⓐ Ⓑ Ⓒ Ⓓ	21 Ⓐ Ⓑ Ⓒ Ⓓ	41 Ⓐ Ⓑ Ⓒ Ⓓ
2 Ⓐ Ⓑ Ⓒ Ⓓ	22 Ⓐ Ⓑ Ⓒ Ⓓ	42 Ⓐ Ⓑ Ⓒ Ⓓ
3 Ⓐ Ⓑ Ⓒ Ⓓ	23 Ⓐ Ⓑ Ⓒ Ⓓ	43 Ⓐ Ⓑ Ⓒ Ⓓ
4 Ⓐ Ⓑ Ⓒ Ⓓ	24 Ⓐ Ⓑ Ⓒ Ⓓ	44 Ⓐ Ⓑ Ⓒ Ⓓ
5 Ⓐ Ⓑ Ⓒ Ⓓ	25 Ⓐ Ⓑ Ⓒ Ⓓ	45 Ⓐ Ⓑ Ⓒ Ⓓ
6 Ⓐ Ⓑ Ⓒ Ⓓ	26 Ⓐ Ⓑ Ⓒ Ⓓ	46 Ⓐ Ⓑ Ⓒ Ⓓ
7 Ⓐ Ⓑ Ⓒ Ⓓ	27 Ⓐ Ⓑ Ⓒ Ⓓ	47 Ⓐ Ⓑ Ⓒ Ⓓ
8 Ⓐ Ⓑ Ⓒ Ⓓ	28 Ⓐ Ⓑ Ⓒ Ⓓ	48 Ⓐ Ⓑ Ⓒ Ⓓ
9 Ⓐ Ⓑ Ⓒ Ⓓ	29 Ⓐ Ⓑ Ⓒ Ⓓ	49 Ⓐ Ⓑ Ⓒ Ⓓ
10 Ⓐ Ⓑ Ⓒ Ⓓ	30 Ⓐ Ⓑ Ⓒ Ⓓ	50 Ⓐ Ⓑ Ⓒ Ⓓ
11 Ⓐ Ⓑ Ⓒ Ⓓ	31 Ⓐ Ⓑ Ⓒ Ⓓ	51 Ⓐ Ⓑ Ⓒ Ⓓ
12 Ⓐ Ⓑ Ⓒ Ⓓ	32 Ⓐ Ⓑ Ⓒ Ⓓ	52 Ⓐ Ⓑ Ⓒ Ⓓ
13 Ⓐ Ⓑ Ⓒ Ⓓ	33 Ⓐ Ⓑ Ⓒ Ⓓ	53 Ⓐ Ⓑ Ⓒ Ⓓ
14 Ⓐ Ⓑ Ⓒ Ⓓ	34 Ⓐ Ⓑ Ⓒ Ⓓ	54 Ⓐ Ⓑ Ⓒ Ⓓ
15 Ⓐ Ⓑ Ⓒ Ⓓ	35 Ⓐ Ⓑ Ⓒ Ⓓ	55 Ⓐ Ⓑ Ⓒ Ⓓ
16 Ⓐ Ⓑ Ⓒ Ⓓ	36 Ⓐ Ⓑ Ⓒ Ⓓ	56 Ⓐ Ⓑ Ⓒ Ⓓ
17 Ⓐ Ⓑ Ⓒ Ⓓ	37 Ⓐ Ⓑ Ⓒ Ⓓ	57 Ⓐ Ⓑ Ⓒ Ⓓ
18 Ⓐ Ⓑ Ⓒ Ⓓ	38 Ⓐ Ⓑ Ⓒ Ⓓ	58 Ⓐ Ⓑ Ⓒ Ⓓ
19 Ⓐ Ⓑ Ⓒ Ⓓ	39 Ⓐ Ⓑ Ⓒ Ⓓ	59 Ⓐ Ⓑ Ⓒ Ⓓ
20 Ⓐ Ⓑ Ⓒ Ⓓ	40 Ⓐ Ⓑ Ⓒ Ⓓ	60 Ⓐ Ⓑ Ⓒ Ⓓ

Section I: Multiple-Choice Questions

Time: 90 minutes

60 questions

Directions: Each of the following questions or statements is followed by four possible answers or sentence completions. Choose the one best answer or sentence completion.

1. The protein tropomyosin is present in a variety of cells, including muscle cells. During RNA processing, the primary RNA transcript that codes for tropomyosin is spliced in a manner specific to cell type. In smooth muscle cells (the type of muscle found in the walls of blood vessels and the digestive tract), exons 1, 2, 4, 5, and 6 are spliced. In contrast, in striated muscle cells (the type of muscle attached to bones that cause movements of the body), exons 1, 3, 4, 5, and 6 are spliced. The following figure illustrates the results of this process.

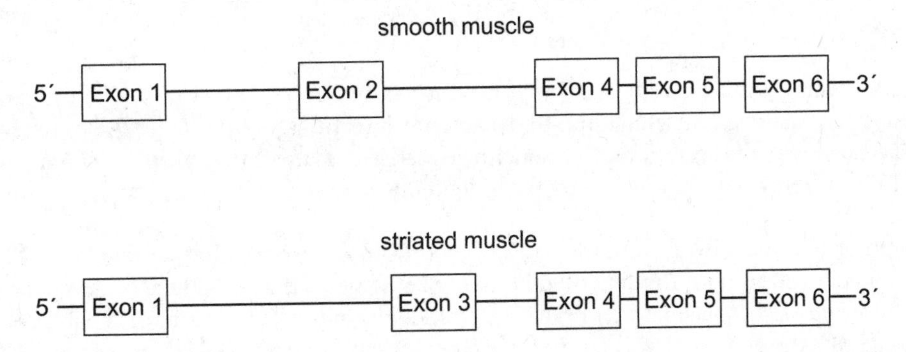

Which of the following describes all possible conditions for exons 2 and 3 if the tropomyosin protein coded for by the mRNA in both cell types is expected to function properly?

A. The nucleotide sequences for exons 2 and 3 must be the same.
B. Exons 2 and 3 must have the same number of nucleotides.
C. The number of nucleotides in exons 2 and 3 must be divisible by 3 with no remainder.
D. The number of nucleotides in exons 2 and 3 must be divisible by 3 with the same remainder (0, 1, or 2).

Questions 2–4 refer to the following figure and description.

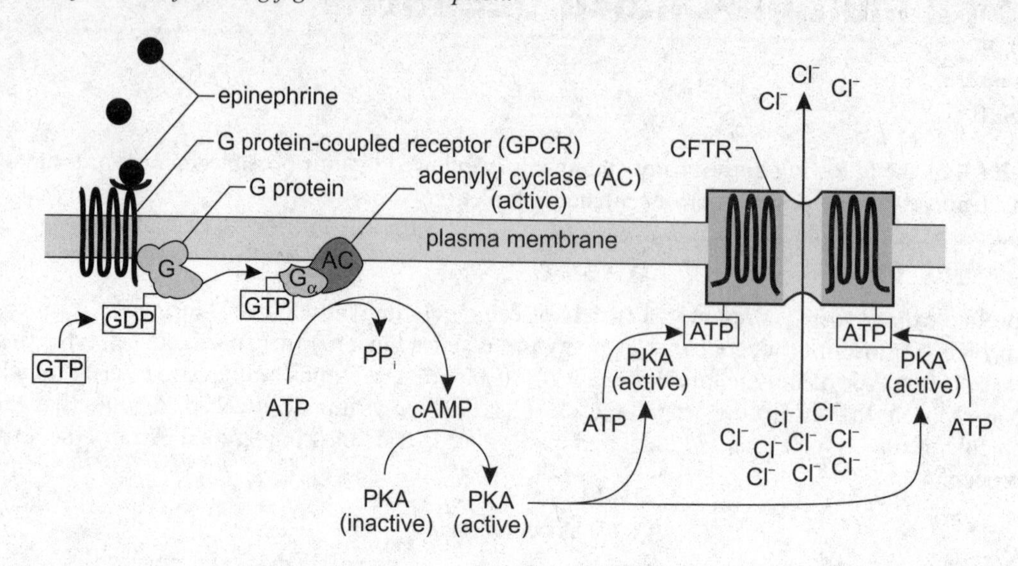

Cystic fibrosis is an inherited disorder that produces a buildup of thick mucus in the lungs that leads to constricted airways, persistent coughing, and bacterial infections of the lungs. Also, mucus buildup from pancreatic cells blocks the secretion of pancreatic digestive enzymes and results in incomplete digestion and diarrhea. Life expectancy is 30 to 40 years for individuals who inherit two copies of the autosomal recessive trait.

The disorder is caused by a mutation in the cystic fibrosis transmembrane regulator (CFTR), an ATP-activated ion channel protein. Binding of ATP to CFTR triggers the opening of a channel that allows Cl^- to passively flow across the plasma membrane. In epithelial lung cells, the channel enables the movement of Cl^- out of the cell, though in other cell types, the channel can be structured so as to promote Cl^- into the cell.

In normal individuals, epinephrine initiates a signal transduction pathway by binding to a G protein-coupled receptor (GPCR) on the plasma membrane of epithelial lung cells. The GPCR then activates an exchange of a GTP for a GDP on a nearby G protein. This G protein, now activated with the GTP, binds to the GPCR. This allows the release of a G subunit (G_α) that binds to and activates adenylyl cyclase. Adenylyl cyclase then catalyzes the conversion of ATP to cyclic AMP (cAMP) and cAMP activates a protein kinase (PKA). PKA enables two ATP molecules to bind to the CFTR protein, triggering the opening of a gated channel in the CFTR and allowing for the passive passage of Cl^- out of the cell.

The movement of Cl^- into the extracellular fluid creates an electrochemical gradient across the plasma membrane that induces water to move out of the cell. An Na^+/K^+ pump, also activated by PKA and ATP, helps regulate the electrochemical gradient to maintain an appropriate surface liquid on the epithelial cells.

2. A signal molecule for this signal transduction pathway is

A. CFTR
B. epinephrine
C. GTP
D. adenylyl cyclase

3. In epithelial cells expressing the cystic fibrosis mutation, the CFTR protein is not functional. How does this contribute to the symptoms of the disorder?

A. Stimulated by a buildup of epinephrine, muscle cells contract and produce coughing.
B. Airways are constricted as a result of the buildup of water in epithelial cells.
C. Cells unable to move Cl^- across the plasma membrane die, and cellular debris produces thick mucus.
D. Thick mucus develops in the lungs because water is not induced to cross the plasma membrane by an electrochemical gradient.

4. Sweat is produced by cells of the sweat glands and transported to the skin surface through ducts. The sweat of individuals with cystic fibrosis is very high in salt (NaCl), so much so that sweat is analyzed for its Cl^- content as a test for cystic fibrosis. This can best be explained by

- **A.** a failure of the CFTR protein to allow the flow of Cl^- into ducts of the sweat glands
- **B.** a failure of the CFTR protein to reabsorb Cl^- from sweat after the sweat is secreted into the ducts
- **C.** a failure of the Na^+/K^+ pump to allow Na^+ to pass through ducts
- **D.** a failure of the G protein-coupled receptor to respond to the signaling molecule

Questions 5–7 refer to the following figure showing four chromosomes (A, B, C, D), each consisting of two chromatids, arranged midway between opposite poles of a cell.

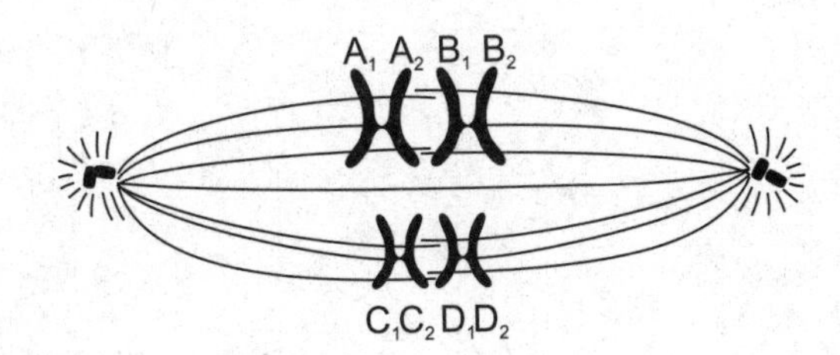

5. Which of the following best describes the process that happens next?

- **A.** Chromatids of all four chromosomes separate and align on a plane midway between the poles of the cell.
- **B.** Chromatids A_1, B_1, C_1, and D_1 move toward one pole of the cell, while chromatids A_2, B_2, C_2, and D_2 move toward the opposite pole.
- **C.** Chromosomes A (A_1 and A_2) and C (C_1 and C_2) move toward one pole of the cell, while chromosomes B (B_1 and B_2) and D (D_1 and D_2) move toward the opposite pole.
- **D.** All four chromosomes align on a plane midway between the poles of the cell.

6. Assuming no mutations during DNA replication, which of the following possesses the same genes?

- **A.** A_1, A_2, B_1, and B_2
- **B.** A_1, B_1, C_1, and D_1
- **C.** A_1 and C_1
- **D.** A_1, A_2, C_1, and C_2

7. Which of the following most closely describes the stage of the cell cycle illustrated in the figure?

- **A.** mitosis
- **B.** meiosis
- **C.** interphase
- **D.** cytokinesis

Questions 8–10 refer to the following text and figure.

Web-building spiders typically eat flies, crickets, and moths. Lizards eat these same foods and also eat spiders. Researchers surveyed the lizards and web-building spiders on 16 Caribbean islands. Half of the islands did not have any lizards; on these islands, they found more spiders and spider species than on islands with lizards. To investigate why this was so, they introduced lizards to four of the eight lizard-free islands and returned every 2 years to resurvey. During each survey, they found lizards and spiders occupying the same areas. The results after 8 years are summarized in the following figure.

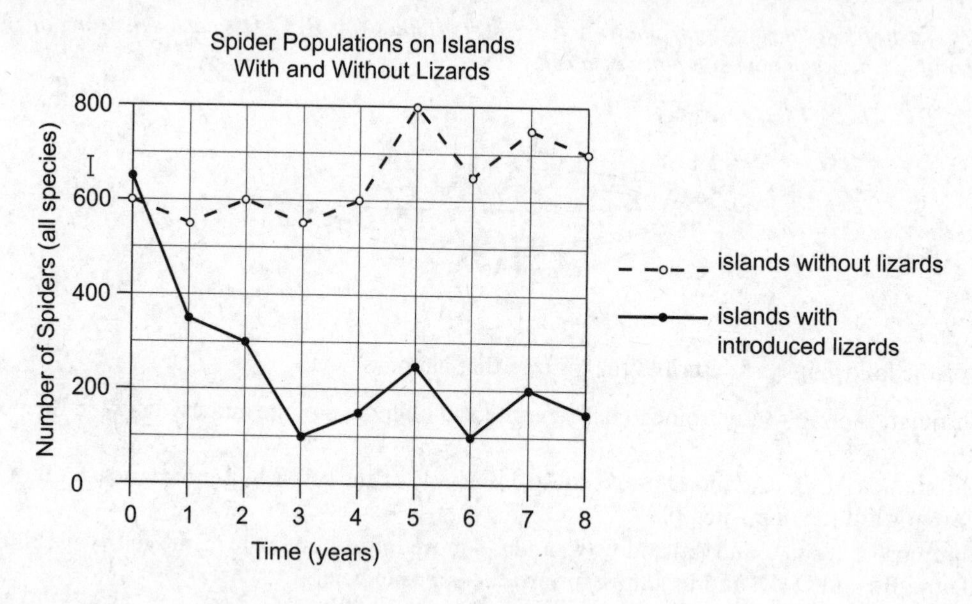

8. Which of the following conclusions is best supported by the data?

A. Spiders cannot escape from predators when they are confined to an island habitat.
B. In the presence of predators, spiders are less likely to emerge from hiding during daylight hours.
C. Where spiders and lizards coexist, sizes of spider populations are reduced by lizard predation.
D. Spiders avoid areas where lizards are present.

9. When the researchers dissected some of the lizards, they found evidence of flies, moths, and crickets in their digestive tracts, but no spiders. How then can the data be explained?

A. The lizards on these islands do not eat spiders.
B. Spiders were capturing and eating lizards.
C. The introduced lizards carried a parasite that infected spiders.
D. Both lizards and spiders compete for the same resources.

10. What does the pattern of variation in the numbers of spiders from year 3 to year 8 on islands both with and without lizards suggest?

A. The numbers of spiders have stabilized.
B. Survival for spiders on islands with and without lizards is increasing.
C. A factor other than lizards is contributing to the numbers of spiders reported.
D. A drought is limiting the amount of food available to spiders.

Questions 11–13 refer to the following equation.

$$\frac{dN}{dt} = r_{max} N\left(\frac{K - N}{K}\right)$$

where

N = population size
r_{max} = intrinsic rate of growth
t = time
K = carrying capacity

11. If $r_{max} > 0$ and $K > 0$, a plot of population size against time produces which of the following graphs?

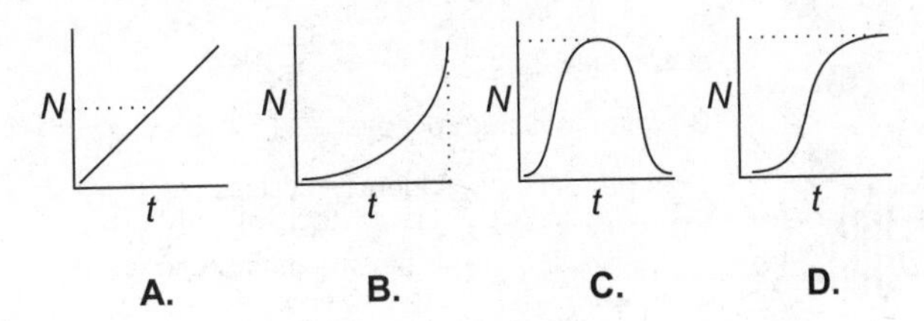

12. When the carrying capacity is equal to the population size, the net rate of increase in the population equals

A. less than 0
B. 0
C. 1
D. the intrinsic rate of growth

13. For which of the following is the net rate of increase in the population greatest?

A. when $N = 0$
B. when $N = 1$
C. when N is small compared to K
D. when $N = K$

Questions 14–16 refer to the following.

Long periods of studying or other kinds of stressful mental activity cause the buildup of adenosine molecules in brain tissue. Adenosine is a ligand that binds to a G protein-coupled receptor on brain cells, activating a G protein by replacing its bound GDP with a GTP. A subunit of the G protein then binds to and activates adenylyl cyclase (AC), a membrane-bound effector protein. Adenylyl cyclase then catalyzes the conversion of ATP to cyclic AMP (cAMP), a second messenger. The cellular response to cAMP varies widely in different types of cells. In brain cells and other cells of the central nervous system, cAMP activates a protein kinase (PKA), which slows brain activity and causes drowsiness. Normally, cAMP concentrations in the cell are kept low by the enzyme cAMP phosphodiesterase (PDE), converting cAMP to regular AMP (not cyclic). But high levels of cAMP can be attained during periods of mental fatigue or other kinds of stress.

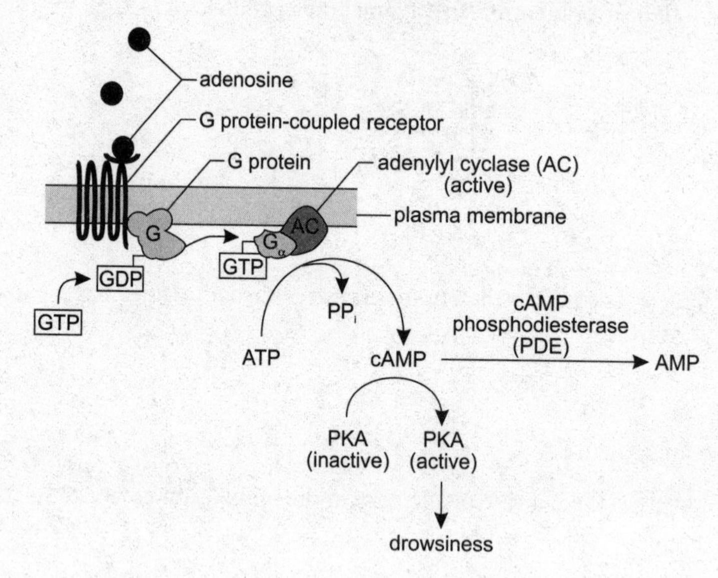

14. What activates adenylyl cyclase (AC)?

- **A.** ATP only
- **B.** ATP and GTP
- **C.** a subunit of the G protein
- **D.** GTP bound to a subunit of the G protein

15. Because caffeine has a structure very similar to adenosine, it can bind to the adenosine receptor but cannot activate it. Which of the following best describes the effect of caffeine when it binds to the adenosine receptor?

- **A.** Caffeine blocks the effect of adenosine by preventing the binding of adenosine to the adenosine receptor.
- **B.** Caffeine enhances the effect of adenosine, overstimulating the adenosine signal transduction pathway and ultimately causing it to fail.
- **C.** Caffeine restores the effect of adenosine by substituting for it as a signaling molecule.
- **D.** Caffeine has no effect on the activity of adenosine.

16. In general, the hormone epinephrine (adrenaline) causes increases in blood pressure, breathing rate, and the rate of cell metabolism—all activities associated with the fight-or-flight response. Like adenosine, epinephrine is also a ligand for a G protein-coupled receptor. Though different from adenosine in other respects, the epinephrine signal transduction pathway, like that of adenosine, produces cAMP and activates PKA. Which of the following is most likely the mechanism that describes how epinephrine works?

- **A.** In muscle cells, PKA triggers the release of glucose from glycogen.
- **B.** In muscle cells, PKA produces muscle fatigue.
- **C.** In brain cells, PKA causes drowsiness.
- **D.** In heart muscle, PKA causes vasoconstriction, or narrowing of the blood vessels.

Questions 17–18 refer to the following graph that plots the rate of a reaction controlled by a human enzyme as a function of temperature.

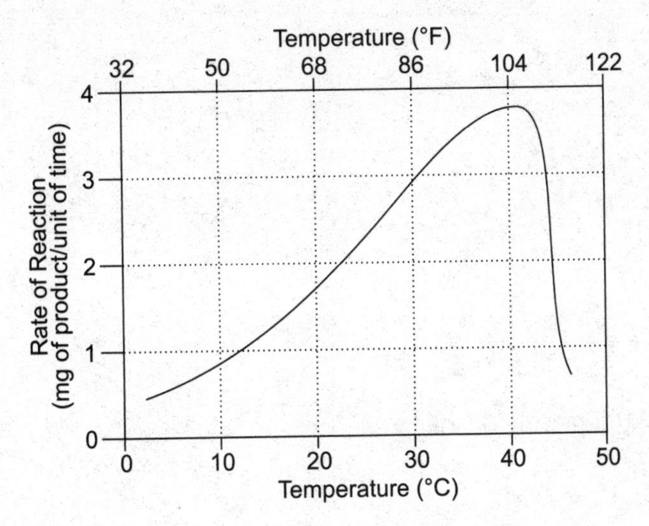

17. What best explains the change in the rate of reaction from 10° to 40°C (50° to 104°F)?

A. With increasing temperature, more substrate molecules collide with sufficient activation energy to react.
B. Because molecules expand as their kinetic energy rises, they become larger targets, increasing their chances of collisions.
C. At higher temperatures, there are more enzyme molecules available to carry out the reactions.
D. As the temperature rises, fewer inhibitory molecules are available to block the active site of the enzyme.

18. What best explains the change in the rate of reaction beyond 45°C (106°F)?

A. At high temperatures, the rapid reaction rate quickly consumes all of the reactants and the reaction is completed.
B. At high temperatures, the substrate becomes locked into the active site of the enzyme, preventing the binding of additional reactants.
C. At high temperatures, the reactants are moving so fast that they cannot stabilize long enough to settle into the active site of the enzyme.
D. At high temperatures, the secondary and tertiary structures of the enzyme are disrupted.

19. The fluidity of the plasma membrane can vary in response to temperature changes. For bacteria and yeast cells, which of the following would be a viable strategy to maintain homeostasis of the plasma membrane in a temperature-varying environment?

A. Increase the concentration of solutes in the cytosol.
B. Change the ratio of phospholipids with double covalent bonds in their tails to those without double covalent bonds in their tails.
C. Replace phospholipids with glucose molecules.
D. Exchange phospholipids from one side of the plasma membrane to the other.

Questions 20–21 refer to the following figure showing a plasma membrane with two embedded proteins that are involved in the movement of materials across the membrane. Each protein is shown in active and inactive states.

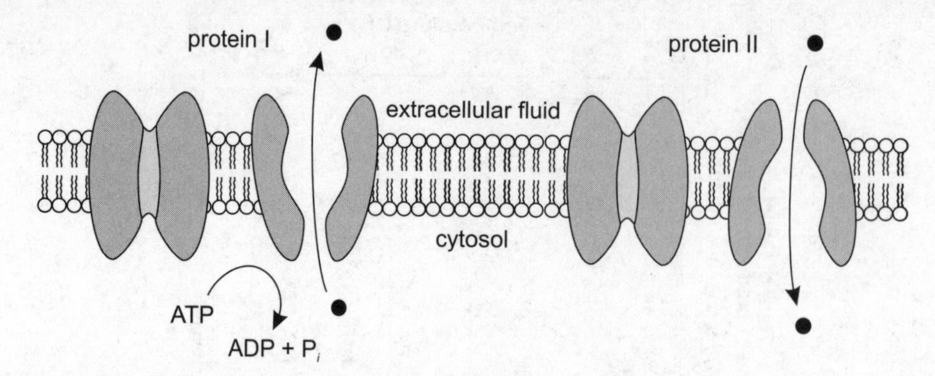

20. The membrane protein II is most likely doing which of the following?

A. transporting glucose *down* a concentration gradient (from a region of higher concentration to one of lower concentration)
B. transporting a small hydrophilic molecule *against* a concentration gradient (from a region of lower concentration to one of higher concentration)
C. transporting Na^+ out of the cell in coordination with the transportation of K^+ into the cell by membrane protein I
D. transporting a steroid hormone out of the cell by active transport

21. The transport of a substance through both protein I and protein II

A. requires ATP
B. requires a gradient from higher concentration to lower concentration in the direction of transport
C. requires that the membrane protein (protein I or protein II) is specific to the substance to be transported
D. requires that the substance to be transported is hydrophobic

22. The structure of ATP synthase in the plasma membranes of bacteria is nearly identical to that in the mitochondria and chloroplasts of eukaryotic cells. This similarity best supports which of the following hypotheses?

A. Eukaryotic and prokaryotic cells share a common ancestor.
B. Mitochondria are derived from ancient aerobic bacteria, and chloroplasts are derived from ancient photosynthetic bacteria.
C. The similarity of ATP synthase in bacteria, mitochondria, and chloroplasts is an example of convergent evolution.
D. Mitochondria and chloroplasts escaped from eukaryotic cells and formed aerobic and photosynthetic prokaryotes.

Question 23 refers to the following diagram of a fragment of a protein. Each amino acid is shown with its R group (side chain) indicated by an R. Amino acids with polar, acidic, or basic side chains are white, while amino acids with nonpolar side chains are shaded.

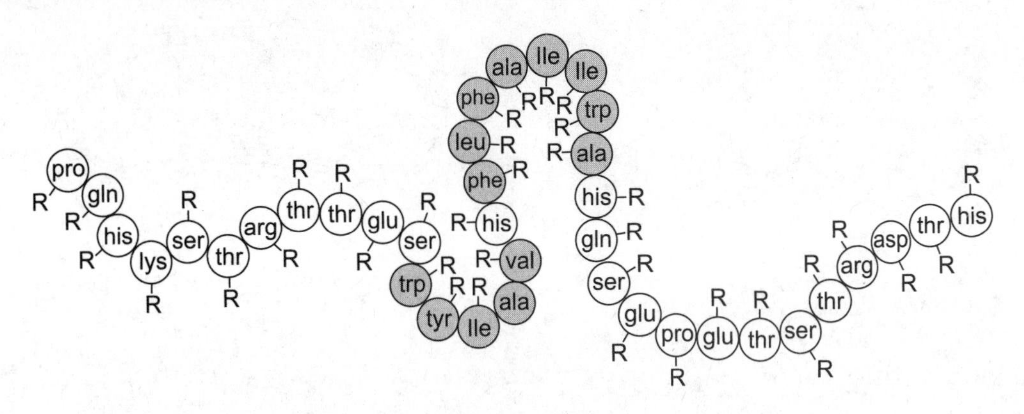

23. The arrangement of the shaded amino acids illustrates how

- **A.** hydrogen bonding between amino acid side chains contributes to the secondary structure of the protein
- **B.** ionic bonding between amino acid side chains contributes to the secondary structure of the protein
- **C.** amino acids with nonpolar side chains contribute to the formation of a helix
- **D.** the hydrophobic response of amino acid side chains contributes to the tertiary structure of the protein

Question 24 refers to the following table showing the population sizes for a community of six species.

Species	Number of Individuals
Black Bear	2
Chipmunk	35
Coyote	15
Ground Squirrel	20
Marmot	10
Pika	18

24. Calculate Simpson's Diversity Index, D, for the data in the table, and round the final answer to the nearest hundredth.

- **A.** 0.23
- **B.** 0.25
- **C.** 0.77
- **D.** 0.79

25. The normal pH of the cytosol is about 7.2. Most enzymes in the lysosome, however, are only active at a pH of 5. This suggests that

- **A.** only acidic substances can undergo enzymatic processing in lysosomes
- **B.** most enzymes in the lysosome are incapable of catalyzing substrates inside the lysosome
- **C.** enzymes in the lysosome are released to the cytosol
- **D.** protons are actively pumped into the lysosome

Question 26 refers to the following figure that shows portions of a mitochondrion and a chloroplast.

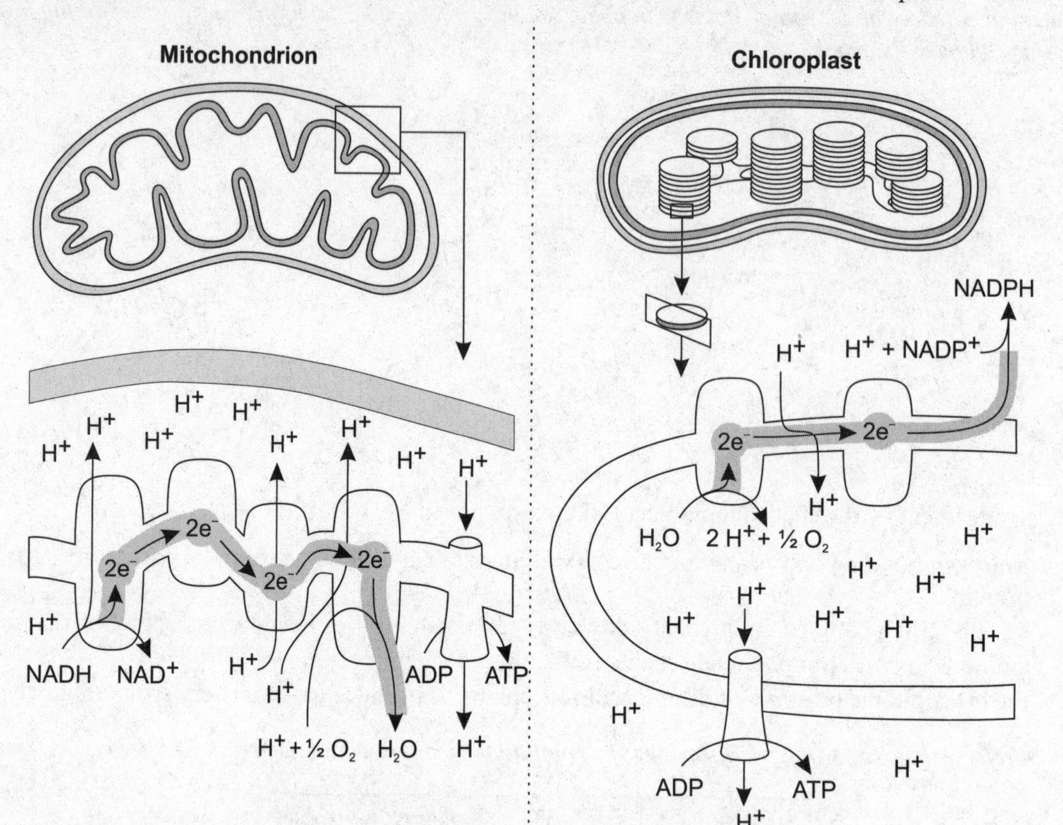

26. A chemiosmotic gradient is generated in both cellular respiration and photosynthesis. The gradient is generated as electrons pass along an electron transport chain in membranes of mitochondria and chloroplasts. Which of the following correctly describes the origin of these electrons as they enter the electron transport chain and the endpoint when they leave the electron transport chain?

A. For respiration, electrons originate from NADH and end in H_2O; for photosynthesis, electrons originate from H_2O and end in NADPH.

B. For respiration, electrons originate from H_2O and end in NADH; for photosynthesis, electrons originate from NADPH and end in H_2O.

C. For both respiration and photosynthesis, electrons originate from NADH or NADPH and end in H_2O.

D. For respiration, the electrons originate from glucose, and for photosynthesis, the electrons originate from CO_2; for both photosynthesis and respiration, the electrons end in ATP.

Questions 27–29 refer to the following paragraph and figure.

To investigate the rate of photosynthesis under different environmental conditions, researchers grew two groups of saltbush (*Atriplex*) plants, one in shade and the other in full sun. Leaves from these plants were then used to determine rates of photosynthesis by measuring CO_2 assimilation while the leaves were exposed to various intensities of photosynthetically active radiation (PAR).

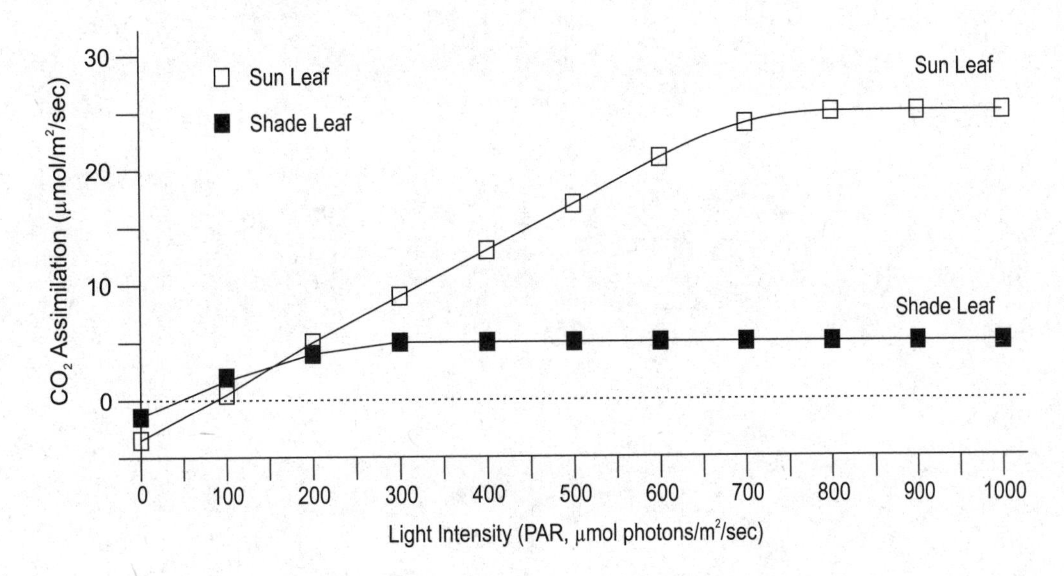

27. As demonstrated by light intensities above 800 μmol photons/m²/sec, leaves from the plant grown in the sun reached a higher maximal photosynthetic rate. Which of the following is the best hypothesis that can explain why plants grown in full sunlight reach a higher maximal photosynthetic rate than plants grown in shade?

A. The sun leaves were exposed to more light than shade leaves.
B. The sun leaves had a greater surface area than shade leaves.
C. Leaves grown in the sun were thicker than shade leaves, with additional layers of photosynthetic cells.
D. The photosynthetic apparatus of leaves grown in the shade were more efficient than those leaves grown in the sun.

28. Which of the following is the best hypothesis that can explain the observed differences in CO_2 assimilation for leaves grown in shade and sun at light intensities between 100 and 150 μmol photons/m²/sec?

A. Shade leaves have a greater surface area than sun leaves.
B. At these low light intensities, high rates of respiration in leaves grown in the shade result in higher rates of CO_2 assimilation than leaves grown in the sun.
C. At these low light intensities, leaves grown in the shade have lower rates of respiration than leaves grown in the sun.
D. At these low light intensities, the photosynthetic apparatus of leaves grown in the sun is more efficient than those grown in the shade.

29. At light intensities below 50 μmol photons/m²/sec, which of the following best explains why values of CO_2 assimilation are negative?

A. Leaf stomata close in response to low light intensities, limiting access to CO_2.
B. Rates of respiration exceed those of photosynthesis at light intensities below 50 μmol photons/m²/sec.
C. The Calvin cycle reverses and releases CO_2 at light intensities below 50 μmol photons/m²/sec.
D. At light intensities below 50 μmol photons/m²/sec, only cyclic photophosphorylation occurs and, thus, no CO_2 is assimilated.

30. Amino acids are joined together by peptide bonds to form proteins. Which of the following groups of amino acid molecules is arranged so that the circled atoms will properly form a peptide bond?

A.

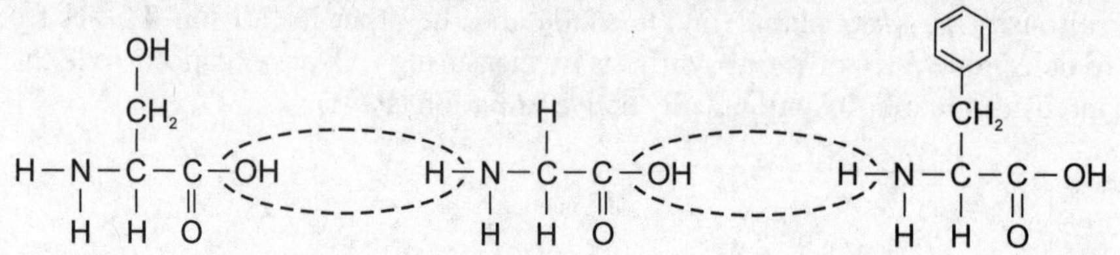

```
          OH                                           (benzene ring)
          |                                                 |
          CH2                  H                           CH2
          |                    |                            |
H—N—C—C—OH            H—N—C—C—OH               H—N—C—C—OH
  |  |  ||               |  |  ||                 |  |  ||
  H  H  O                H  H  O                  H  H  O
```

B.

```
                               COOH                        CH3
                                |                           |
          CH3                  CH2                     HO—C—H
          |                     |                           |
H—N—C—C—OH            H—N—C—C—OH              HO—C—C—N—H
  |  |  ||               |  |  ||                  ||  |  |
  H  H  O                H  H  O                   O   H  H
```

C.

```
          COOH                  CH3
          |                     |
          CH2                  CH2
          |                     |
          CH2                H—C—CH3                     H
          |                     |                         |
H—N—C—C—OH            HO—C—C—N—H             H—N—C—C—OH
  |  |  ||                 ||  |  |                |  |  ||
  H  H  O                  O   H  H                H  H  O
```

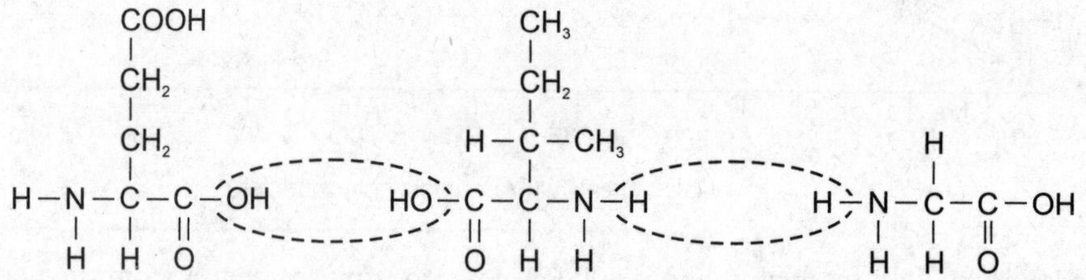

D.

```
          CH3
          |
          CH2
          |
       H—C—CH3               H                      O  H  H
          |                  |                      ||  |  |
HO—C—C—N—H            H—N—C—C—OH            HO—C—C—N—H
    ||  |  |              |  |   ||                  |
    O   H  H              H  CH2 O                  CH3
                             |
                             OH
```

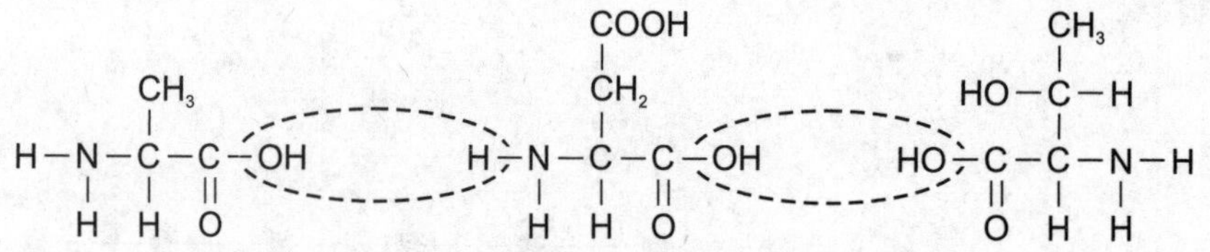

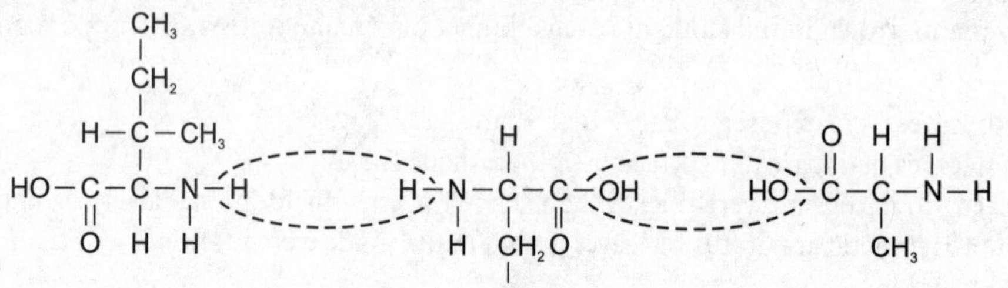

Questions 31–32 refer to the following figure of the glycophorin A protein, a protein passing through the plasma membrane, extending from the cytosol to the extracellular fluid. The protein consists of two identical polypeptides, each forming a helix within the membrane. Circles represent amino acids, and squares represent carbohydrate side chains.

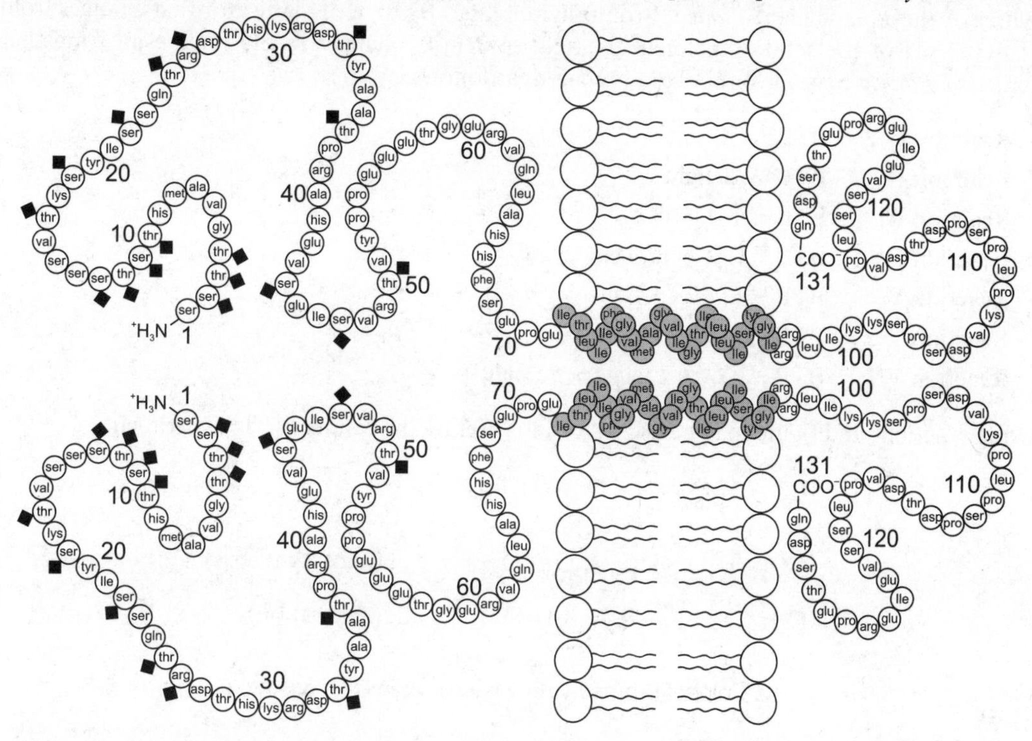

31. The parts of the protein represented by the shaded circles (numbers 73–95) are most likely dominated by amino acids with

A. hydrophilic side chains
B. hydrophobic side chains
C. polar side chains
D. charged side chains

32. The protein shown in the figure illustrates characteristics of which levels of protein structure?

A. only primary structure
B. only primary and secondary structures
C. only primary, secondary, and tertiary structures
D. primary, secondary, tertiary, and quaternary structures

Questions 33–38 refer to the following.

A student experiment investigates the effects that the presence and absence of light have on the metabolic activities of the aquatic plant *Elodea*. Bromothymol blue (BTB), a pH indicator that changes from yellow to blue as the pH of the solution changes, is used to identify changes in pH that result from the metabolic activities of *Elodea*. Seven test tubes are prepared, as follows:

Reaction I:	BTB + CO_2
Reaction II:	BTB + no light
Reaction III:	BTB + CO_2 + light
Reaction IV:	BTB + CO_2 + *Elodea* + light
Reaction V:	BTB + *Elodea* + light
Reaction VI:	BTB + *Elodea* + no light
Reaction VII:	BTB + CO_2 + *Elodea* + no light

The following reactions and figure may be useful for interpreting the results of the experiment:

$$CO_2 + H_2O \leftrightarrow H_2CO_3 \leftrightarrow H^+ + HCO_3^-$$

$$CO_2 + H_2O \xrightarrow{\text{light}} C_6H_{12}O_6 + O_2 \quad \text{(photosynthesis)}$$

$$C_6H_{12}O_6 + O_2 \rightarrow CO_2 + H_2O \quad \text{(respiration)}$$

The Color of Bromothymol Blue (BTB) in Response to pH

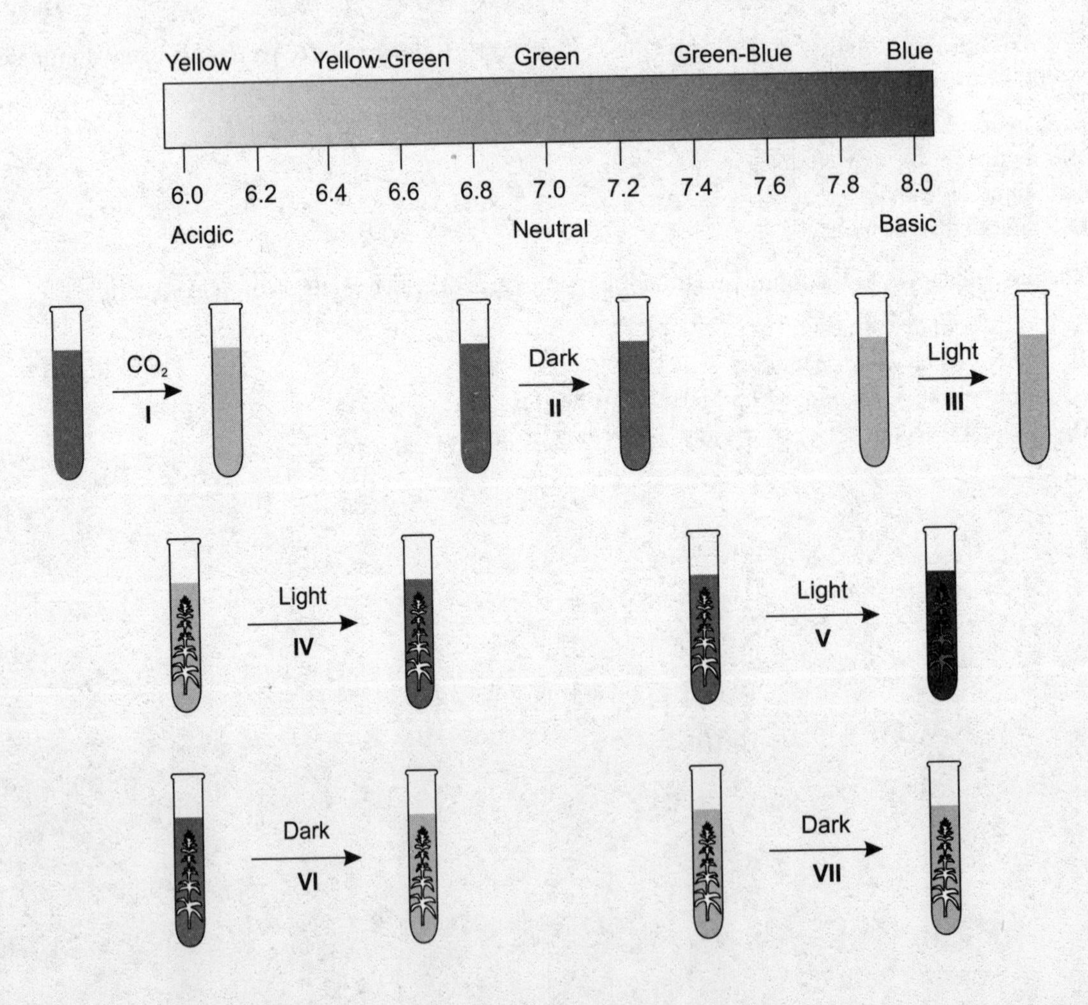

33. The addition of carbon dioxide (CO_2) to BTB has which of the following effects?

A. It causes the BTB solution to become basic.
B. It increases the concentration of hydrogen ions (H^+) in the BTB solution.
C. It turns BTB green.
D. It has no effect on the BTB solution.

34. In Reaction IV, the BTB solution with *Elodea* changes from yellow to blue when exposed to light. Which of the following is the most likely cause of this change?

A. Light is causing the BTB to become more acidic.
B. *Elodea* is releasing CO_2 in response to the exposure to light.
C. Photosynthesis is occurring in *Elodea*.
D. The *Elodea* plant is dying.

35. Which reaction supports a conclusion that the change in the color observed in Reaction IV is caused by the influence of light on *Elodea* and not by some acidic or basic substance in the *Elodea* plant?

A. Reaction II
B. Reaction III
C. Reaction V
D. Reaction VII

36. What metabolic process is responsible for the change observed in Reaction VI?

A. When light exposure ends, CO_2 stored in the chlorophyll of the *Elodea* plant is released.
B. The *Elodea* plant is carrying out cellular respiration.
C. The absence of light is causing the *Elodea* plant to die and to release O_2 into the BTB solution.
D. In the absence of light, the Calvin cycle reverses and releases CO_2.

37. Which of the following reactions is a control for Reaction VI?

A. Reaction II
B. Reaction III
C. Reaction IV
D. Reaction V

38. Which of the following can be used to demonstrate that a metabolic activity rather than a substance stored in the *Elodea* plant is responsible for the change observed in Reaction VI?

A. Reaction IV
B. Reaction V
C. Reaction VII
D. a dead plant in a solution of blue BTB

39. The following map shows the distribution of the *B* allele across Europe and Africa.

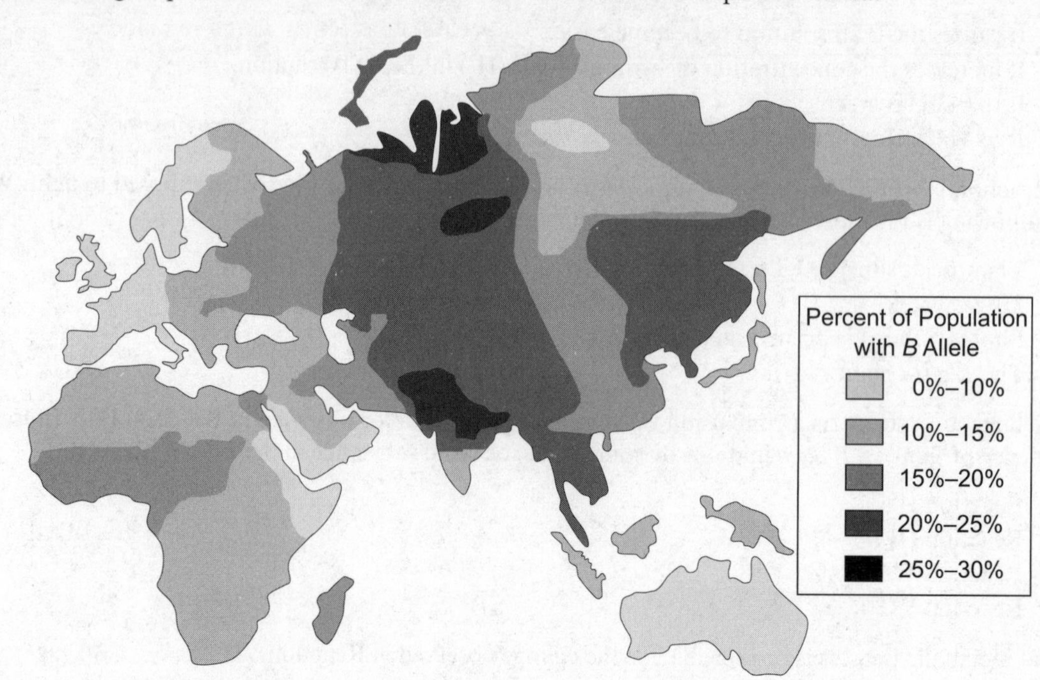

Which of the following best explains the distribution of the *B* allele across Europe and Africa?

A. A mutation for the *B* allele occurred independently in various locations throughout Europe and Africa.
B. Variations in environmental conditions selected for different frequencies of the *B* allele throughout Europe and Africa.
C. A mutation for the *B* allele occurred in western Asia and spread through Europe and Africa as individuals carrying the *B* allele migrated westward.
D. The variations in the *B* allele observed throughout Europe and Africa occurred as a result of genetic drift.

40. The following table describes the distribution of O, A, B, and AB blood types among four populations of Native Americans.

		Phenotypic Percentages			
Population	**Location**	**O**	**A**	**B**	**AB**
Utes	Utah	97	3	0	0
Navajos	Northern Mexico	77	23	0	0
Eskimos	Alaska	41	53	4	2
Blackfoot	Montana	24	76	0	0

Historically, the Blackfoot of Montana were a small, isolated population. Which of the following best explains their distinctive phenotypic percentages as compared to the other three Native American populations?

A. Various groups of Native Americans assembled in Montana to create a diverse genetic mix.
B. The Blackfoot population began as a group of individuals whose genetic makeup with respect to the ABO blood group was not representative of the population from which they originated.
C. The genetic makeup of the Blackfoot population with respect to the ABO blood group began with phenotypic percentages similar to that observed in other Native American populations but, due to chance, became distinctive over time.
D. Natural selection selected against the *B* allele.

Questions 41–43 refer to the following.

Oxygen is transported throughout the body by hemoglobin molecules within red blood cells. The blood types A, B, and O identify red blood cells distinguished by specific sugars attached to membrane-bound proteins. There are three alleles (*A*, *B*, and *O*), two of which (*A* and *B*) code for enzymes that attach a specific sugar to the membrane-bound protein. The combination of any two of the three alleles results in six genotypes, producing four phenotypes, A, B, AB, and O.

The following figure of a phylogenetic tree shows the nucleotide substitutions, deletions, and insertions for one possible explanation for the evolution of the ABO allele group.

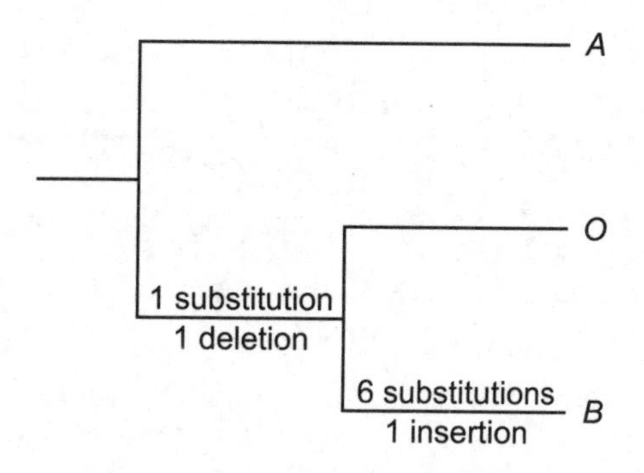

41. According to the phylogenetic tree, which of the alleles is the most likely ancestral?

- **A.** *A* allele
- **B.** *B* allele
- **C.** *O* allele
- **D.** unable to determine from data provided

42. If the exon of the *A* allele contains 1,065 nucleotides, how many nucleotides code for the *B* allele?

- **A.** 1,063
- **B.** 1,064
- **C.** 1,065
- **D.** 1,072

43. Unlike the *A* and *B* alleles, the enzyme produced by the *O* allele does not attach any sugars to the membrane-bound protein. Considering the phylogenetic tree above, which of the following provides the best explanation for why this might be?

- **A.** Multiple nucleotide substitutions during evolution most likely led to a nonfunctional gene.
- **B.** A nucleotide substitution in the *A* allele during evolution resulted in an *O* allele that, when transcribed, produced an mRNA with a new codon.
- **C.** Because of a nucleotide deletion, the *O* allele produced a new protein that enzymatically terminates the membrane-bound protein without attaching a sugar.
- **D.** The evolution of the *O* allele introduced a deletion into the nucleotide sequence that created a frameshift mutation.

Questions 44–47 refer to the following:

Two color morphs of a beetle are controlled by a single gene with two alleles. The *B* allele produces black beetles and is dominant over the *b* allele, which produces brown beetles. The life cycle of this beetle is 1 year, and each year the population is replaced by its descendants. Data for one large population over a period of 50 years are shown here.

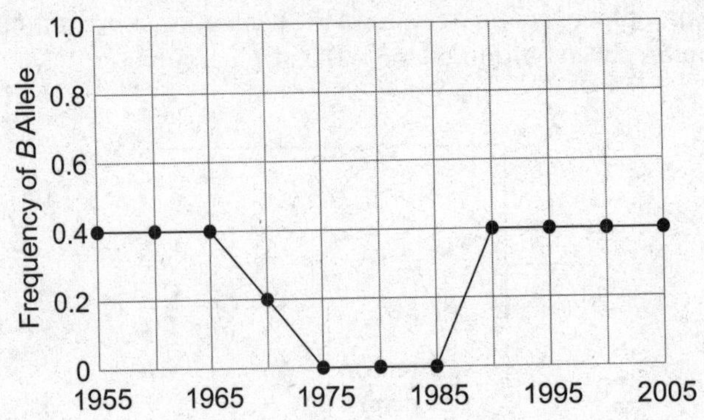

44. Assuming that the population was in Hardy-Weinberg equilibrium from 1955 to 1965, what was the frequency of the *b* allele?

A. 0
B. 0.2
C. 0.4
D. 0.6

45. From 1955 to 1965, what was the percentage of black beetles?

A. 4%
B. 16%
C. 48%
D. 64%

46. Which of the following is most likely responsible for the change in allele frequencies from 1965 to 1975?

A. selection against the brown-colored beetle
B. selection against the black-colored beetle
C. genetic drift
D. gene flow

47. Which of the following is most likely responsible for the change in allele frequencies from 1985 to 1990?

A. selection against the brown-colored beetle
B. mutation
C. genetic drift
D. gene flow

48. The data in the following table compare the DNA sequences of a short segment of DNA from five species (J, K, L, M, and N). The number of nucleotide differences between each pair of species was determined and then divided by the total number of nucleotides in the segment and multiplied by 100.

Species	J	K	L	M	N
J	0	20	6	24	14
K		0	28	8	30
L			0	18	12
M				0	34
N					0

Which of the following phylogenetic trees best represents the evolutionary relationships of the five species presented in the table?

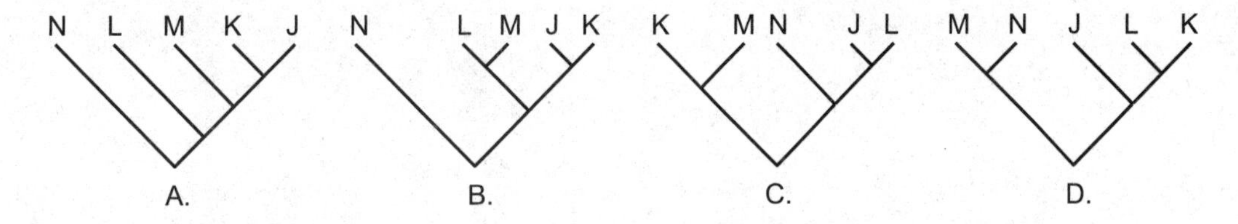

Question 49 refers to the following figure.

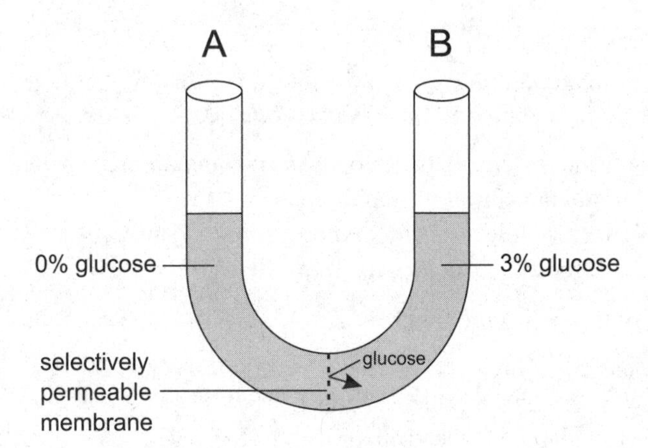

49. A glass tube, shaped into the form of a U, has a selectively permeable membrane inserted in its center. The ends of the tube, labeled A and B in the figure, are open. The membrane is permeable to water but not to glucose. Side A is filled with water and side B is filled with a 3% aqueous solution of glucose. Both sides are filled to the same height. After several hours, what happens to the height of the solutions in each side of the tube?

A. The height of the solution in side A rises, and the height in side B falls.
B. The height of the solution in side B rises, and the height in side A falls.
C. The heights of the solutions rise in both sides A and B.
D. The heights of the solutions in both sides remain unchanged.

50. The end product of the alcohol fermentation pathway is the alcohol ethanol. A considerable amount of energy still remains in ethanol, yet ethanol is a waste product for the yeast cells that produce it. What, then, is the purpose of fermentation?

A. Fermentation provides CO_2 for photosynthesis.
B. Fermentation provides CO_2 for cellular respiration.
C. Yeast cells excrete ethanol in order to provide a defense against invading bacteria.
D. Fermentation replenishes the NAD^+ necessary for glycolysis to continue.

Questions 51–53 refer to the following.

Enzyme E regulates the reaction J → K + L. The following graph shows the progress of this reaction after the enzyme is mixed with J.

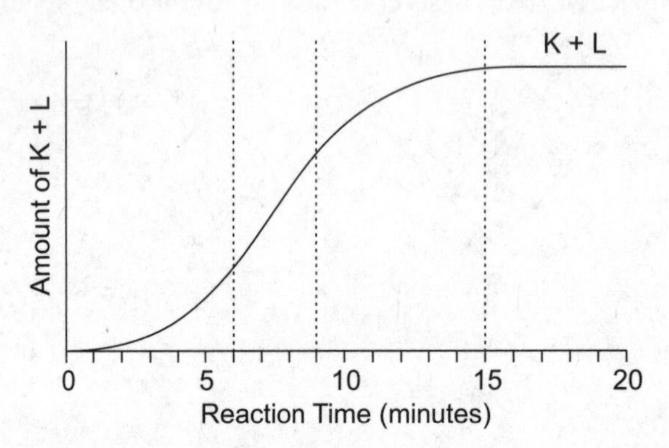

51. The reaction maintains its maximum rate during the interval from 6 to 9 minutes. Which of the following best explains why the reaction is limited to the observed rate?

A. The cleft formed by the active site of the enzyme is at its maximal aperture.
B. The concentration of reactant J is at its maximum.
C. All of the available enzyme molecules are bound to substrate molecules.
D. The temperature of the reaction is at its maximum value.

52. Which of the following best explains the progress of the reaction after 15 minutes?

A. The enzyme is depleted.
B. All of reactant J is converted to products K and L.
C. The reaction is completed and all chemical activity has ended.
D. For every additional molecule of K and L generated, a molecule of K and a molecule of L combine to form a molecule of J.

53. Which of the following actions is most likely to cause the reaction to generate additional amounts of product K and L?

A. Add more of enzyme E.
B. Add enzymes that will convert K and L to other substances.
C. Add an enzyme that will convert J to a different molecule, M.
D. Add energy to the reaction by raising the temperature of the solutions.

Questions 54–56 refer to the following.

The pink mold *Neurospora* obtains all of its nutritional requirements when it grows on bread. In the laboratory, a minimal culture medium can be prepared that also allows *Neurospora* to grow. When wild-type *Neurospora* is exposed to x-rays, however, the resulting induced mutants can grow only on media that have been supplemented with additional nutrients. Various nutritional supplements to the minimal medium for each of four mutants are shown in the following table. Each of the mutants possesses a single mutation that inactivates a single biochemical conversion. Growth occurs when any one of the supplements indicated by a plus sign is added to the minimum culture medium.

		Supplements to Minimal Medium			
	Minimal Medium	**Ornithine**	**Citrulline**	**Argininosuccinate**	**Arginine**
Wild-type	+	+	+	+	+
Mutant I		+	+	+	+
Mutant II			+	+	+
Mutant III				+	+
Mutant IV					+

54. Which of the following describes the metabolic pathway for the synthesis of arginine?

A. substance in minimal medium → ornithine → citrulline → argininosuccinate → arginine
B. substance in minimal medium → arginine → argininosuccinate → citrulline → ornithine
C. substance in minimal medium → citrulline → argininosuccinate → arginine → ornithine
D. substance in minimal medium → argininosuccinate → arginine → citrulline → ornithine

55. Mutant II will grow if citrulline, argininosuccinate, or arginine is added to the minimal medium. If a small amount of citrulline is added to allow growth to begin, what product is expected to have accumulated after all the added citrulline is used up?

A. ornithine
B. citrulline
C. argininosuccinate
D. arginine

56. Which of the following mutant cultures can grow if ornithine, citrulline, and argininosuccinate are all added to the minimal medium?

A. only mutant I
B. only mutants I and II
C. only mutants I, II, and III
D. all of them

Questions 57–60 refer to the following.

A plant produces fruits that are either red or white and that have either a smooth or a spiny surface. Table I shows the phenotypes of the F_1 offspring that result from crossing true-breeding parents with these traits.

Three groups of students then used computer simulations to predict the numbers of each phenotype produced when F_1 progeny were test crossed to individuals homozygous recessive for both traits. The results of the computer simulations are shown in Table II.

Table I: Cross of True-Breeding Parents

P	F_1
Red × White	All Red
Smooth × Spiny	All Smooth
Red and Smooth × White and Spiny	All Red and Smooth

Table II: Test Cross of F_1

Phenotypes Produced from Test Cross of F_1	Number of Offspring Generated		
	Simulation 1	Simulation 2	Simulation 3
Red and Smooth	394	155	272
Red and Spiny	405	0	124
White and Smooth	415	0	116
White and Spiny	386	145	288
TOTALS	1,600	300	800

57. The results of the cross of true-breeding parents in Table I indicate that no F_1 individuals express the white or spiny traits. How can this best be explained?

- **A.** White individuals and spiny individuals are always carriers when the white and spiny traits are not expressed.
- **B.** The allele for red-colored fruits is dominant to the allele for white-colored fruits, and the allele for smooth fruits is dominant to the allele for spiny fruits.
- **C.** Only one of two alleles of a gene can be inherited at the same time.
- **D.** Individuals with smooth fruits in the F_1 generation are homozygous.

58. The results shown in Table II for each of the computer simulations are different, presumably because the students made different assumptions about how the traits of color and texture are inherited. Which of the following best explains the results generated by simulation 1?

- **A.** The students assumed that homozygous recessive zygotes are unable to survive or develop normally.
- **B.** The students assumed that the alleles for these two traits are codominant; that is, the red and white alleles are codominant and the smooth and spiny alleles are codominant.
- **C.** The genes for color and texture were assumed to assort independently during cell division.
- **D.** Only the results of simulation 1 can be considered correct because it is the only simulation that generated a large enough number of offspring to establish reliability.

59. In simulation 2, no individuals were generated with red and spiny or white and smooth fruits. Which of the following best explains these results?

A. The total number of offspring generated by simulation 2 is not large enough to produce reliable results.
B. The students assumed that the genes for color and texture are found very close together on the same autosome.
C. Recombination between a chromosome containing a dominant allele and a chromosome containing a recessive allele produces results that can cancel each other out in some individuals.
D. A mutation occurred that blocked recombination of alleles.

60. Although all four phenotypes are generated by simulation 3, the numbers are different from those generated by simulation 1. How can this best be explained?

A. The students assumed that the alleles for red and white are codominant and that the alleles for smooth and spiny are also codominant.
B. The students requested that the number of offspring to be generated by simulation 3 be only half that of simulation 1.
C. The students assumed that crossing over of linked genes occurred during cell division.
D. The students assumed that the genes for color and texture assort independently.

Section II: Free-Response Questions

Time: 90 minutes

2 long free-response questions: Questions 1–2 (allow about 20 minutes each)
4 short free-response questions: Questions 3–6 (allow about 10 minutes each)

Directions: In this part of the exam, the first two questions are long free-response questions worth 8 to10 points each. These questions are followed by four short free-response questions. Each short free-response questions has four parts, each worth 1 point for a total of 4 points per question.

Answer all questions as completely as possible within the recommended timeframe. Use complete sentences (NOT outline form) for all your answers. When questions have multiple parts, separate your answers to each part and identify them with the letter for that part. You may use diagrams to supplement your answers, but a diagram alone without appropriate discussion is inadequate.

1. The endothelial cells that form the walls of blood vessels are surrounded by smooth muscle. Contraction of the smooth muscle constricts the blood vessels and reduces blood flow. In contrast, relaxation of the smooth muscle dilates blood vessels and increases blood flow. To control this smooth muscle activity, neurons (nerve cells) secrete the neurotransmitter acetylcholine (ACh) at their terminal ends. Acetylcholine activates a G protein-coupled receptor on the plasma membrane of endothelial cells, which, in turn, initiates a signal transduction pathway that ultimately opens channel proteins of the endoplasmic reticulum (ER), releasing Ca^{2+} into the cytoplasm. The Ca^{2+} activates the enzyme nitric oxide synthase (NOS), which, in turn, catalyzes the release of nitric oxide (NO) from the amino acid arginine. NO, a gas, acts as a signaling molecule for the adjacent smooth muscle cell, activating the enzyme guanylyl cyclase. Guanylyl cyclase catalyzes the production of cyclic GMP (cGMP) from the nucleotide GTP. Cyclic GMP then triggers relaxation of smooth muscle. The following figure summarizes the signal transduction pathways involved in these processes.

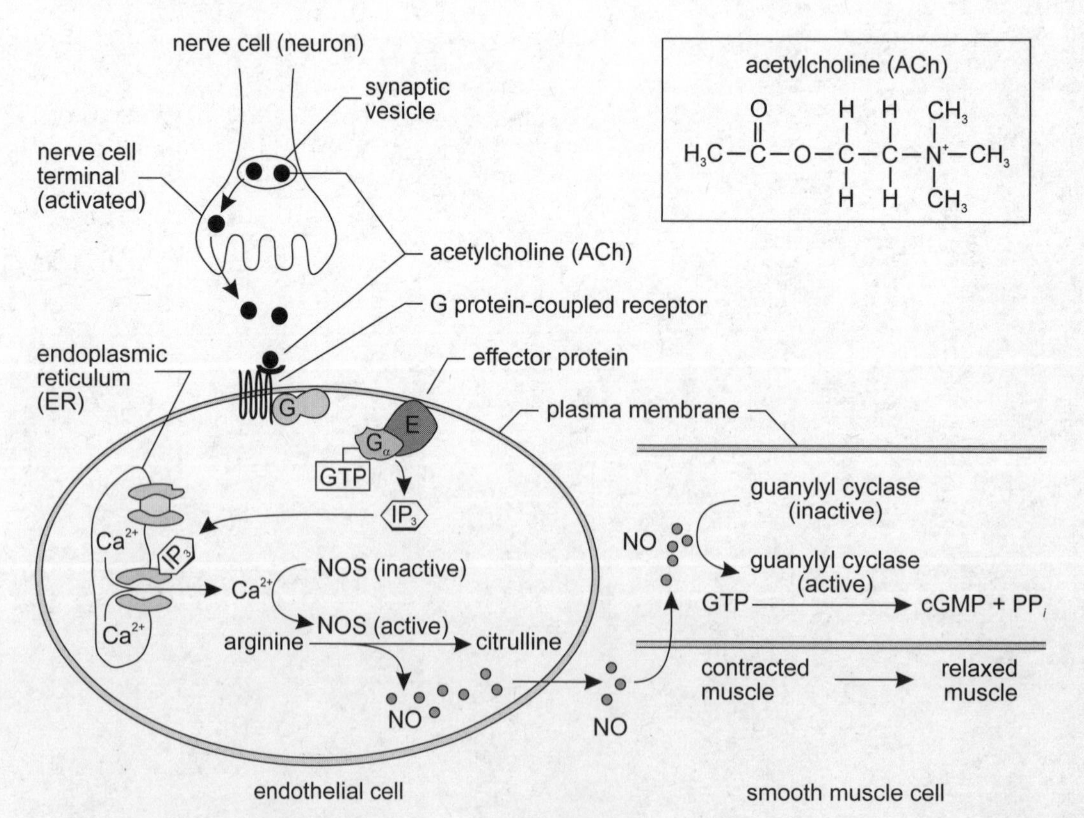

a. Acetylcholine does not directly initiate the signal transduction pathway itself. Instead, an intermediary mechanism, the G protein-coupled receptor, is required to transmit the signal of acetylcholine into the cell. **Explain** why acetylcholine does not enter the cell directly.

b. **Explain** how the signal transduction pathway transmits a signal from the G protein-coupled receptor to the endoplasmic reticulum.

c. Nitroglycerine is often used to alleviate chest pains in patients suffering from heart conditions where blood vessels that supply blood to heart muscle are constricted or blocked. Nitroglycerine tablets are placed on the tongue, where they are rapidly absorbed into the circulatory system. Nitroglycerine is broken down in the cells by the mitochondrial enzyme, mitochondrial aldehyde dehydrogenase (mtALDH), as follows.

$$\text{nitroglycerin} \xrightarrow{\text{mtALDH}} \text{glyceryl dinitrate} + \text{NO}$$

Explain why nitroglycerin alleviates chest pains.

d. Mountain climbers at high altitudes sometimes suffer from high-altitude sickness. One form of high-altitude sickness is pulmonary edema. The condition develops in the lungs where blood vessels constrict and blood pressure rises. One of the following drugs is most likely expected to help this condition. **Choose** ONE drug and **justify** your choice:

- Drug 1 competes with acetylcholine for binding to the G protein-coupled receptors on endothelial cells.
- Drug 2 binds to IP_3 and prevents its attachment to channel proteins in the endoplasmic reticulum.
- Drug 3 blocks the enzyme that degrades cGMP.
- Drug 4 releases citrulline into endothelial cells.

2. As a result of hunting, poisoning, and bounty programs, wolves were eliminated from Yellowstone National Park by 1925. With the elimination of this major natural predator, park management was concerned that the elk population would grow and exceed the carrying capacity of the park environment. In response to this concern, park management culled (removed) elk from the herds. Remaining selective pressures on elk survival included food shortages, disease, weather, and predation by grizzly bears, black bears, cougars, and coyotes.

In 1968, park management decided to stop culling the elk (except for diseased animals) and allowed nature to proceed with minimal human influence. In response to the 1973 Endangered Species Act and in an effort to allow Yellowstone to return to its most natural state, 11 wolves obtained from Canada were introduced in 1995.

The following figure plots population changes for elk, bison, and wolves.

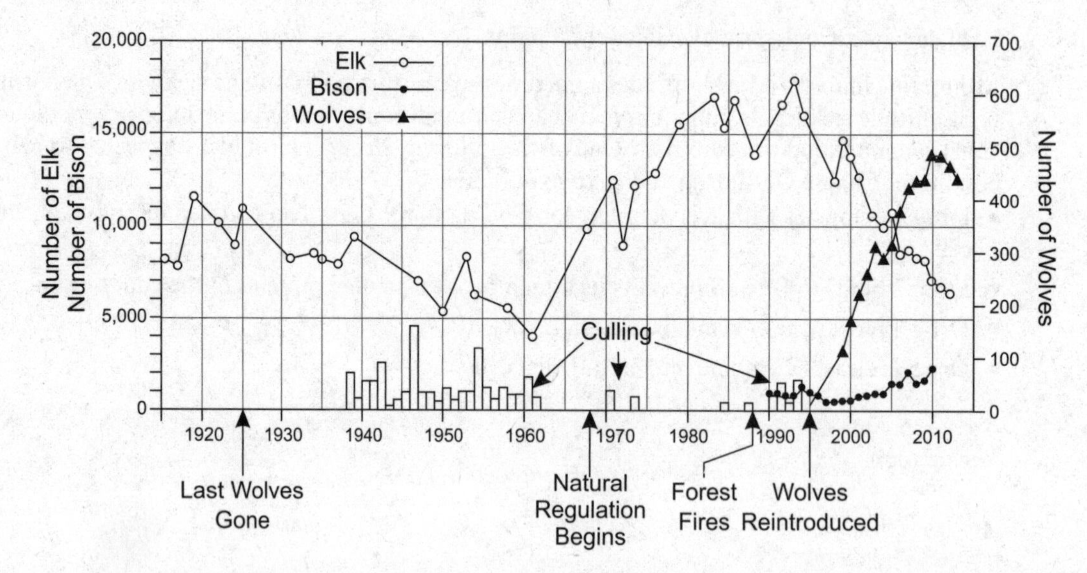

a. From 1925 to 1995, the elk population size experienced an increase followed by a decrease. Based on the information in the introductory paragraphs and in the figure,

- **state** a hypothesis for the decline in the elk population from 1925 to 1960.
- **state** a hypothesis for the increase in the elk population from 1960 to 1995 and **describe** a method for testing your hypothesis.

b. Data in the figure show that the elk population decreased from 1995 to 2010. One explanation for this decrease is predation by the introduced wolves. **State** and **explain** THREE other possible causes for this decrease.

c. **Predict** changes, if any, in the vegetation coverage from 1995 to 2010. **Justify** your prediction.

d. Beginning in 2010, the data indicated a decline in the wolf population. **Propose** an explanation for this decline.

3. The mermaid's wineglass, *Acetabularia,* is a one-celled alga. The cell is large (2 to 10 cm) and consists of a cap, a slender stalk, and a base that attaches to the substrate. A single nucleus resides in the base. Various *Acetabularia* species differ by the shape of their caps. To investigate cell regeneration, researchers cut off the caps and stalks of two species, *A. mediterranea* and *A. crenulata*. The stalks and caps died, but the bases survived and regenerated new stalks with new caps. Refer to the following figure for parts a and b.

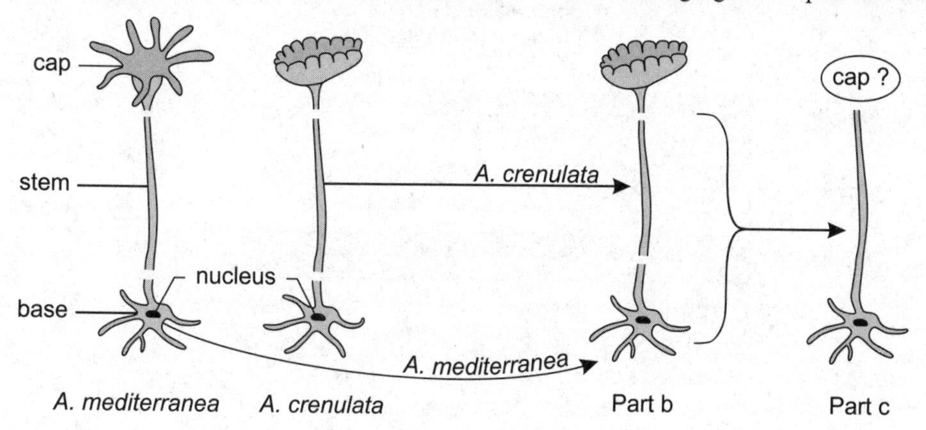

a. **Explain** why only the bases survived and were able to regenerate new stalks and caps.
b. When the base of an *A. mediterranea* was removed and grafted to a stalk of an *A. crenulata,* the stalk grew an *A. crenulata* cap. **Pose** an explanation for this regeneration.
c. Consider the regenerated *A. crenulata* cap in part b. If this cap were removed, **predict** the kind of cap that would be regenerated a second time. **Justify** your answer.
d. **Predict** the kind of cap that would be regenerated if the bases of *A. mediterranea* and *A. crenulata* were removed from their stalks and grafted together. **Justify** your answer.

4. Pheromones are chemicals that animals use for communication. When a female beetle of the species *Popilia japonica* is ready to mate, she broadcasts to males her location by releasing the volatile pheromone japonilure. One chemical form of japonilure is an attractor that signals to males that a female is receptive to mate. A second form is an enantiomer, or structural mirror image, of the first form; it acts as an inhibitor to signal that the female is no longer receptive. Females of another beetle species, *Anomala osakana*, use the same two forms of japonilure, except that the messages of the two forms are reversed.

a. **Provide** an evolutionary advantage for why the two forms of japonilure are reversed for these two beetle species.
b. **Propose** an example for why using a volatile chemical to attract mates may be a disadvantage for a species.

Plants use colors, shapes, scents, and sweet rewards to attract insect pollinators. Certain orchids are only pollinated by the males of the solitary bee species *Andrena nigroaenea*.

c. **Explain** why only males are involved in the pollination of these orchids.
d. **Explain** why only members of *Andrena nigroaenea* are attracted to the orchid.

5. Researchers collected eggs from three fish populations and recorded the dates that the eggs hatched. When the hatchlings reached sexual maturity, they identified the sex of the fish. The following graph shows hatch-time data for males and females of the three fish populations. The horizontal error bars represent two standard errors of the mean (SEM).

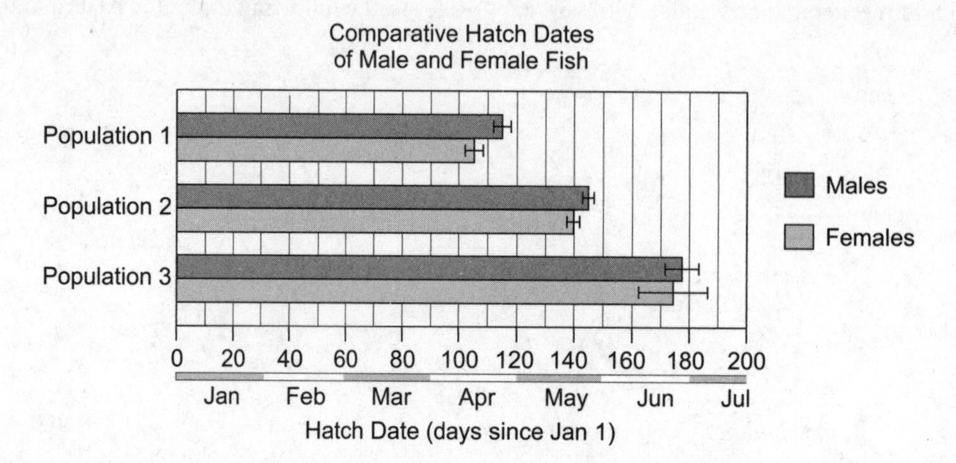

a. **Identify** the fish population that hatches earliest in the year.
b. **Identify** any differences in hatch time between males and females for the populations.
c. **Describe** an evolutionary advantage for your answer to part b.
d. **Propose** an explanation for why the three populations hatch at different times of the year.

6. Each of four bags made from dialysis tubing was filled with the same sucrose solution. Water, but not sucrose, can pass through dialysis tubing membrane. Each bag was then immersed in four beakers containing sucrose solutions of 0.2 *M*, 0.4 *M*, 0.6 *M*, and 0.8 *M*. After 30 minutes, each bag was weighed and its change in weight was calculated. All solutions were at 25°C. The results are shown in the following graph:

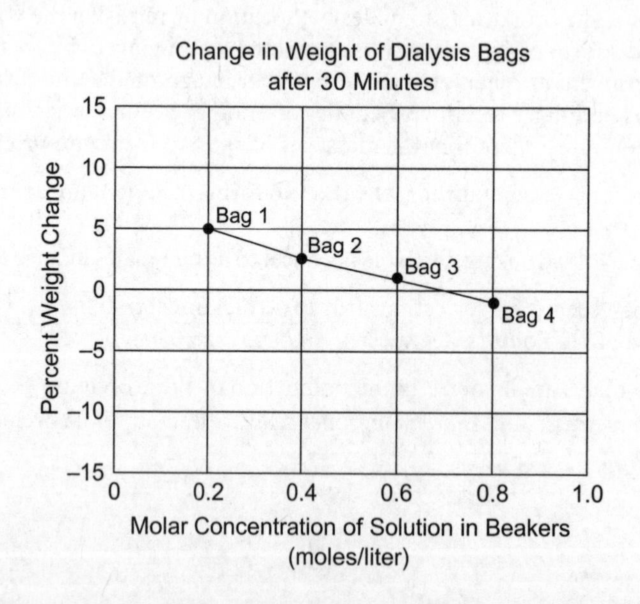

a. **Describe** the movements of solute and solvent between the bags and the beakers.
b. **Determine** the concentration of sucrose in the bags.
c. **Determine** the pressure potential (Ψ_p) for the solutions in the beakers.
d. **Determine** the water potential of the solution in the dialysis bag, rounded to the nearest tenth.

Answer Key for Practice Exam 1

Section I: Multiple-Choice Questions

1. D
2. B
3. D
4. B
5. C
6. A
7. B
8. C
9. D
10. C
11. D
12. B
13. C
14. D
15. A
16. A
17. A
18. D
19. B
20. A
21. C
22. B
23. D
24. C
25. D
26. A
27. C
28. C
29. B
30. A
31. B
32. D
33. B
34. C
35. D
36. B
37. A
38. D
39. C
40. B
41. A
42. C
43. D
44. D
45. D
46. B
47. D
48. C
49. B
50. D
51. C
52. D
53. B
54. A
55. A
56. C
57. B
58. C
59. B
60. C

Scoring Your Practice Exam

Section I: Multiple-Choice Questions

TOTAL number of multiple-choice questions out of 60 you answered correctly	

Section II: Free-Response Questions

Each long free-response question can be worth 8 to 10 points, depending upon the question and the exam. Each short free-response question is worth 4 points.

Score for free-response question 1 (0–8 points)	
Score for free-response question 2 (0–10 points)	
Score for free-response question 3 (0–4 points)	
Score for free-response question 4 (0–4 points)	
Score for free-response question 5 (0–4 points)	
Score for free-response question 6 (0–4 points)	
TOTAL for Section II (0–34 points)	

Combined Score (Section I + Section II)

Total for Section I (from above)		× 0.83 =	
Total for Section II (from above)		× 1.47 =	
TOTAL combined score		=	

AP Grade Estimate	
60–100	5
50–59	4
41–49	3
33–40	2
0–32	1

Answers and Explanations for Practice Exam 1

Section I: Multiple-Choice Questions

1. **D.** Both mRNAs begin with exon 1. Also, the exon that follows the alternative exons (exons 2 and 3) for both mRNAs is the same (exon 4). Exon 1 may not necessarily end with a complete codon of three nucleotides, and exon 4 may not necessarily begin at the start of a new codon. Therefore, both exons 2 and 3 must pick up where exon 1 ends and add to where exon 4 begins. This requires, at a minimum, that both exons 2 and 3 must have sufficient nucleotides to code for complete codons (that is, 3 nucleotides) and must have, in addition, any extra nucleotides to complete the codons at the end of exon 1 and at the beginning of exon 4. Only answer choice D allows for properly completing any codons at the end of exon 1 or at the beginning of exon 4. If these codons 1 and 4 are not completed by either exon 2 or 3, the resulting processed mRNAs will have a frameshift mutation, resulting in a radically different, and most likely nonfunctional, protein. On the other hand, either exons 2 or 3 can have additional codons (3 nucleotides) that would result in the insertion of one or more additional amino acids into the tropomyosin protein (and that might influence its function). Answer choices B and C could accurately describe some splices, but neither addresses the possibility that the codon that terminates exon 1 and the codon that begins exon 4 might be incomplete with 1 or 2 nucleotides.

2. **B.** The signaling molecule, or ligand, is epinephrine. It initiates the signal transduction pathway when it binds to the G protein-coupled receptor. At the end of the pathway, a protein kinase (PKA) is activated, which, in turn, activates the binding of an ATP to the CFTR protein, opening its gated Cl^- channel.

3. **D.** The function of the CFTR protein is to generate an electrochemical gradient that allows the passive flow of Cl^- across the plasma membrane. In response, water follows, moving across the membrane in response to the gradient. When water does not flow, the surface liquid on the cells is deficient and thick mucus develops.

4. **B.** The sweat of individuals with cystic fibrosis is salty because it contains high amounts of Cl^- and Na^+. In the sweat-producing cells of normal individuals, Cl^- is reabsorbed by CFTR. So, in sweat-producing cells with a mutated form of the CFTR protein, the movement of Cl^- by the CFTR protein must be in the opposite direction from its movement in epithelial cells of the lungs. As a result, in sweat glands that express a mutated form of the CFTR protein, Cl^- does not return to cells and remains in the sweat. Answer choices A and C are incorrect because both would result in decreasing the amount of Cl^- or Na^+ in sweat—opposite from that occurring in individuals with cystic fibrosis. Answer choice D is incorrect because cystic fibrosis is caused by mutations of the gene that code for CFTR, not the G protein-coupled receptor.

5. **C.** The chromosomes in the figure (each consisting of two identical chromatids, labeled as 1 and 2) are arranged in pairs. Chromosome A (A_1 and A_2) paired with chromosome B (B_1 and B_2) represents a homologous pair of chromosomes. Chromosome C (C_1 and C_2) paired with chromosome D (D_1 and D_2) represents another homologous pair. This arrangement of homologous pairs between the two poles of a cell is distinctive of metaphase (alignment) of meiosis I. The next step in the process is for one member of each homologous pair to migrate to one pole and the other member of each homologous pair to migrate to the opposite pole, as described in answer choice C.

6. **A.** A homologous pair of chromosomes consists of one chromosome derived from each parent paired together because they have the same genes (although alleles of each chromosome may differ). In addition, each chromosome consists of two chromatids that are identical, assuming that no copying errors occurred when the second chromatid was replicated from the first. Thus, the chromatids A_1, A_2, B_1, and B_2 all contain the same genes (but only A_1 and A_2 are identical, as are B_1 and B_2). Although not listed among the answer choices, note that chromatids C_1, C_2, D_1, and D_2 also contain genes that are the same.

7. **B.** Homologous chromosomes pair as illustrated in the figure only during meiosis. In contrast, in mitosis all chromosomes individually (not paired) align along an equatorial plane between the poles. During cytokinesis, chromosomes in both mitosis and meiosis have migrated toward the poles and are no longer positioned midway between the poles. During interphase, chromosomes are not visible; instead, the chromosomes unwind from their condensed structure and appear as chromatin—long, thin filaments of DNA and protein.

8. **C.** Because the graph shows a decline in the sizes of spider populations only on islands where lizards were introduced, one can conclude that the introduction of lizards was responsible for that decline. Of the choices given, answer choice C provides a simple, straightforward explanation consistent with the data—that lizards were eating the spiders. Answer choice A cannot be justified because no data were presented for spider populations on habitats other than islands. Answer choices B and D suggest that spiders pursue concealment to avoid getting eaten when lizards are present, but no observations are presented to support that conclusion and the data indicate that they were being eaten in spite of any elusive behaviors.

9. **D.** Although it is clear that the dissected island lizards were not eating spiders (answer choice A), that explanation does not explain why the spider populations were smaller on the islands with the introduced lizards. Instead, the lizards and spiders were competing for the same resources—food, in particular. When the lizards were introduced to the islands, the amount of food available for the spiders decreased, reducing the sizes of their populations. Answer choice C suggests that the size of the spider populations declined as a result of an introduced parasite. Although that is a possible explanation, no data are provided to support this conclusion. In contrast, the introductory paragraph makes it clear that spiders and lizards eat the same food.

10. **C.** From years 3 to 8, the numbers of reported spiders on both islands with and without lizards appear to be increasing and decreasing in parallel. This suggests that some other factor (rain, drought, or temperature, for example) is affecting the number of surviving spiders from year to year. None of these factors in particular, however, can be identified with the data provided in the figure. Thus, answer choice D cannot be accepted.

11. **D.** For this, the logistic equation, early stages of growth are depicted by a rapid (logarithmic) rise in the population size, followed by a gradual tapering off of the growth rate as the carrying capacity is approached. When the population size reaches the carrying capacity, the net rate of growth is zero, as expressed by the horizontal line (slope of the line is zero) of the plotted equation.

12. **B.** As the population size approaches the carrying capacity, the growth rate decreases and becomes zero when $N = K$.

13. **C.** The net rate of growth is greatest early in the plotted equation, when N is still small compared to K. To see this mathematically, rearrange the equation like this:

$$\frac{dN}{dt} = r_{\max} N\left(\frac{K-N}{K}\right) = r_{\max} N\left(1 - \frac{N}{k}\right)$$

When N is small compared to K, the fraction $\frac{N}{k}$ is small compared to 1. Thus, $1 - \frac{N}{k}$ is approximately equal to 1, and the value of the equation is maximum and equal to $r_{\max}N$.

14. **D.** When adenosine binds to the G protein-coupled receptor, it causes a GTP to replace the GDP attached to the inactive G protein. This exchange, a GTP for a GDP, activates the G protein. Activation causes a subunit of the G protein (Gα), together with the GTP, to bind to and activate the membrane-bound effector protein, adenylyl cyclase (AC).

15. **A.** Caffeine is an adenosine signaling antagonist, blocking the effect of adenosine. When caffeine binds to the adenosine receptor, it blocks the G protein-coupled receptor from initiating the adenosine signaling pathway. As a result, PKA is not activated and drowsiness does not occur.

16. **A.** It is important to understand that a single signaling molecule, initiating the same signal transduction pathway, can have different effects in different cell types. Also, there are only three types of membrane receptors (G protein-coupled receptors, ligand-gated ion receptors, and protein kinase receptors), but there are more than 700 known signaling molecules for G protein-coupled receptors alone. Each signaling molecule binds to a specific receptor, initiating a signal transduction pathway that is specific to the signaling molecule and cell type. The PKA produced by the epinephrine signal transduction pathway, then, is going to be different from that of the PKA produced by the adenosine pathway. Since epinephrine is associated with promoting the fight-or-flight response, PKA triggers the release of glucose from glycogen. The release of glucose provides a source of energy for fight-or-flight activities. Muscle fatigue, drowsiness, and reducing blood flow to the heart (as a result of vasoconstriction) do not promote the fight-or-fight response.

17. **A.** Temperature is a measure of the average kinetic energy of the molecules in a system—the higher the temperature, the greater the average kinetic energy. Kinetic energy is a measure of how fast an object is moving—the faster a molecule is moving, the greater its kinetic energy. When the temperature increases, molecules move faster, colliding more frequently with the necessary activation energy (the energy necessary for a reaction to progress to completion).

18. **D.** At some point as the temperature rises, the bonds that hold the enzyme together and maintain its three-dimensional shape break down. When the three-dimensional shape is lost, the enzyme can no longer function and is said to be denatured.

19. **B.** A phospholipid with an unsaturated fatty acid tail (one with one or more double covalent bonds) creates a kink in the tail that spreads adjacent phospholipids apart, making the membrane more fluid. In contrast, phospholipids with saturated fatty acids pack more closely together and make the membrane more viscous. Answer choice A is incorrect because an increase in cytosol solute concentration will help the cytosol tolerate chilling, but it will not keep the plasma membrane fluid. Answer choice C is incorrect because glucose and other sugar molecules do not occur in the plasma membrane except as attached to phospholipids (glycolipids) or to proteins (glycoproteins). Furthermore, replacing phospholipids entirely with glucose molecules would destroy the integrity of the plasma membrane. Answer choice D is incorrect because the two phospholipid layers of the plasma membrane are asymmetric, each side specific to the direction it faces. Exchanging the phospholipids from one side to the other would damage that integrity.

20. **A.** The activity of protein II is an example of facilitated transport. Since no ATP is required (as is for protein I), the transport must be passive and *down* a concentration gradient. In contrast, the transport of Na^+ and K^+ is carried out by a single cotransporter protein (the Na^+/K^+ pump) and transports both Na^+ and K^+, each against a concentration gradient, and requires ATP. In addition, the Na^+/K^+ pump moves Na^+ out of the cell and K^+ into the cell, not the other way around. Steroids traverse the plasma membrane without the assistance of protein carriers because steroids are nonpolar, and the nonpolar characteristic of the plasma membrane does not present a barrier.

21. **C.** A protein that transports a substance across the plasma membrane, either actively (like protein I) or passively (like protein II), is specific to the substance to be transported. In that way, the plasma membrane controls which substances to be transported. Furthermore, different kinds of cells have specific proteins that serve the needs of that kind of cell. Only protein I requires ATP, and only protein II requires a downhill gradient. Proteins are necessary for the transport of ions and large, uncharged polar substances. But small, nonpolar (hydrophobic) substances (such as O_2 and CO_2) can diffuse across the membrane without the assistance of a transport protein.

22. **B.** The sole function of mitochondria is to carry out the same energy-generating metabolic pathway that occurs in aerobic bacteria. Similarly, the sole function of chloroplasts is to carry out the energy-generating metabolic pathway that occurs in photosynthetic bacteria. In addition, all use the same molecular machinery (ATP synthase) to accomplish these processes. These observations best support the hypothesis that mitochondria and chloroplasts are derived from bacteria. The remaining answer choices do not support the observed similarities as well: eukaryotic and prokaryotic cells may share a common ancestor, but there are no observations that establish that such an ancestor used ATP synthase; convergent evolution would result in similar functions (as observed), but through different mechanisms; and there are no observations that suggest that mitochondria or chloroplasts can survive independently outside a eukaryotic cell.

23. **D.** The shaded amino acids have nonpolar side chains. Nonpolar side chains are hydrophobic; when these amino acids occur together, they cluster so that the side chains are facing away from the surrounding aqueous solution.

24. **C.** Retrieve the formula for Simpson's Diversity Index from the Equations and Formulas pages at the beginning of this exam:

$$\text{Simpson's Diversity Index} = 1 - \sum\left(\frac{n}{N}\right)^2$$

Then calculate $\frac{n}{N}$ for each species, then square each one. Add together the squares for all six species. Finally, subtract the sum from 1.

Species	Number of Individuals (n)	$\frac{n}{N}$	$\left(\frac{n}{N}\right)^2$
Black Bear	2	0.02	0.0004
Chipmunk	35	0.35	0.1225
Coyote	15	0.15	0.0225
Ground Squirrel	20	0.2	0.04
Marmot	10	0.1	0.01
Pika	18	0.18	0.0324
TOTALS	$N = 100$		$\sum\left(\frac{n}{N}\right)^2 = 0.2278$

Simpson's Diversity Index $= 1 - \sum\left(\frac{n}{N}\right)^2 = 1 - 0.2278 = 0.7722$. Rounded to the nearest hundredth, $D = 0.77$.

25. **D.** The enzymes in the lysosome are enzymatically active because the pH inside the lysosome is 5. The pH is maintained by pumping H^+ ions (protons) into the lysosome by active transport proton pumps in the lysosome membrane.

26. **A.** In cellular respiration, glucose is broken down and energy is extracted and stored in NADH, $FADH_2$, and ATP. During the oxidative phosphorylation stage of respiration, electrons from NADH (and $FADH_2$) enter the electron transport chain (ETC) in the mitochondrion inner membrane. As this occurs, protons (H^+) move across the membrane from the matrix into the intermembrane space, generating a proton gradient. At the end of the ETC, the electrons combine with a proton and oxygen to form H_2O. In contrast, the electrons in photosynthesis originate in H_2O from inside the thylakoids of chloroplasts and enter an ETC chain in the thylakoid membrane. As the electrons pass along the ETC, protons move into the thylakoids from the stroma, generating a proton gradient. At the end of the ETC, the electrons combine with $NADP^+$ and a proton (H^+) to form NADPH. Ultimately, the NADPH is used to generate glucose. For both respiration and photosynthesis, electrons pass through an ETC and generate a proton gradient. But for the origins and endpoints of the electrons involved, the two processes are the reverse of one another.

27. **C.** This question evaluates your ability to read and understand data in graphical form. Answer choices A, B, and D are not supported by the data presented in the figure. Answer choice A is not supported by the data in the figure because the data compare shade and sun leaves exposed to the same amount of light at various intensities (800, 900, and 1,000 μmol photons/m^2/sec, for example). Answer choice B is not supported by the data in the figure because the data compare equal surface areas of plant leaves (light intensity is per m^2). Answer choice D implies that shade leaves have a higher maximal photosynthetic rate than sun leaves, but the data for light intensities above 200 μmol photons/m^2/sec indicate the opposite. Because the hypothesis presented in answer choice C cannot be refuted by the data in the graph, it is the best of the hypotheses presented among the four answer choices.

28. **C.** Like question 27, this question evaluates your ability to read and understand data in graphical form. According to the figure, at light intensities between 100 and 150 μmol photons/m²/sec, CO_2 assimilation in shade leaves is greater than that in sun leaves. Two hypotheses are plausible: shade leaves are *assimilating more* CO_2 than sun leaves because they are doing more photosynthesis than sun leaves or shade leaves are *releasing less* CO_2 because they are doing less cellular respiration than sun leaves. This latter hypothesis occurs because under low light conditions, some cells may be doing photosynthesis while other cells may be doing cellular respiration. The hypothesis in answer choice C describes this scenario where the shade leaves are releasing less CO_2 because more cells are photosynthesizing than respiring compared to sun leaves. For these leaves, apparently, shade leaves are more efficient than sun leaves under low light conditions. Answer choice A is not supported by the data because the CO_2 assimilation data in the figure is independent of leaf surface area (CO_2 assimilation is per m²). Answer choice B is incorrect because respiration releases CO_2, the opposite of CO_2 assimilation. Answer choice D contradicts the data presented in the figure, as shade leaves are more efficient than sun leaves for light intensities between 100 and 150 μmol photons/m²/sec. Therefore, only answer choice C supports one of the two plausible hypotheses that could explain the data in the figure.

29. **B.** At light intensities below 50 μmol photons/m²/sec, the rate of photosynthesis is so low that cells require additional energy from respiration. When respiration increases, the production of CO_2 from respiration exceeds assimilation of CO_2 by photosynthesis. Although other answer choices may explain why CO_2 assimilation is low, only answer choice B answers why CO_2 assimilation is negative below 50 μmol photons/m²/sec.

30. **A.** The AP exam does not expect you to recognize the exact structure of each amino acid. Instead, this question evaluates your understanding of how amino acids bond together. In particular, two amino acids form a peptide bond as a result of hydrolysis. That is, a water molecule is formed and released as a covalent bond is created between two adjacent amino acid molecules. More specifically, the OH in the carboxyl group (COOH) of one amino acid and the H of the amino group (NH_2) of an adjacent amino acid combine to form a water molecule, as shown below:

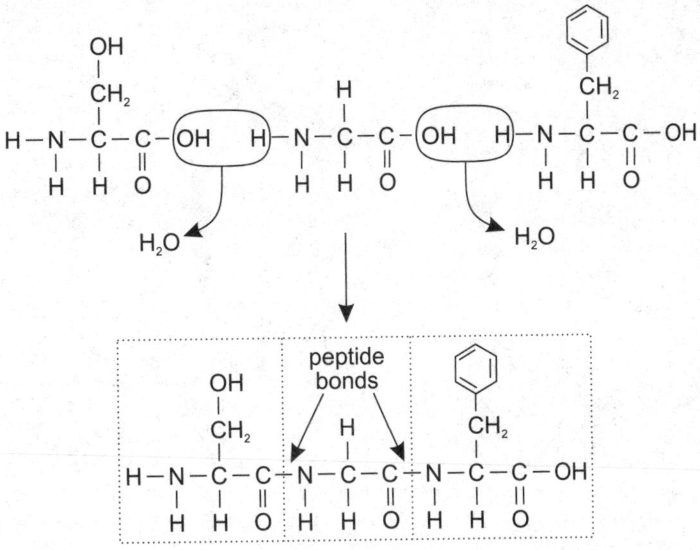

31. **B.** The amino acids represented by shaded circles are inside the plasma membrane among the hydrophobic fatty acid chains of the phospholipids. As a result, that part of the protein is made up of hydrophobic amino acids, that is, amino acids with hydrophobic side chains (R groups). These side chains project outward into the fatty acids so that hydrophobic-hydrophilic incompatibilities are averted. In contrast, the amino acids that penetrate the aqueous environments of the cytosol and extracellular fluid (white circles) are dominated by amino acids with hydrophilic side chains (polar or charged R groups).

32. **D.** The circles show the primary amino acid sequence (amino acid names and their sequential order), and the α-helices (as the protein passes through the membrane) illustrate the protein's secondary structure (hydrogen bonding between amino acids). Tertiary structure describes interactions of the hydrophobic and hydrophilic side chains of the amino acids (as shown by the aggregation of hydrophobic R groups with adjacent membrane phospholipids and the folding of the protein inside and outside of the membrane), hydrogen bonding or ionic bonding between side chains (as shown by protein folding), and disulfide bonds of cysteine (none occurs in this protein). Because glycophorin is a protein that consists of two separate polypeptide strands, it is a protein with quaternary structure.

33. **B.** Carbon dioxide combines with water, forming HCO_3^- and H^+. The increase in concentration of H^+ makes the BTB solution acidic, causing a blue solution of BTB to turn yellow.

34. **C.** When *Elodea* is exposed to light, electrons are energized and noncyclic and cyclic photophosphorylation begins, generating ATP, NADPH, and O_2. In the Calvin cycle phase of photosynthesis, the CO_2 in the BTB solution that the ATP and NADPH generated in photophosphorylation is used to generate glucose. When the CO_2 is consumed, the solution becomes basic (as H^+ is converted to H_2CO_3) and the yellow BTB turns blue. The *reverse* reaction is observed in Reaction I, where the addition of CO_2 causes the blue BTB to become more acidic and turns the blue BTB to yellow.

35. **D.** Reaction VII, occurring in the dark, shows that *Elodea* does not cause the yellow BTB solution to turn blue. Thus, light must be present for the observed color change (from yellow to blue) to occur in Reaction IV. In addition, Reaction III shows that light alone does not cause the color change observed in Reaction IV.

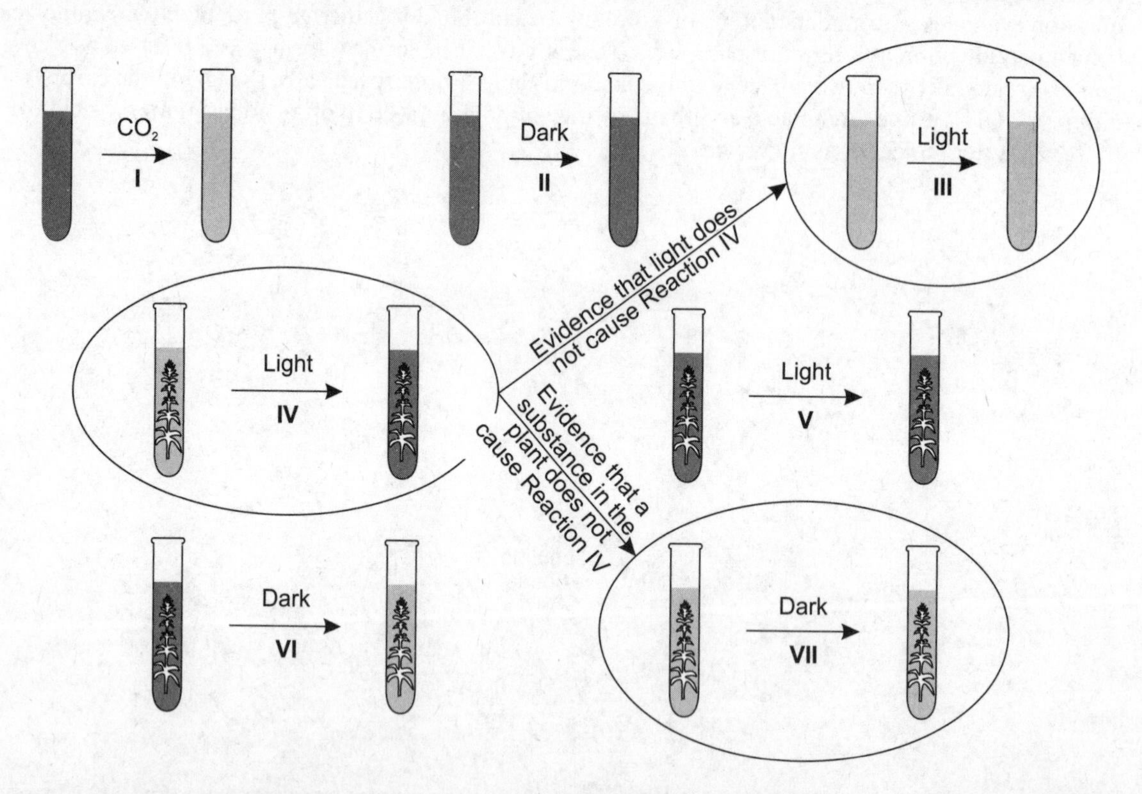

36. **B.** When light is not available to carry out photosynthesis, cellular respiration becomes the dominant metabolic process. Cellular respiration breaks down glucose to generate ATP and CO_2. The blue BTB turns yellow as the solution becomes more acidic from the addition of CO_2 (as observed in Reaction I in the figure). If, as in answer choice C, the *Elodea* plant were decaying, organisms responsible for the decay would be releasing CO_2 (from respiration), not O_2.

37. **A.** Reaction II is a control that demonstrates that darkness does not cause a blue BTB solution to yellow. Thus, the plant, in some way, is responsible for the change.

38. **D.** To demonstrate that there is no substance in *Elodea* that causes the change observed in Reaction VI, a dead plant, unable to carry out cellular respiration, can be put in a solution of BTB and kept in the dark.

39. **C.** Because the frequency of the *B* allele is highest in western Asia and gradually decreases moving westward through Europe and Africa, the *B* allele most likely extended its range westward (and eastward) from a point in western Asia where the mutation for the *B* allele originated. Movement of alleles from one population to another, as observed for this allele, is an example of gene flow. In contrast, many independent mutations (answer choice A) would show multiple frequency peaks of the *B* allele. Also, because the map encompasses such a large area (two continents plus part of a third one!) representing so many environmental, geographical, and ecological variations, it is unlikely that the environment could be responsible for a process of natural selection that would generate the observed gradual declining pattern of the *B* allele as it moves westward (answer choice B). Genetic drift (answer choice D), or changes in gene frequencies as a result of chance, is also not likely responsible for the observed pattern because the pattern shows a steady decline (that is, not random, or by chance) in the frequency of the *B* allele moving westward.

40. **B.** The distinctive characteristic of the Blackfoot population relative to other Native American populations is its very high percentage of the A phenotype. This is best explained by the founder effect, in which a group of individuals, genetically atypical of the larger population of which they were a member, left the population to create a new population. In this scenario, the new population, by chance, had a much higher frequency of the *A* allele than the population from which they emigrated.

41. **A.** Because the *A* allele appears in the tree unaltered by substitutions, deletions, or insertions, it is the most ancestral allele.

42. **C.** The *B* allele differs from the *A* allele by 7 substitutions, 1 deletion, and 1 insertion. Thus, the total number of nucleotides remains the same as the *A* allele with 1,065 nucleotides (1,065 – 1 + 1 = 1,065). Substitutions are replacements for existing nucleotides and do not increase or decrease the total number.

43. **D.** The phylogenetic tree indicates that the *O* allele differs from the *A* allele by 1 nucleotide substitution and 1 nucleotide deletion. A deletion would result in an mRNA with a sequence shifted by one nucleotide. This generates a frameshift mutation, where all the codons following the shift are now different from those in the original mRNA sequence. In fact, additional analyses have discovered that one of those codon changes creates a stop codon, terminating the translation of the mRNA long before the end of the mRNA is reached. This early termination, in addition to the frameshift errors introduced in other codons, is responsible for producing a nonfunctional protein. As a result, no sugar is attached to the membrane-bound protein.

44. **D.** The ordinate (vertical axis) of the graph represents p, the frequency of the dominant allele *B*. The frequency of q, the recessive allele *b*, is calculated from $p + q = 1$, or $q = 1 - p$, which is equal to $1 - 0.4 = 0.6$.

45. **D.** The frequency of phenotypes is calculated using the equation $p^2 + 2pq + q^2 = 1$, where p^2 individuals are homozygote black, $2pq$ are heterozygote black, and q^2 are homozygote brown. Black-colored beetles, then, are expressed by both p^2 and $2pq$. So, $p^2 + 2pq = (0.4)^2 + (2)(0.4)(0.6) = 0.16 + 0.48 = 0.64 = 64\%$.

46. **B.** For the 10-year period from 1965 to 1975, there was a dramatic shift in the frequency of p and q. When p became 0, q became fixed at 1.0. Selection (perhaps a disease) against the black phenotype (both *BB* and *Bb* phenotypes) is the most likely cause. Genetic drift is usually significant only in small populations (this is a large population), and a sudden bottleneck resulting from some catastrophic event would not likely have eliminated all the black beetles. Similarly, gene flow through emigration can be eliminated as the correct answer as it is extremely unlikely that only brown beetles remained while all black beetles emigrated.

47. **D.** The change in p from 1985 to 1990 can best be explained by gene flow. Immigrants carrying the dominant allele, *B*, entered the population. This implies that the selection pressure against the black-colored beetles (perhaps a disease) was removed sometime between 1975 and 1985.

48. C. Here is a rough way to determine the best fitting phylogenetic tree. First determine which two species are most closely related (have the fewest nucleotides differences). Because J and L have the fewest at 6, call that cluster 1. J and L will share a common node (point where two lines meet at a common ancestor). Second, find the next closest pair of species; that will be K and M at 8. Call that cluster 2. Now, try to determine if these two clusters can fit on one of the provided trees. Only tree C provides a pairing for these two clusters.

There are other ways to create a tree, but they require a little math and often too much time. Instead, try to find pairs in the data that will fit one of the provided trees.

49. B. Glucose cannot pass through the selectively permeable membrane, so consider what is happening with the water. Side A has a 100% water concentration, and side B has a 97% water concentration. The concentration of water is higher in side A, so the water will move, by diffusion, from side A to side B, or from higher concentration to lower concentration. As a result, the solution will rise in side B. The movement of water across a selectively permeable membrane is osmosis.

50. D. Because there is a finite amount of NAD^+ in the cytoplasm, glycolysis would stop once all of the NAD^+ is converted to NADH. In the presence of oxygen, there would always be enough NAD^+ because NADH is converted back to NAD^+ during oxidative phosphorylation, when electrons from NADH are used to generate ATP from ADP. But when oxygen is absent, as it is during fermentation, oxidative phosphorylation (and cellular respiration) cannot occur because oxygen is not available as the final acceptor of the electrons in the electron transport chain. Under conditions where oxygen is absent, then, fermentation serves to replenish the NAD^+, allowing glycolysis to continue and to generate 2 ATP. Note that yeasts are fungi and do not do photosynthesis.

51. C. Once all of the available enzyme molecules are engaged with the reactants, no additional reactants can be catalyzed. As a result, the reaction rate remains constant during the interval between 6 and 9 minutes. Note that the reaction rate, the amount of product formed per unit of time, is the slope of the plotted curve (a straight line between 6 and 9 minutes).

52. D. Enzymes catalyze both forward and backward reactions. The concentrations of reactants and products determine which direction is dominant. When the reaction reaches equilibrium, each reaction in the forward direction ($J \rightarrow K + L$) is matched by one in the reverse direction ($K + L \rightarrow J$), and the net activity is zero.

53. B. When substances K and L are removed by subsequent reactions, the reaction $J \rightarrow K + L$ is no longer in equilibrium. As a result, more of K and L will be produced by the forward reaction ($J \rightarrow K + L$) than that of J in the reverse reaction ($K + L \rightarrow J$).

54. A. Mutant I cannot grow unless ornithine is added to the minimal medium. This suggests that an enzyme that converts a substance in the minimal medium to ornithine is either absent or nonfunctional. Thus, the induced mutation occurs in the DNA that codes for this enzyme. Other enzymes in the metabolic pathway remain functional because adding any one of the supplements ornithine, citrulline, or argininosuccinate will result in the production of arginine. Similarly, the enzyme in mutant II that converts ornithine to citrulline is absent or nonfunctional; adding citrulline or argininosuccinate allows arginine to be produced because the enzymes that convert citrulline to argininosuccinate and the enzyme that converts argininosuccinate to arginine remain functional. In a similar way, mutant III has a nonfunctional enzyme that converts citrulline to argininosuccinate, and mutant IV has a nonfunctional enzyme that converts argininosuccinate to arginine. Because each mutant has only one mutation that inactivates a single enzyme, adding ornithine, citrulline, or argininosuccinate (or all three supplements) to mutant IV does not lead to the production of arginine, as the required enzyme, the one that converts argininosuccinate to arginine, is nonfunctional. Only answer choice A describes a metabolic pathway that is consistent with the data.

55. A. In mutant II, the enzyme that converts ornithine to citrulline is missing or nonfunctional. If some citrulline is added to the medium, growth will occur and substances in the minimal medium will be converted to ornithine. But this newly generated ornithine cannot be converted to citrulline because of the mutation in the enzyme that converts ornithine to citrulline. Thus, once all of the added citrulline is used up (converted to argininosuccinate), growth stops and the newly generated ornithine remains unconverted to citrulline.

56. C. Mutant IV cannot grow because it cannot convert argininosuccinate to arginine.

57. B. If the allele for the red-colored fruit is dominant (*R*) to that of the white-colored (*r*) fruit, a plant that is true-breeding for red fruit is homozygous dominant, or *RR*, and a plant that is true-breeding for white-colored fruit is homozygous recessive, or *rr*. When these two genotypes are crossed, *RR* × *rr*, the progeny are all *Rr*, or red. Similarly, when plants with smooth fruits are crossed with plants with spiny fruits (*SS* × *ss*), all the progeny are *Ss*, or smooth. When both traits are considered together, the F_1 individuals are *RrSs*. Answer choice A is incorrect because individuals that are carriers for a trait do not express that trait. In this question, it is the plants with red and smooth fruits that are carriers for the white and spiny traits. Answer choice C is incorrect because each parent contributes one of each chromosome to every offspring and, as a result, every offspring receives two alleles for every gene. Answer choice D is incorrect because the F_1 plants bearing smooth fruits must be heterozygous in order to satisfy the results.

58. C. When genes for different traits assort independently, it means that the genes are on different chromosomes, and a Punnett square can be used to show all possible combinations of gametes. When plants from the F_1 generation (*RrSs*) are test-crossed with plants homozygous recessive for both traits (*rrss*), one of each of the possible genotypes is produced for a 1:1:1:1 ratio, as shown in the following table.

Gametes from homozygote recessive individuals (*rrss*)	Gametes from F_1 individuals (*RrSs*)			
	RS	*Rs*	*rS*	*rs*
rs	*RrSs* (red and smooth)	*Rrss* (red and spiny)	*rrSs* (white and smooth)	*rrss* (white and spiny)

Answer choice A is incorrect because homozygous recessives are, indeed, produced (white and spiny, for example). Answer choice B is incorrect because if codominance were the mode of inheritance, there would be a blending of the traits, perhaps pink phenotypes from *Rr* genotypes. Answer choice D is incorrect for two reasons: First, the question asks for the best explanation of the results, not if the results are accurate. Second, simulations 2 and 3 both have sufficient numbers of offspring to produce reliable results.

59. B. When genes are on the same chromosome, they cannot assort independently, so they are inherited together. Recombination can still occur if crossing over ensues. But when the genes are very close together, crossing over is not likely, as there is little space between them to break and exchange with a similar break on the homologous chromosome. This is the case in simulation 2: When the genes are linked and no crossing over occurs because they are very close, each parent can provide only two kinds of gametes. But when one of the parents is homozygous recessive for both traits, that parent has only one kind of gamete. For the parents in this question, then, one parent is donating two kinds of gametes that combine (after fertilization) with only one kind of gamete from the second parent. Thus, there are only two kinds of offspring produced. *Watch for these kinds of questions: When the phenotypes of offspring are produced in numbers skewed toward the phenotypes of certain parents, suspect linkage with crossing over. When skewing is dramatic, as in this question, suspect linkage with no crossing over.*

60. **C.** If genes are on the same chromosome, they do not assort independently, but they can exchange (cross over) with like genes on the homologous chromosome. Generally, the farther the genes are separated on the chromosome, the more often they are likely to cross over (because there are more places between the genes for the chromosome to break and recombine). As in this question, where the ratio of phenotypes is about 7:3:3:7, crossing over generates phenotypes among offspring in ratios that may not look like typical crosses (for example, 9:3:3:1 or 1:1:1:1). *Watch for these kinds of questions: When the phenotypes of offspring are produced in numbers skewed toward the phenotypes of certain parents, suspect linkage with crossing over. When skewing is dramatic, as in the previous question, suspect linkage with no crossing over.* Although this question does not request it, the crossing-over frequency in simulation 3 is 30%, as the following figure illustrates.

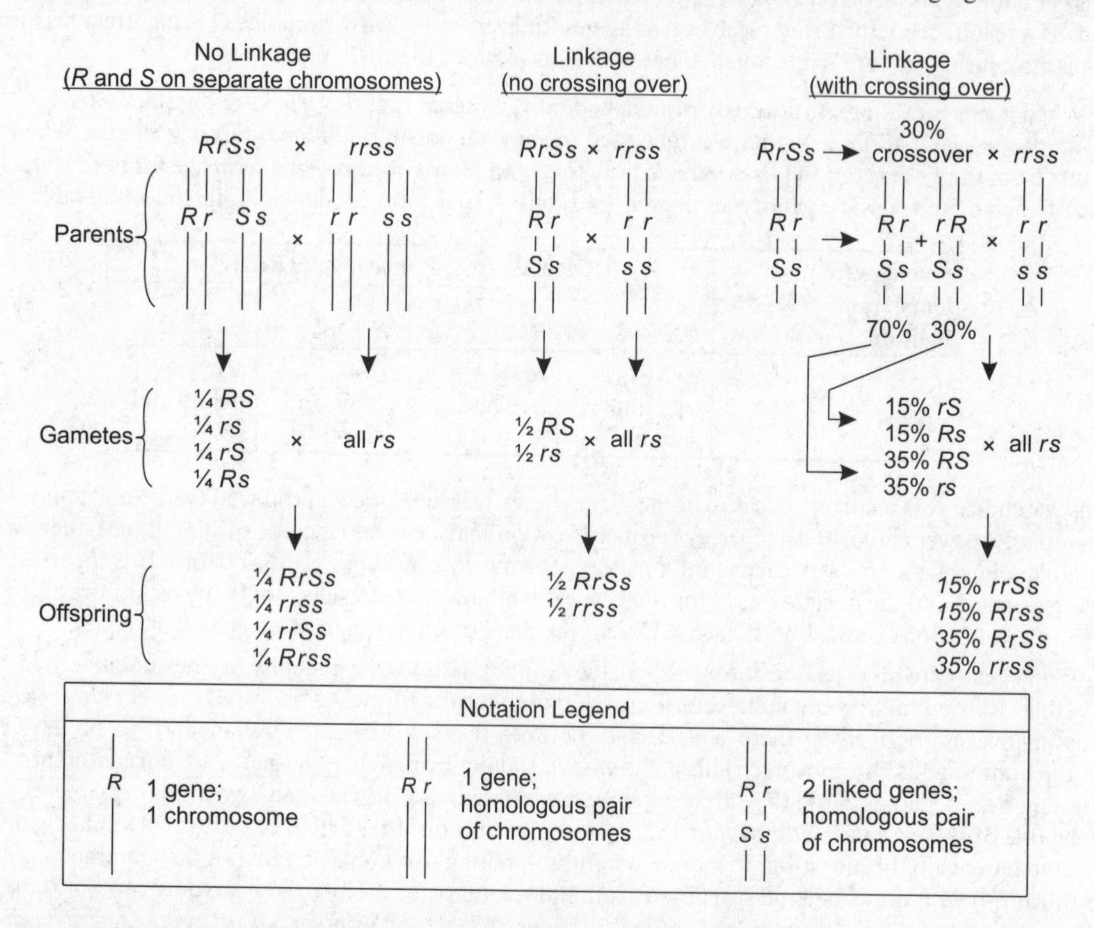

Section II: Free-Response Questions

Scoring Standards for the Free-Response Questions

To score your answers, award points to your response using the standards given below. For each item listed below that matches the content and vocabulary of a statement or explanation in your response, add the indicated number of points to your score (to the maximum allowed for each section). Your score for each question will range from 0 to 10 points.

Words appearing in parentheses in answers represent alternative wording.

Question 1 (8 points maximum)

a. (1 point maximum)
1 pt: Acetylcholine is a polar molecule.
1 pt: Acetylcholine cannot pass across the membrane.

b. (1 point maximum)
1 pt: IP_3 communicates the message to the endoplasmic reticulum.
1 pt: A second messenger, IP_3, signals the endoplasmic reticulum to open Ca^+ channels.

c. (3 points maximum)
1 pt: Dilating blood vessels in heart muscle increases the delivery of blood (oxygen).
1 pt: Nitroglycerin releases NO when broken down.
1 pt: NO activates guanylyl cyclase.
1 pt: Guanylyl cyclase converts GTP to cGMP (cyclic GMP).
1 pt: cGMP (cyclic GMP) relaxes (dilates, opens up) blood vessels.

d. (3 points maximum)
1 pt: Drug 3
2 pts: By blocking the breakdown of cGMP, its effect is prolonged, allowing it to maintain a relaxed state of smooth muscle.

Question 2 (10 points maximum)

a. (3 points maximum)
1 pt: The elk population declined as a result of park management culling.
1 pt: The elk population increased as a result of the absence of its major predator (wolves).
1 pt: Reintroducing wolves would be a test for this hypothesis.

b. (3 points maximum) (statement / explanation)
1 pt: Competition for food / from increasing numbers of bison.
1 pt: Food shortages / caused by a decline in vegetation productivity resulting from changes in the weather.
1 pt: Destruction of habitat / by fires.
1 pt: An increase in disease / resulting from an increase in herd densities.

c. (2 points maximum)
1 pt: The vegetation upon which the elk browse will decrease.
1 pt: Increasing numbers of elk require increasing amounts of food.
OR Vegetation availability declines as a result of excessive browsing.

d. (2 points maximum) (proposal + justification)
2 pts: The wolf population declined because of the declining numbers of elk.
2 pts: The wolf population declined to the carrying capacity of its environment.
2 pts: The wolf population declined because of disease.

Question 3 (4 points maximum)

a. (1 point maximum)

1 pt: Only the bases retained a nucleus.

1 pt: The nucleus possesses the genetic information that directs the growth of new stalks and caps.

b. (1 point maximum)

1 pt: The new cap is that of *A. crenulata* because the stalk still contained *A. crenulata* mRNA that coded for an *A. crenulata* cap.

c. (1 point maximum)

1 pt: The new cap would be the species type of the base (*A. mediterranea*) because the nucleus would have replaced the old mRNA in the grafted stalk with its own mRNA.

1 pt: The regenerated cap would be *A. mediterranea* because the base contains the nucleus of an *A. mediterranea,* which will produce mRNA that codes for regenerating an *A. mediterranea* cap.

d. (1 point maximum)

2 pts: The new cap will be a mixture, or intermediate combination, of both caps.

2 pts: The mRNA produced by both nuclei will mix and together determine a new cap type.

Question 4 (4 points maximum)

a. (1 point maximum)

1 pt: The different messages promote reproductive isolation.

1 pt: The different messages keep the species from intermating.

b. (1 point maximum)

1 pt: A volatile chemical can attract predators to the females as well as male suitors.

1 pt: The same volatile chemical can be released by predators to attract males into traps.

c. (1 point maximum)

1 pt: These orchids attract male bees by releasing a pheromone that mimics the mating pheromone of the female bees.

d. (1 point maximum)

1 pt: Only males of *Andrena nigroaenea* are attracted to the orchid because the orchid's pheromone mimic is species-specific.

Question 5 (4 points maximum)

a. (1 point maximum)

1 pt: Eggs from population 1 have the earliest hatch date.

b. (1 point maximum)

1 pt: Females hatch first in populations 1 and 2. There is no significant difference in the hatch times for females and males in population 3.

c. (1 point maximum)

1 pt: Earlier hatch time for females may be advantageous for some populations because it allows more time for females to feed and grow and, thus, provides for the additional nutritional needs for producing eggs.

d. (1 point maximum)

1 pt: The populations may be at different locations where they experience different environmental factors such as water temperature, food, predators, or shelter.

Question 6 (4 points maximum)

a. (1 point maximum)

1 pt: As described in the question, the solute, sucrose, cannot pass through the dialysis bag membrane. In bags where the concentration of solute is higher than that in the beaker, water, the solvent, moves from higher water concentration (in the beaker) to lower concentration (in the bag). That occurs in bags 1, 2, and 3. In bag 4, water moves out of the bag and into the beaker.

b. (1 point maximum)

1 pt: When the concentration of sucrose inside a bag is the same as the concentration outside the bag, there is no net movement of solutes between the bag and the beaker and, therefore, no change in the weight of the bag after it is immersed into the beaker. None of the bags has a net change in weight of 0, but, according to the graph, a weight change of zero would occur at 0.7. Therefore, the sucrose solution in the bags is 0.7 moles/liter.

c. (1 point maximum)

1 pt: The pressure potential (Ψ_p) for a solution in an open beaker is 0. There are no external forces on the solutions.

d. (1 point maximum)

1 pt: At 0.7 moles/liter, the water potentials (Ψ) of the solutions in the bag and the beaker are equal. To calculate Ψ, use the formulas for water potential and solute potential provided on the Equations and Formulas pages at the beginning of this exam:

$$\Psi = \Psi_p + \Psi_s$$

and

$$\Psi_s = -iCRT$$

where

$i = 1$ (ionization constant is 1 because sucrose does not ionize)

$C = 0.7$, molar concentration obtained from graph where the percent weight change is zero

$R = 0.0831$ liter-bar/mole °K (from Equations and Formulas pages)

$T = (273 + 25)$°K (25°C provided in question)

$$\begin{aligned}\Psi(\text{beaker}) = \Psi(\text{bag}) = \Psi_s + \Psi_p = \Psi_s + 0 &= -iCRT \\ &= -1\left(\frac{0.7\text{ moles}}{\text{liter}}\right)\left(\frac{0.0831\text{ liter-bars}}{\text{mole-°K}}\right)\left(\frac{(273+25)\text{°K}}{1}\right) \\ &= -(0.7)(0.0831)(298)\text{ bars} \\ &= -17.33466 \\ &\approx -17.3\text{ bars}\end{aligned}$$

Chapter 14

Practice Exam 2

Answer Sheet for Practice Exam 2

(Remove This Sheet and Use It to Mark Your Answers)

Multiple-Choice Questions

1 Ⓐ Ⓑ Ⓒ Ⓓ	21 Ⓐ Ⓑ Ⓒ Ⓓ	41 Ⓐ Ⓑ Ⓒ Ⓓ
2 Ⓐ Ⓑ Ⓒ Ⓓ	22 Ⓐ Ⓑ Ⓒ Ⓓ	42 Ⓐ Ⓑ Ⓒ Ⓓ
3 Ⓐ Ⓑ Ⓒ Ⓓ	23 Ⓐ Ⓑ Ⓒ Ⓓ	43 Ⓐ Ⓑ Ⓒ Ⓓ
4 Ⓐ Ⓑ Ⓒ Ⓓ	24 Ⓐ Ⓑ Ⓒ Ⓓ	44 Ⓐ Ⓑ Ⓒ Ⓓ
5 Ⓐ Ⓑ Ⓒ Ⓓ	25 Ⓐ Ⓑ Ⓒ Ⓓ	45 Ⓐ Ⓑ Ⓒ Ⓓ
6 Ⓐ Ⓑ Ⓒ Ⓓ	26 Ⓐ Ⓑ Ⓒ Ⓓ	46 Ⓐ Ⓑ Ⓒ Ⓓ
7 Ⓐ Ⓑ Ⓒ Ⓓ	27 Ⓐ Ⓑ Ⓒ Ⓓ	47 Ⓐ Ⓑ Ⓒ Ⓓ
8 Ⓐ Ⓑ Ⓒ Ⓓ	28 Ⓐ Ⓑ Ⓒ Ⓓ	48 Ⓐ Ⓑ Ⓒ Ⓓ
9 Ⓐ Ⓑ Ⓒ Ⓓ	29 Ⓐ Ⓑ Ⓒ Ⓓ	49 Ⓐ Ⓑ Ⓒ Ⓓ
10 Ⓐ Ⓑ Ⓒ Ⓓ	30 Ⓐ Ⓑ Ⓒ Ⓓ	50 Ⓐ Ⓑ Ⓒ Ⓓ
11 Ⓐ Ⓑ Ⓒ Ⓓ	31 Ⓐ Ⓑ Ⓒ Ⓓ	51 Ⓐ Ⓑ Ⓒ Ⓓ
12 Ⓐ Ⓑ Ⓒ Ⓓ	32 Ⓐ Ⓑ Ⓒ Ⓓ	52 Ⓐ Ⓑ Ⓒ Ⓓ
13 Ⓐ Ⓑ Ⓒ Ⓓ	33 Ⓐ Ⓑ Ⓒ Ⓓ	53 Ⓐ Ⓑ Ⓒ Ⓓ
14 Ⓐ Ⓑ Ⓒ Ⓓ	34 Ⓐ Ⓑ Ⓒ Ⓓ	54 Ⓐ Ⓑ Ⓒ Ⓓ
15 Ⓐ Ⓑ Ⓒ Ⓓ	35 Ⓐ Ⓑ Ⓒ Ⓓ	55 Ⓐ Ⓑ Ⓒ Ⓓ
16 Ⓐ Ⓑ Ⓒ Ⓓ	36 Ⓐ Ⓑ Ⓒ Ⓓ	56 Ⓐ Ⓑ Ⓒ Ⓓ
17 Ⓐ Ⓑ Ⓒ Ⓓ	37 Ⓐ Ⓑ Ⓒ Ⓓ	57 Ⓐ Ⓑ Ⓒ Ⓓ
18 Ⓐ Ⓑ Ⓒ Ⓓ	38 Ⓐ Ⓑ Ⓒ Ⓓ	58 Ⓐ Ⓑ Ⓒ Ⓓ
19 Ⓐ Ⓑ Ⓒ Ⓓ	39 Ⓐ Ⓑ Ⓒ Ⓓ	59 Ⓐ Ⓑ Ⓒ Ⓓ
20 Ⓐ Ⓑ Ⓒ Ⓓ	40 Ⓐ Ⓑ Ⓒ Ⓓ	60 Ⓐ Ⓑ Ⓒ Ⓓ

Section I: Multiple-Choice Questions

Time: 90 minutes

60 questions

Directions: Each of the following questions or statements is followed by four possible answers or sentence completions. Choose the one best answer or sentence completion.

Questions 1–3 refer to the following bar graphs, which show the variation in beak depth of two species of Darwin's finches, Geospiza fuliginosa *and* G. fortis, *both seed-eating ground finches. Beak depths are highly correlated with the seed size in the diet. The graphs show these birds on islands where they occur separately on Los Hermanos and Daphne Major and where they coexist on Santa Cruz, all islands of the Galápagos archipelago.*

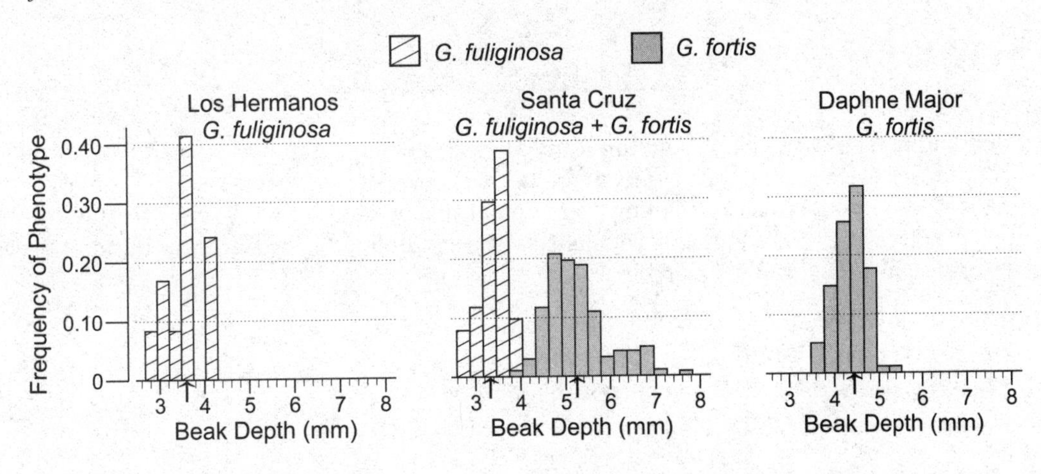

1. In which species and on which island is the beak depth variation greatest?

- **A.** *G. fuliginosa* on Los Hermanos
- **B.** *G. fuliginosa* on Santa Cruz
- **C.** *G. fortis* on Santa Cruz
- **D.** *G. fortis* on Daphne Major

2. At one time, *G. fuliginosa* was the only ground finch on Santa Cruz, but some *G. fortis* arrived 22 years ago. Upon their arrival, beak depths for both species on Santa Cruz measured the same as their respective beak depths on Los Hermanos and Daphne Major. Thus, the beak depths on Santa Cruz indicate changes after 22 years. The average beak depth for each species is indicated by an arrow on the *x*-axis. Which of the following accurately describes the greatest average change in beak depth on Santa Cruz after 22 years?

- **A.** an increase in beak depth for *G. fuliginosa*
- **B.** a decrease in beak depth for *G. fuliginosa*
- **C.** an increase in beak depth *G. fortis*
- **D.** a decrease in beak depth for *G. fortis*

3. Which of the following best explains the observed changes on Santa Cruz?

- **A.** not enough mating sites
- **B.** competition for food
- **C.** genetic drift
- **D.** mutation

Question 4 refers to the following figure.

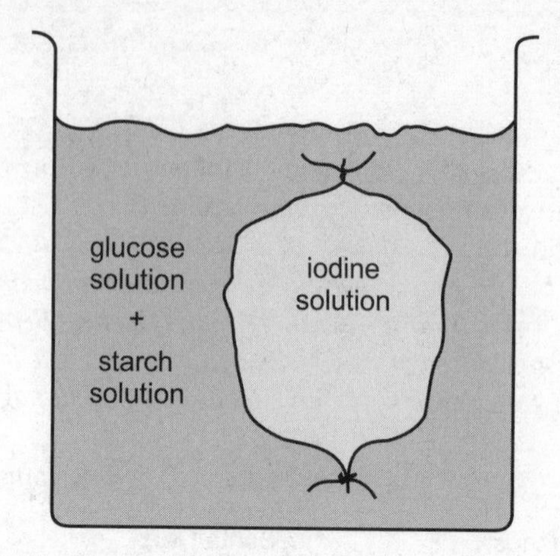

4. An iodine solution (IKI or Lugol's solution) turns from yellow-brown to blue when it reacts with starch, but not when it reacts with glucose. A bag made from a selectively permeable material and containing iodine solution is placed into a beaker containing a solution of glucose and starch. Using only this information, which of the following can be correctly concluded if the solution in the beaker turns blue but the contents of the bag remain unchanged in color?

A. Glucose moved from the beaker into the bag.
B. The bag is permeable to IKI, but not to starch.
C. The bag is permeable to IKI and glucose, but not to starch.
D. The bag is permeable to IKI, but not to starch and glucose.

Questions 5–7 refer to the following figure, which shows a mitogen-activated protein (MAP) signaling cascade. Mitogens are substances that stimulate cell division.

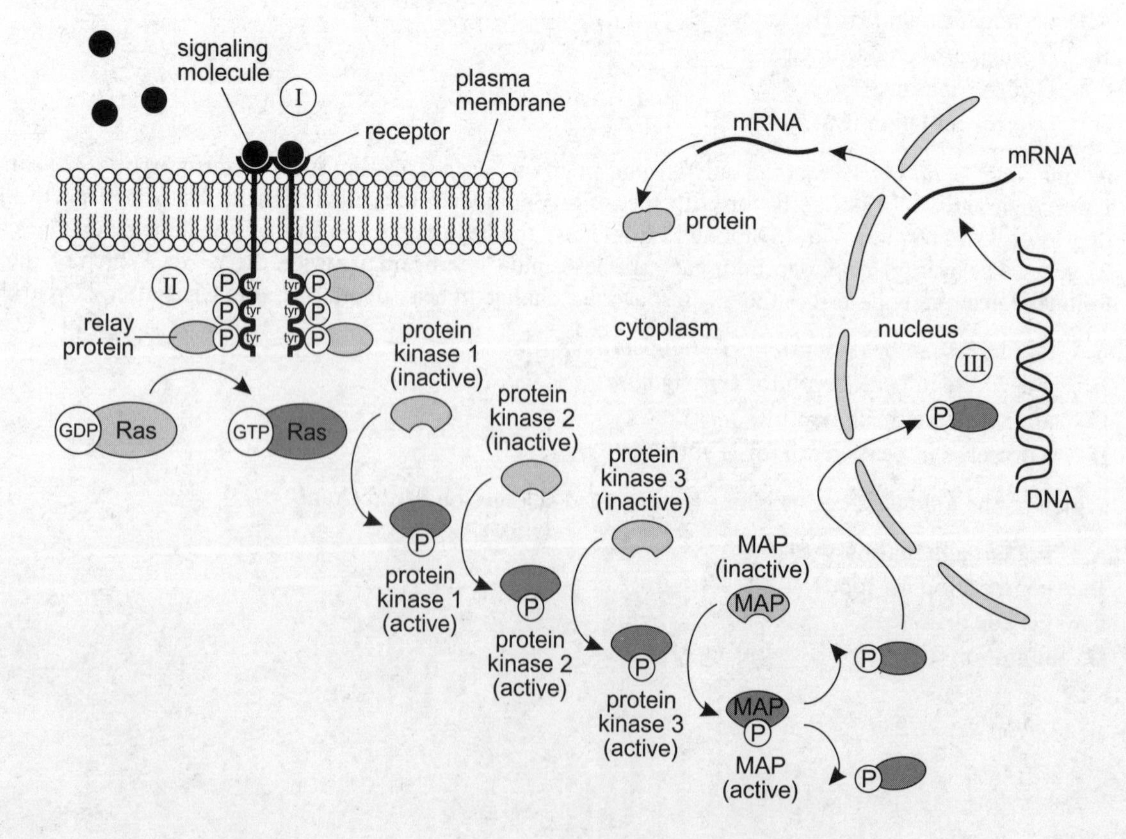

5. The figure shows two signaling molecules binding to a transmembrane receptor protein (shown at I). What do these signaling molecules do?

- **A.** They provide energy to the receptor molecules that will be used to activate relay proteins and cascade proteins.
- **B.** They disable a signal transduction pathway by causing two adjacent receptor proteins to lock together.
- **C.** They activate the receptor proteins, causing them to initiate a signal transduction pathway.
- **D.** They activate the receptor proteins, causing them to open a gated ion channel, allowing the passage of specific molecules across the membrane.

6. In the figure, the activated receptor proteins allow the binding of a relay protein (shown at II). In turn, the relay protein triggers the activation of the G protein called Ras. How is the Ras protein activated?

- **A.** The relay molecule binds to Ras.
- **B.** A GDP molecule on Ras is replaced with a GTP molecule.
- **C.** Ras is enzymatically split to form two active proteins.
- **D.** Ras binds to a second relay protein, forming the active form of the protein.

7. A signaling cascade is illustrated by protein kinase 1, protein kinase 2, protein kinase 3, and MAP. What purpose does this cascade serve?

- **A.** The cascade makes the target of the signal more specific.
- **B.** The cascade amplifies the signal.
- **C.** The cascade reduces the necessary activation energy for the overall reaction.
- **D.** The cascade generates the necessary energy for activating transcription.

8. The final step shown for this pathway is the transcription of an mRNA (shown at III). What is the most likely mechanism that initiates this step?

- **A.** The signaling cascade produces RNA polymerases.
- **B.** The signaling cascade generates the ATP required for transcription.
- **C.** The mRNA is activated by a product of the signaling cascade.
- **D.** A product of the signaling cascade activates transcription factors.

9. Mutations in the Ras protein are found in humans. One of these mutations leads to a form of Ras that does not respond to the relay protein but is permanently activated. For such a mutated Ras, which of the following is the most likely effect on its mitogen functions?

- **A.** uncontrolled cell growth, cell division, and cancer
- **B.** apoptosis, or programmed cell death
- **C.** an increase in metabolic activities and, for cells that secrete substances, an increase in secretions
- **D.** inability to undergo cell division

Questions 10–13 refer to the following figure showing the replicative cycle of an RNA retrovirus such as HIV, the AIDS virus.

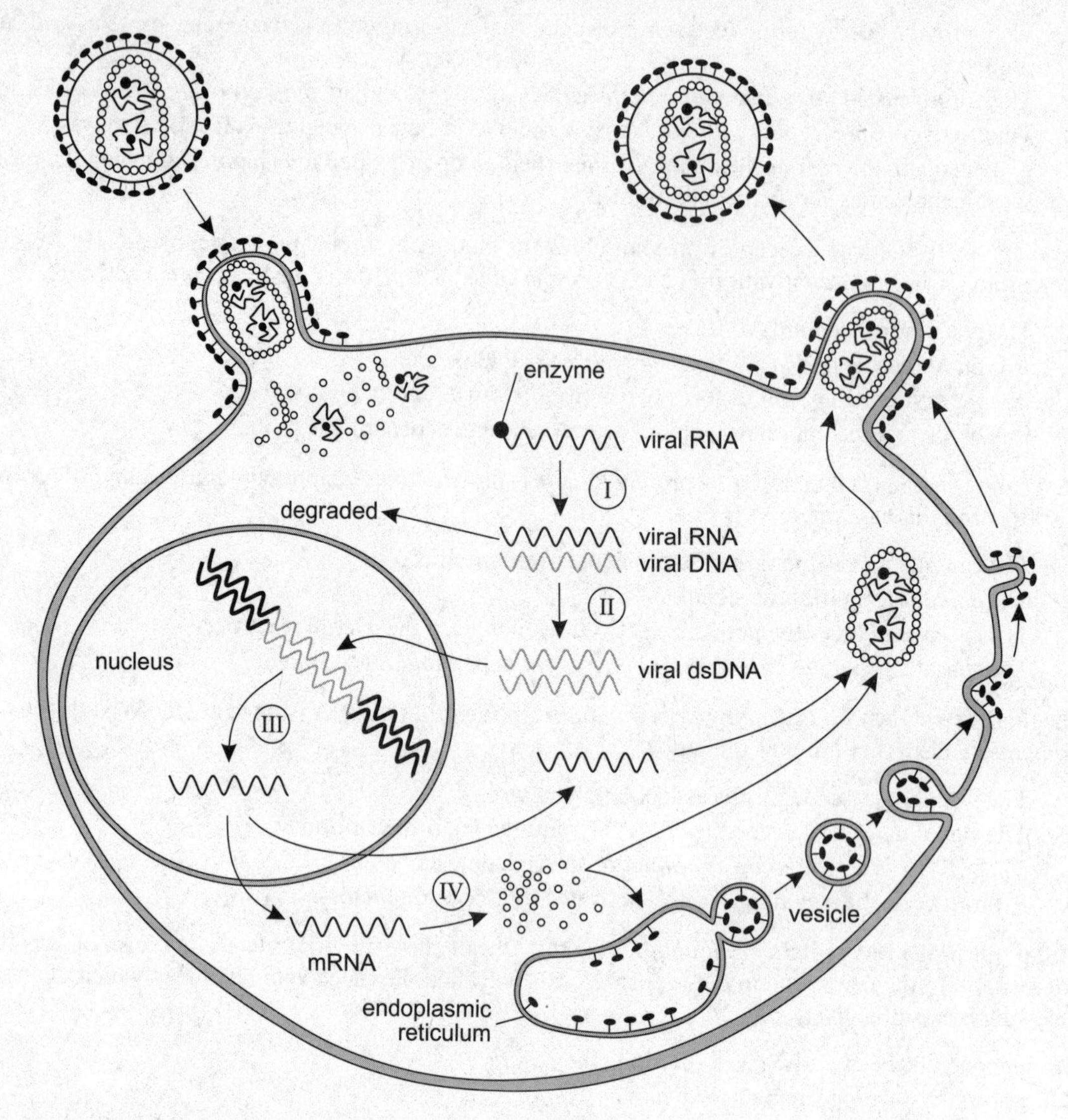

10. The process indicated by process I in the figure is catalyzed by

A. DNA polymerase
B. RNA polymerase
C. reverse transcriptase
D. ligase

11. The process indicated by process III in the figure is catalyzed by

A. DNA polymerase
B. RNA polymerase
C. reverse transcriptase
D. ligase

12. To complete the process of viral replication, the two processes following the arrow at process IV in the figure produce

A. RNA and DNA
B. proteins and RNA
C. proteins and glycoproteins
D. DNA and glycoproteins

13. Where are viral genes transcribed into mRNA?

A. cytoplasm
B. nucleus
C. endoplasmic reticulum
D. vesicle

14. Compared to DNA viral replication, RNA viral replication is more susceptible to mutation because

A. RNA replication takes place in the cytoplasm, whereas DNA replication takes place in the nucleus.
B. RNA viruses replicate much faster than DNA viruses.
C. RNA viruses use uracil instead of thymine.
D. RNA replication lacks the proofreading functions characteristic of DNA replication.

Questions 15–16 refer to the following table that indicates the presence or absence of traits X, Y, and Z in species I, II, and III. The number 1 indicates that a character is present; a 0 indicates that it is absent.

Species	Trait		
	X	Y	Z
I	1	1	0
II	1	1	1
III	1	0	0
Outgroup	0	0	0

15. Which of the following trees correctly reflects the data shown in the table?

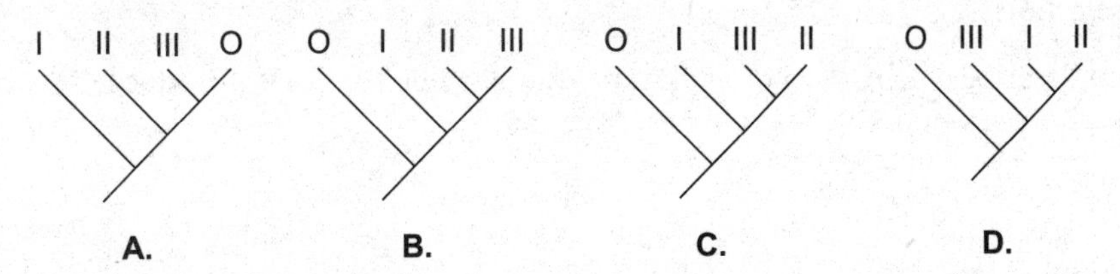

16. With respect to traits X, Y, and Z, which of the following is true?

A. Species II is most closely related to species I.
B. Species II is most closely related to species III.
C. Species II is most closely related to the outgroup.
D. Species II is related to both species I and III equally.

17. The reaction A + B → C is catalyzed by enzyme *K*. If the reaction is in equilibrium, which of the following would allow more product C to be produced?

A. removing some of reactant A
B. removing some of reactant C
C. adding more enzyme *K*
D. increasing the temperature of the system

18. Most animal cells, regardless of species, are relatively small and about the same size. Relative to larger cells, why is this?

A. Smaller cells avoid excessive osmosis and subsequent lysis.
B. Smaller cells have a smaller surface-to-volume ratio.
C. Smaller cells have a larger surface-to-volume ratio.
D. Smaller cells fit together more tightly.

Questions 19–21 refer to the following figure that illustrates the evolution of nine species from a single ancestral species. Roman numerals I through V identify different areas of the figure.

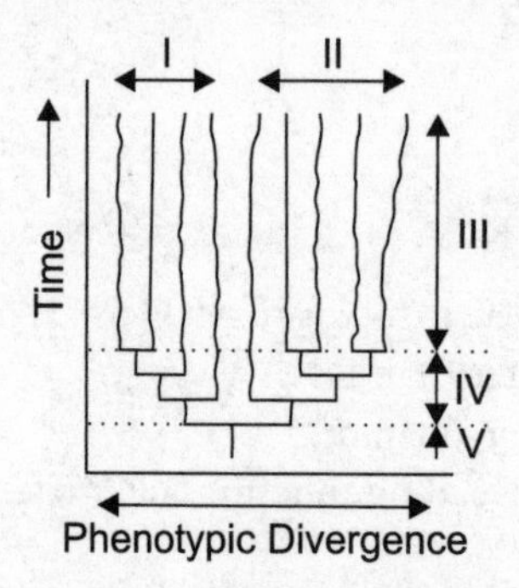

19. During which of the indicated periods is the rate of evolution most rapid?

A. II
B. III
C. IV
D. V

20. Which of the following could be responsible for the evolutionary pattern indicated by area III?

A. directional selection
B. stabilizing selection
C. disruptive selection
D. sexual selection

21. If the diagram describes the pattern of evolution after a single species is introduced to a remote, newly formed island, the pattern in the diagram best suggests

A. adaptive radiation
B. allopatric speciation
C. coevolution
D. multiple occurrences of gene flow

Questions 22–23 refer to the following.

A clear plastic chamber containing a cylindrical cage was used as an animal chamber. An opening at one end of the chamber was sealed with a stopper. Passing through the stopper was a graduated burette. Gas production or consumption inside the chamber could be measured by movements of a solution rising or falling inside the burette. Potassium hydroxide (KOH) was added to the chamber to absorb any CO_2 produced.

Twenty hamsters were weighed and put into one of 20 chambers. Half of the chambers were maintained at 10°C, and half were maintained at 25°C. Oxygen consumption in ml was recorded for each hamster every 30 seconds and plotted as ml per gram of hamster weight. The data are shown in the graph that follows.

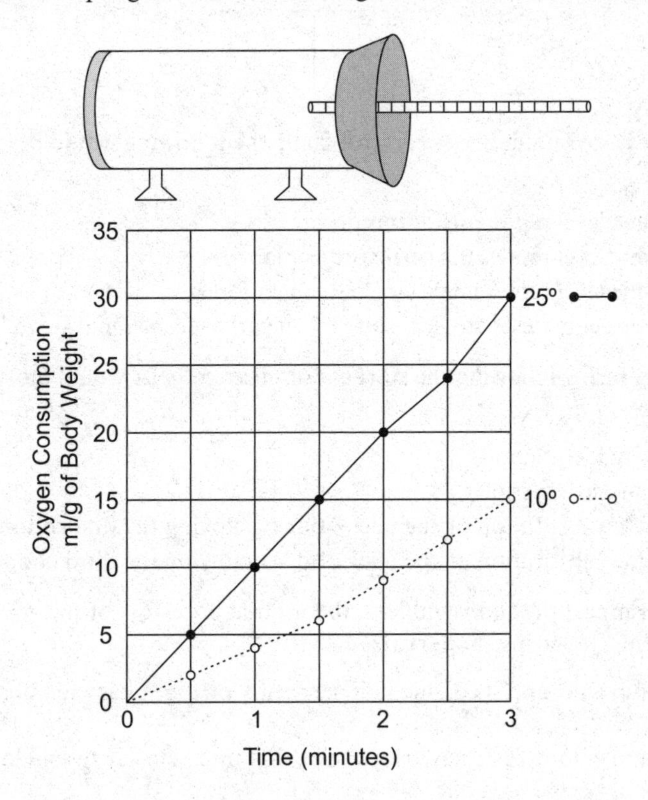

22. The data in the graph above are consistent with which of the following conclusions?

A. As time progresses, respiration rate increases in hamsters.
B. The rate of respiration for hamsters increases with increases in environmental temperature.
C. Temperature increases with increases in the weights of hamsters.
D. The larger the hamster, the greater its rate of respiration.

23. All of the following must remain unchanged during data collection EXCEPT:

A. atmospheric pressure
B. the volume of the animal chamber
C. the amount of CO_2 in the animal chamber
D. the amount of oxygen in the animal chamber

Question 24 refers to the following ecological pyramid.

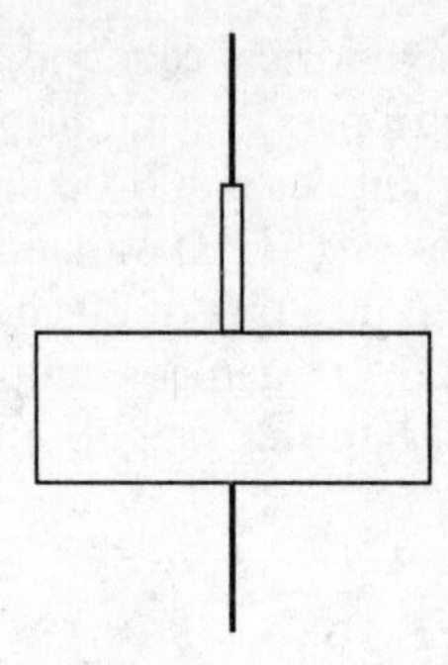

24. If the figure illustrates a pyramid of *numbers,* which of the following would best explain the relative sizes of the trophic levels?

A. The lowest trophic level represents decomposers.
B. The lowest trophic level represents carnivorous plants.
C. The lowest trophic level represents a single tree in a forest.
D. The pyramid represents the ecological state of an early successional stage.

25. Which of the changes below following the start codon in an mRNA would most likely have the greatest deleterious effect?

A. a deletion of a single nucleotide
B. a deletion of a nucleotide triplet
C. a single nucleotide substitution of the nucleotide occupying the first codon position
D. a single nucleotide substitution of the nucleotide occupying the third codon position

26. All of the following support the endosymbiotic theory that ancestors of mitochondria and chloroplasts were once independent, free-living prokaryotes EXCEPT:

A. Mitochondria and chloroplasts divide independently of the eukaryotic host cell by a process similar to binary fission.
B. Mitochondria and chloroplasts have ribosomes that more closely resemble those of bacteria than of eukaryotic cells.
C. Mitochondria and chloroplasts function independently of the eukaryotic host cell.
D. Mitochondria, chloroplasts, and bacteria have a single, circular chromosome without histones or proteins.

Questions 27–28 refer to the following.

Mitochondria isolated from cells can be induced to carry out respiration if an appropriate substrate is provided and a low temperature is maintained to prevent enzyme degradation. The following experiment uses isolated mitochondria in this manner to investigate respiration.

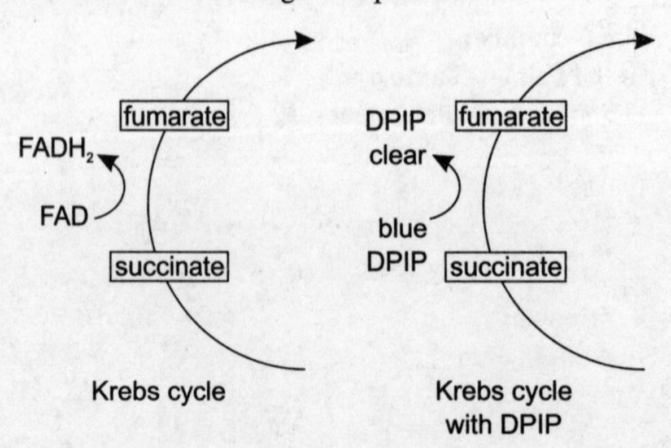

During the Krebs cycle, succinate is oxidized to fumarate as electrons are transferred to FAD for its reduction to $FADH_2$ (see figure above). In this experiment, DPIP is provided as a substitute electron acceptor for FAD. DPIP, which is blue in its oxidized state, accepts electrons from succinate. After accepting the electrons, DPIP turns clear, its reduced state. A spectrophotometer is used to quantify the degree of color change by measuring the amount of light that is transmitted through a cuvette. A cuvette with a higher transmittance percentage indicates a solution with more reduced DPIP.

Three cuvettes are prepared at low temperatures with an appropriate buffer. The contents of a fourth cuvette are similarly prepared except that the mitochondrial suspension is first preheated to 100°C and then returned to a low temperature. Distilled water was added to bring the contents of all cuvettes to the same volume. A summary of the contents of the cuvettes follows:

Cuvette 1: 0.2 ml succinate added to isolated mitochondria

Cuvette 2: 0.1 ml succinate added to isolated mitochondria

Cuvette 3: 0 ml succinate added to isolated mitochondria

Cuvette 4: 0.1 ml succinate added to isolated mitochondria *preheated* to 100°C

Transmittance in each cuvette is measured every 10 minutes. The results of the experiment are summarized in the following graph.

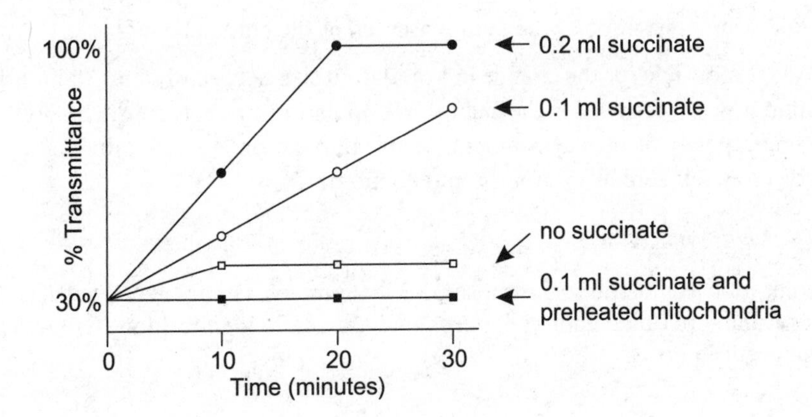

27. The results of the experiment indicate that the respiratory *rate* increases as

A. the concentration of substrate increases
B. the concentration of DPIP increases
C. the concentration of enzyme increases
D. the time increases

28. Which of the following is true with respect to the availability of electrons from other steps of the Krebs cycle?

A. No other steps of the Krebs cycle are involved in the reduction of electron acceptors.
B. Electrons from other steps could also reduce DPIP, but the substrates for these steps are consumed before succinate is added.
C. FAD is the only electron acceptor for the Krebs cycle.
D. NAD^+ is the only electron acceptor for the Krebs cycle.

29. In reference to a segment of DNA, which of the following molecules contains the *fewest* number of nucleotides?

A. a single strand of the original DNA segment
B. a single strand of the original DNA segment after a point mutation
C. the primary RNA transcript (before splicing) from the original DNA segment
D. a single strand of complementary DNA (cDNA) made from an mRNA, which was made from the original DNA segment

Question 30 refers to the following graph.

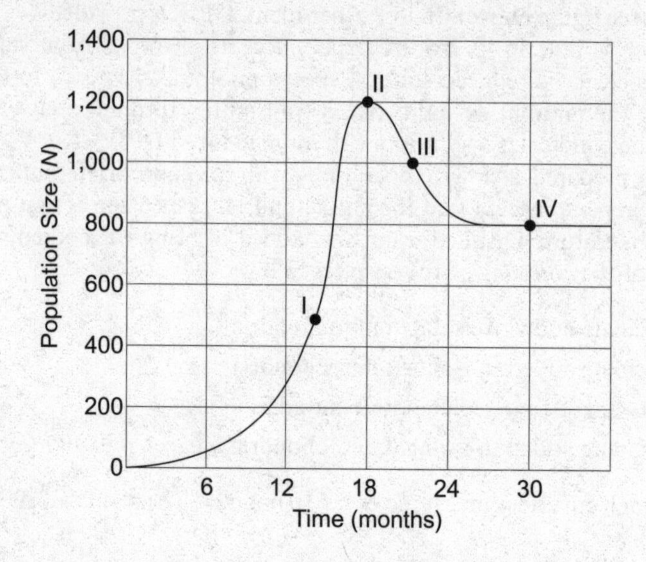

30. Which of the following best supports the data presented in the graph above?

A. Disease was responsible for the change in population size between 18 and 30 months.
B. Competition was responsible for the change in population size between 18 and 30 months.
C. The carrying capacity of the environment for this population is 1,200 individuals.
D. The rate of growth is zero at 18 months and 30 months.

Question 31 refers to the following.

The figure that follows illustrates chromosome 3 for humans, chimpanzees, gorillas, and orangutans (left to right). Boxes 1 and 4 outline banding patterns that are nearly identical for all four species, but differences between species occur in boxes 2 and 3.

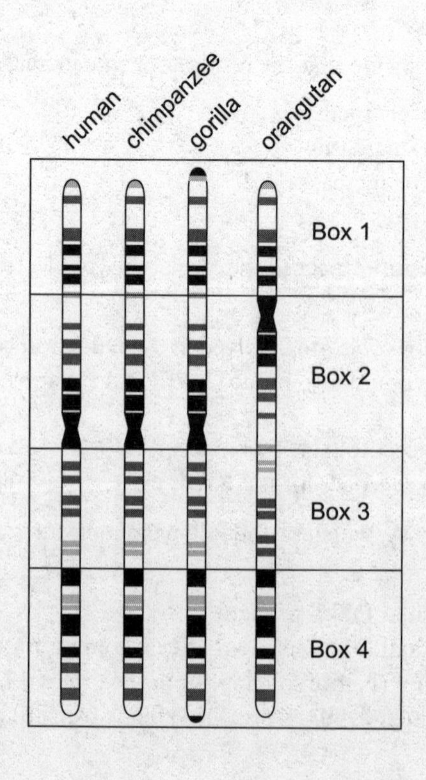

31. With respect to the chromosome segments in boxes 2 and 3, which of the following best explains how orangutans differ from the other species?

A. chromosomal deletion
B. chromosomal addition
C. chromosomal translocation
D. chromosomal inversion

32. Which of the following would most likely describe the fate of a vesicle formed as a result of phagocytosis?

A. The vesicle merges with a mitochondrion.
B. The vesicle merges with a lysosome.
C. The vesicle is shuttled to the nucleus, and its contents become part of the nucleolus.
D. The vesicle releases its contents to the cytoplasm to be digested.

33. More than a dozen species of Darwin's finches with various specialized adaptations inhabit the Galápagos Islands. Each species possesses a specialized adaptation for obtaining food. Why is it that similar adaptations do not exist among the finches of the South American mainland, the presumed origin of the Galápagos finch ancestor?

A. The various foods available on the mainland are different from those on the Galápagos Islands.
B. South American predators limited evolution of the mainland finch.
C. Reproductive isolation is not possible on the mainland.
D. The available niches that Darwin's finches exploited on the Galápagos Islands were already occupied by other species of birds on the mainland.

34. In addition to light, aquatic plants require various nutrients. To identify the nutrient whose absence is most limiting to growth, researchers compared the effect on phytoplankton growth by the addition of four different nutrients, one at a time. The results of the comparisons are shown in the following figure. Numbers in parentheses are the number of experiments, and error bars represent 95% confidence intervals.

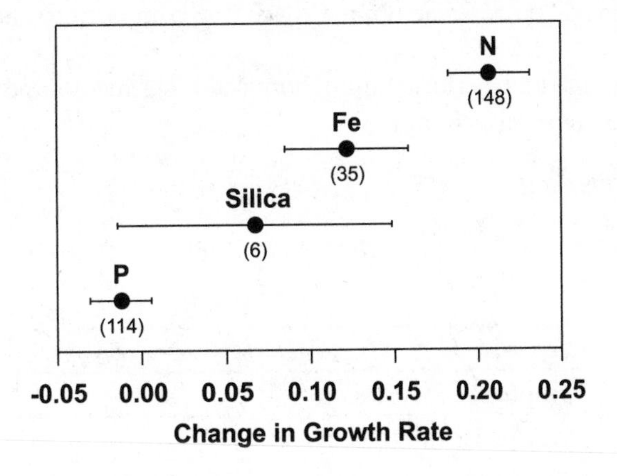

According to the data in the figure, which of the nutrients was most limiting on phytoplankton growth?

A. Nitrogen
B. Iron
C. Silica (SiO_2)
D. Phosphorus

35. In the pedigree shown below, circles indicate females and squares indicate males. A horizontal line connecting a male and female indicates that these two individuals produced offspring. Offspring are indicated by a descending vertical line that branches to the offspring. A filled circle or filled square indicates that the individual has a particular trait.

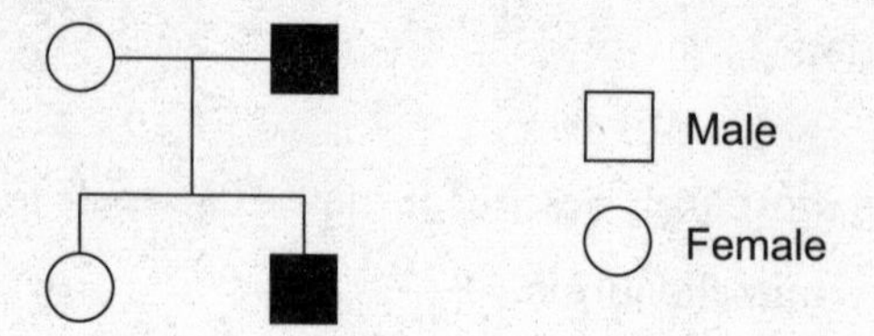

The pedigree can be explained by all of the following inheritance patterns EXCEPT:

A. autosomal dominant allele
B. autosomal recessive allele
C. X-linked dominant allele
D. Y-linked trait

36. When water evaporates from the surface of a pond, what happens to the remaining liquid water?

A. The surface water cools.
B. The surface water warms.
C. The surface water temperature remains unchanged.
D. The pH of the remaining water decreases.

37. Oligomycin is an antibiotic that blocks proton channels in the cristae of mitochondria by binding to ATP synthase. Which of the following would be the first expected response after the application of oligomycin to cells?

A. Water production would increase.
B. H^+ would increase inside the intermembrane space (the compartment between inner and outer membranes).
C. H^+ would increase in the matrix (inside mitochondrial inner membrane).
D. H^+ would increase outside mitochondria.

Question 38 refers to the following data.

Phases of mitosis in a randomly dividing section of a root tip were observed through a microscope in the following numbers:

Phase	Number of Cells
prophase (condensation)	20
metaphase (alignment)	10
anaphase (separation)	15
telophase (restoration)	15
interphase	40

38. Which of the following can be correctly concluded from this data?

A. DNA replication occurs during interphase.
B. Interphase takes longer than all other phases combined.
C. Prophase consumes 10% of the time required to complete a cell cycle.
D. Metaphase consumes the least amount of time during the cell cycle.

39. In the inherited disorder called Pompe disease, glycogen breakdown in the cytosol occurs normally and blood glucose levels are normal, yet glycogen accumulates in lysosomes. This suggests a malfunction with

A. enzymes in the lysosomes
B. enzymes in the mitochondria
C. membrane transport during exocytosis
D. membrane transport during endocytosis

40. Which of the following best summarizes the process of allopatric speciation?

A. differential changes in allele frequencies → reproductive isolation → reproductive barriers → new species
B. catastrophic event → reproductive isolation → reproductive barriers → new species
C. reproductive isolation → differential changes in allele frequencies → reproductive barriers → new species
D. geographic isolation → reproductive isolation → differential changes in allele frequencies → reproductive barriers → new species

Questions 41–42 refer to the following:

A piece of potato is dropped into a beaker of pure water.

41. Which of the following describes the initial condition when the potato enters the water?

A. The water potential of the pure water is negative.
B. The water potential of the pure water is positive.
C. The water potential of the potato is positive.
D. The water potential of the potato is negative.

42. Which of the following describes the activity after the potato is immersed into the water?

A. Water moves from the potato into the surrounding water.
B. Water moves from the surrounding water into the potato.
C. Potato cells plasmolyze.
D. Solutes in the water move into the potato.

Questions 43–45 refer to the following.

The graph that follows shows the absorption spectra for individual pigments found inside a chloroplast and the action spectrum for photosynthesis.

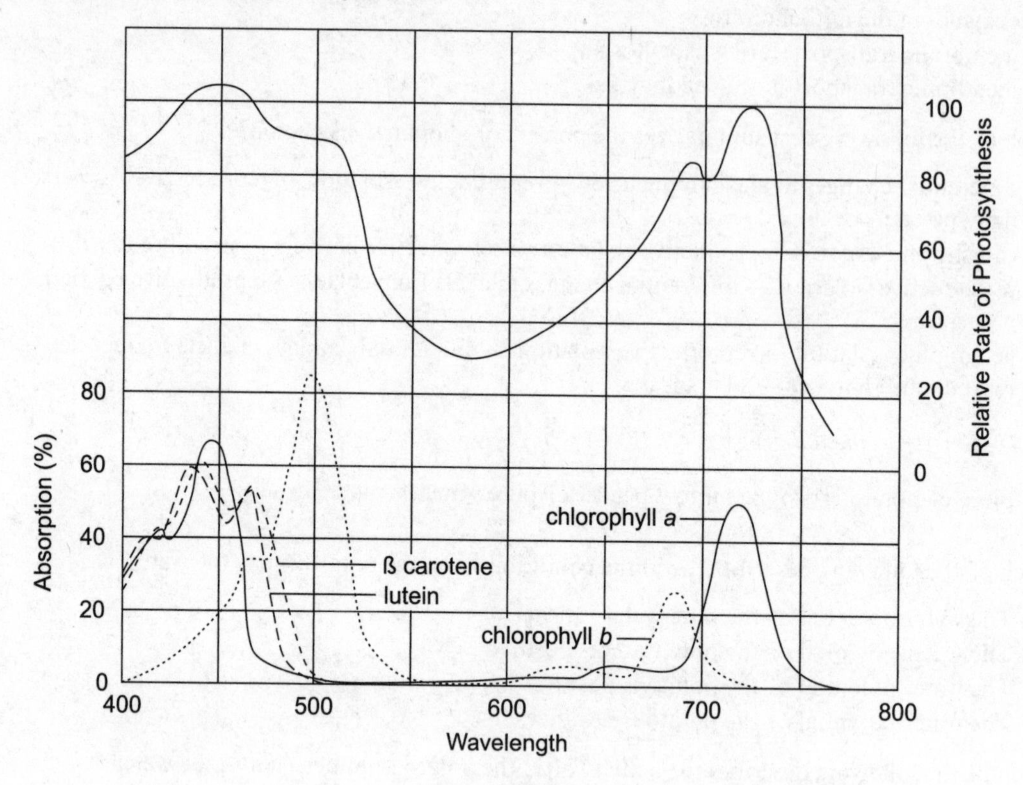

43. Which of the pigments absorbs the most light energy at the *longest* wavelength?

A. β carotene
B. lutein
C. chlorophyll *a*
D. chlorophyll *b*

44. At what wavelength of light is the rate of photosynthesis *greatest?*

A. 450 nm
B. 500 nm
C. 550 nm
D. 650 nm

45. What wavelengths would most contribute to the composition of light *reflected* from a chloroplast?

A. 400–450 nm
B. 450–500 nm
C. 525–575 nm
D. 650–700 nm

46. Which of the following reactions occurs in the forward direction during glycolysis but in the reverse direction during fermentation?

A. pyruvate → lactate
B. pyruvate → ethanol
C. $NAD^+ + H^+ + 2e^- \rightarrow NADH$
D. ADP + P → ATP

Questions 47–49 refer to the following structural formulas.

A.

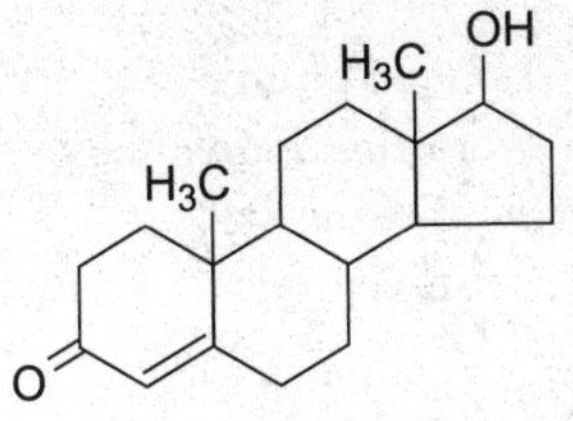

B.

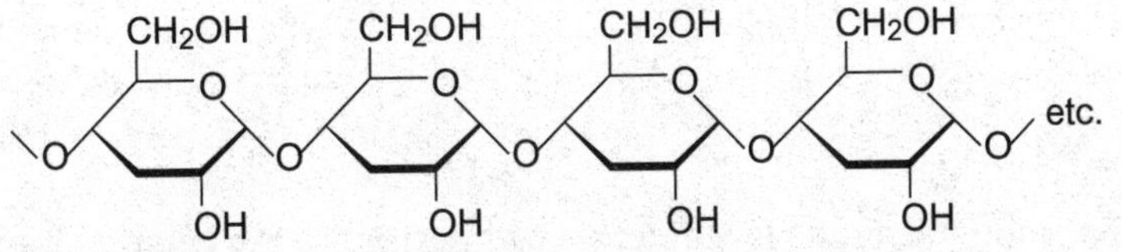

C.

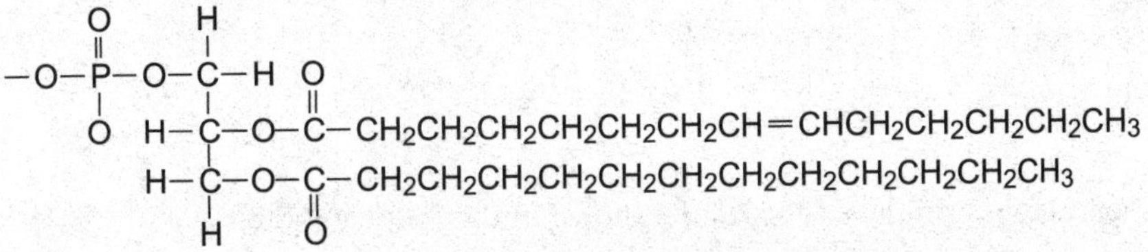

D.

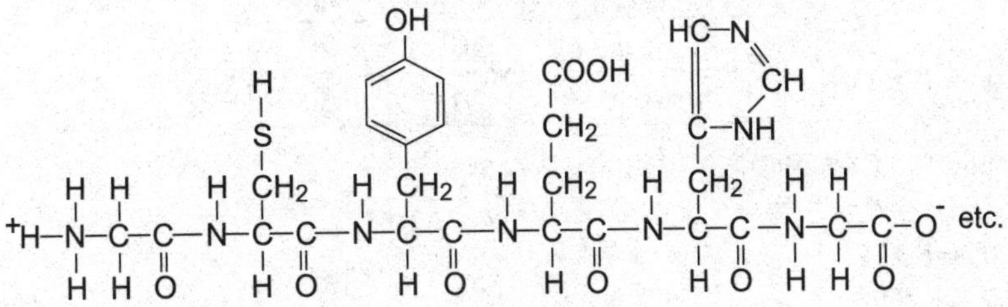

47. Which molecule is drawn to show only its primary structure?

48. Which molecule forms a hydrophilic head and a hydrophobic tail?

49. For which molecule is energy storage a major function?

Questions 50–51 refer to the following.

The following graph shows changes in species abundance with time from early to late successional stages of a temperate forest community typical of the eastern United States.

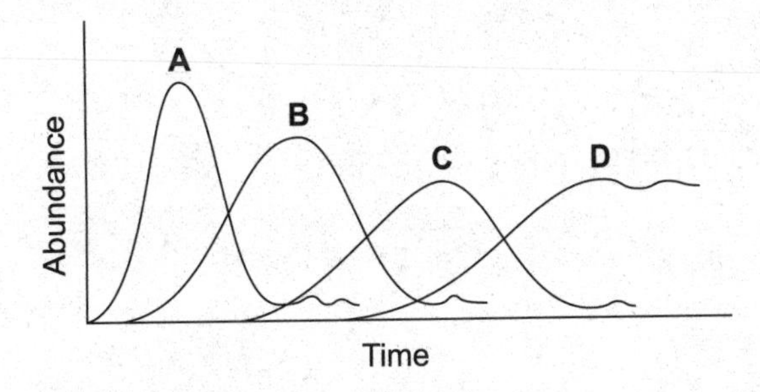

50. Which successional stage would contain the greatest abundance of plants with an *r*-selected life history?

51. Which successional stage would contain the greatest abundance of hardwood trees?

52. Which of the following is most likely to promote the formation of two species in which only one species existed before?

A. the separation of the population into two reproductively isolated groups
B. a mutation in one individual that provides an advantage over other individuals
C. a catastrophic event that reduces the population size to 50 individuals
D. the introduction of a disease to which many individuals are susceptible

Questions 53–56 refer to the following.

In fruit flies, a gene for eye color is X-linked. The dominant wild type allele produces red eyes (*R*), and the recessive mutant allele produces white eyes (*r*). A female who is heterozygous for red eyes is crossed with a white-eyed male.

53. What are the expected phenotypic results for this cross with respect to eye color and sex?

A. 2:1:1:1
B. 1:1:1:1
C. 2:1:2:1
D. 0:2:0:2

54. An experimental cross produced the results shown in the following table.

Phenotype	Females	Males
red-eyed	30	20
white-eyed	20	30

What is the chi-square (χ^2) statistic for the observed results?

A. between 0 and 1.99
B. between 2.0 and 3.99
C. between 4.0 and 5.99
D. 6.0 or greater

55. How many degrees of freedom (*df*) are there for this experiment?

A. 1
B. 2
C. 3
D. 4 or more

56. What can you conclude from the results of the χ^2 statistical analysis?

A. Accept the null hypothesis.
B. Reject the null hypothesis.
C. Conclude that the inheritance of this gene is not X-linked.
D. Conclude that the mixing of genes was not random.

Questions 57–60 refer to the following.

A researcher is investigating the success of a procedure that transfers plasmids to bacteria. She uses a recombinant plasmid that contains a gene that gives resistance to the antibiotic tetracycline.

To determine the success of the procedure, she prepares two batches of bacteria: one with untreated bacteria, the other with bacteria that have been treated with the recombinant gene. She also prepares two kinds of agar plates: one with tetracycline and one without. She transfers each of the two kinds of bacteria to each of the two kinds of agar plates, yielding a total of four plates.

The results of the experiment are shown here, with the number of bacterial colonies indicated.

	Number of Colonies on Agar Plate	
	no tetracycline	**with tetracycline**
normal bacteria	result I: numerous	result III: 0
treated bacteria	result II: numerous	result IV: 15

Use the following answer choices for questions 57–60.

A. Only the treated bacteria are resistant to tetracycline.
B. Some untreated bacteria are resistant to tetracycline.
C. Both normal and treated bacteria are able to grow in the absence of tetracycline.
D. Not all treated bacteria absorbed the recombinant plasmid.

57. What does a comparison of results I and II indicate?

58. What does a comparison of results III and IV indicate?

59. Why is the number of colonies on the tetracycline agar for the treated bacteria limited to 15 instead of showing numerous colonies?

60. What could explain the appearance of one bacterial colony (instead of zero colonies) for result III?

Section II: Free-Response Questions

Time: 90 minutes

2 long free-response questions: Questions 1–2 (allow about 20 minutes each)

4 short free-response questions: Questions 3–6 (allow about 10 minutes each)

Directions: In this part of the exam, the first two questions are long free-response questions worth 10 points each. These questions are followed by four short free-response questions. Each short free-response question has four parts, each worth 1 point for a total of 4 points per question.

Answer all questions as completely as possible within the recommended timeframe. Use complete sentences (NOT outline form) for all your answers. When questions have multiple parts, separate your answers to each part and identify them with the letter for that part. You may use diagrams to supplement your answers, but a diagram alone without appropriate discussion is inadequate.

1. The following results are measurements of dissolved oxygen obtained from water samples collected at various depths in a freshwater lake. Temperatures of samples do not differ significantly.

Depth (meters) (ml O_2/L)	Daytime Measurements (ml O_2/L)		Nighttime Measurements (ml O_2/L)	
	8 a.m.	4 p.m.	8 p.m.	4 a.m.
0	6.0	18.0	16.0	8.0
0.5	4.0	14.0	13.0	6.0
1	3.0	11.0	10.0	4.0
2	2.0	7.0	6.0	1.0
2.5	1.0	5.0	5.0	0.5

a. Using the graph grid provided, **draw** a graph showing how the amount of oxygen gained during the period from 8 a.m. to 4 p.m. varies with the depth of the collected sample. On the same set of axes, **draw** a graph showing how the amount of oxygen lost during the period from 8 p.m. to 4 a.m. varies with the depth of the collected sample.

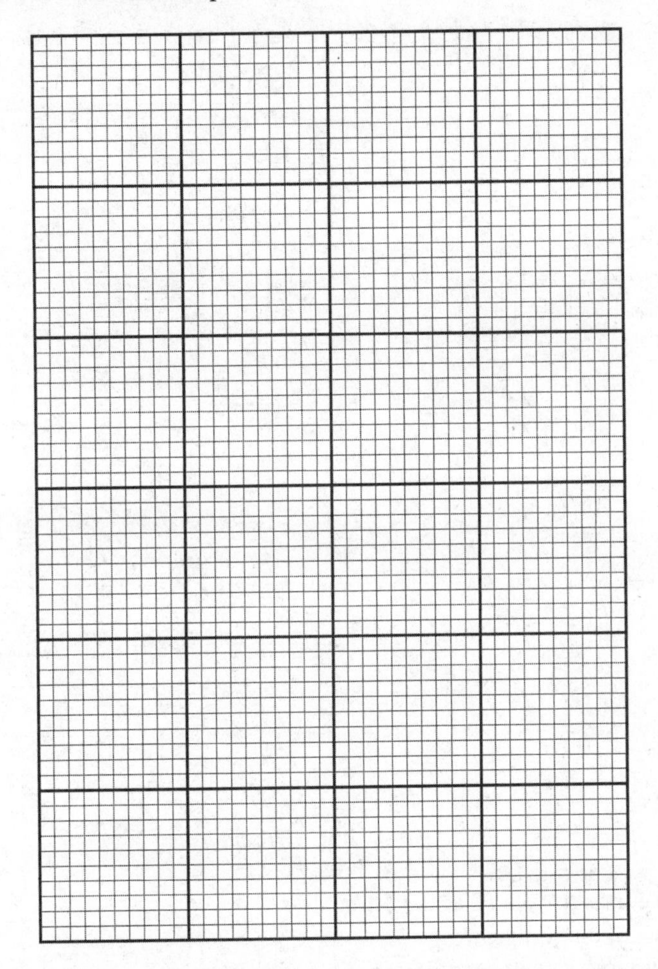

b. **Identify** three abiotic variables that would change the results reported for this lake and **explain** the change or changes that would result.

c. **Explain** why there are differences in oxygen levels between daytime and nighttime measurements made at the same lake depth.

d. **Explain** why there are differences in oxygen levels measured at different lake depths.

2. The graph that follows shows the early growth of a population, labeled I, with two possible outcomes, labeled II and III.

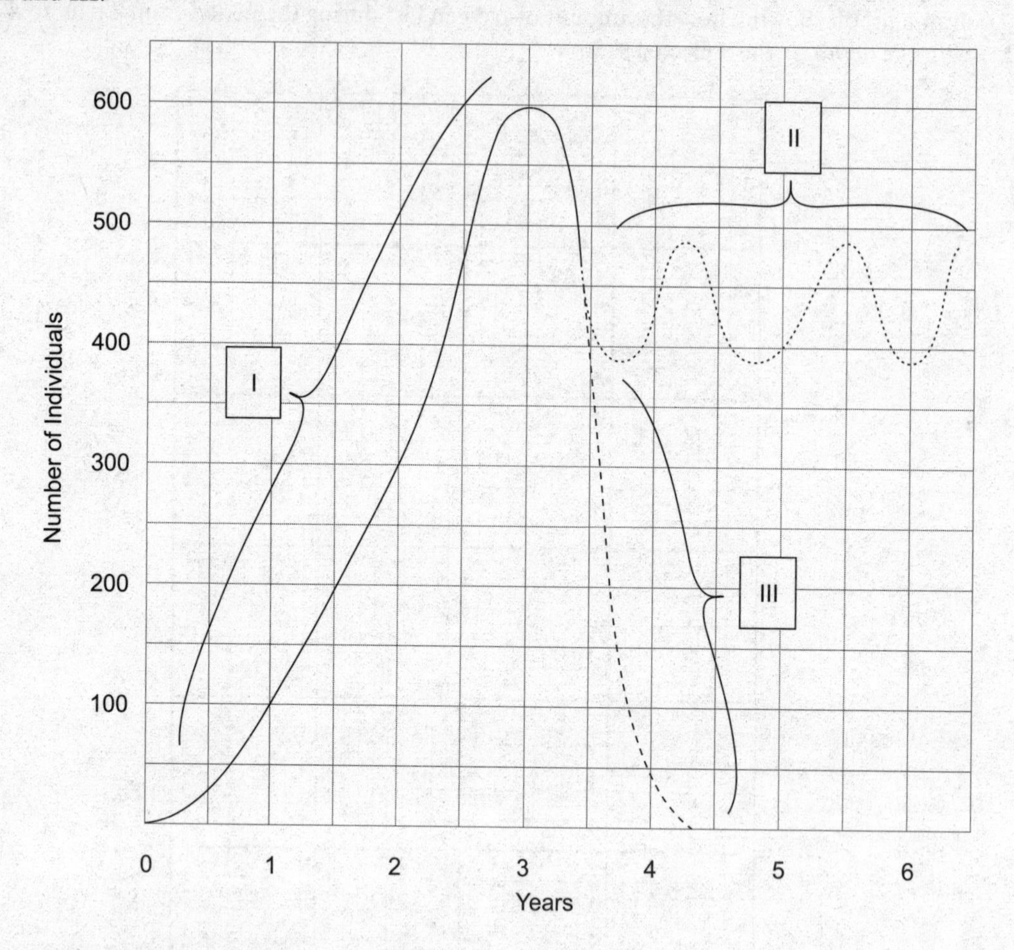

a. For part I of the curve, between 0 and 3.5 years,

(1) Describe the population growth.

(2) Describe factors that may be influencing the rate of growth.

b. Determine the change in population size and **calculate** *r*, the rate of population increase, during the period between the end of the first year and the end of the second year.

c. For part II of the curve,

(1) Describe the population growth.

(2) Describe factors that may be influencing the rate of growth.

(3) Describe the life-history strategy for this population.

(4) Describe an example of a population (or type of organism) that could be experiencing this type of growth.

d. For part III of the curve,

(1) Describe the population growth.

(2) Describe factors that may be influencing the rate of growth.

(3) Describe the life-history strategy for this population.

(4) Describe an example of a population (or type of organism) that could be experiencing this type of growth.

3. Protein structure is described at multiple levels by the interaction of its constituent amino acids with each other and with their environment.

- **a.** **Describe** one kind of amino acid interaction that contributes to the three-dimensional shape of a protein.
- **b.** **Describe** how a protein functions as an enzyme.
- **c.** **Describe** how an enzyme may change when it is moved from its natural aqueous environment to a nonpolar solvent.
- **d.** Ceviche is a Mexican dish that is not cooked with heat. Instead, it is prepared by soaking raw fish in a strong citric acid solution made from limes or other citric juices. **Explain** how this method of preparation is similar to cooking with heat.

4. Sometime before the sixteenth century, a form of deafness, inherited as an autosomal recessive allele, appeared in a town in southern England. In the middle 1600s, several families from the town migrated to North America and settled on Martha's Vineyard, an island off the coast of Massachusetts. During the years from 1700 to 1900, deafness among the 3,000 descendants of the original families was unusually high (as high as 1 in 25) compared with other populations in the United States (1 in 5,700). During this same period, the population on Martha's Vineyard subsisted without interaction with the mainland. Marriages occurred between individuals of the families that lived on the island. During the early 1900s, however, some families moved away from the island, while many new families arrived on the island. By 1950, few individuals with deafness remained.

- **a.** **Describe** the most likely cause for the original appearance of deafness in England.
- **b.** **Determine,** by **constructing** a Punnett square, the probability that two individuals, each with one allele for this deafness, will have a deaf child.
- **c.** **Describe** the evolutionary mechanism most likely responsible for the *increase* in the frequency of deafness from 1700 to 1900.
- **d.** **Describe** the evolutionary mechanism most likely responsible for the *decline* in the frequency of deafness after 1900.

5. A portion of a DNA molecule with a mutation is shown below its wild (normal) type. The genetic code is shown below for reference. Note: The *template* strand is transcribed.

wild type

coding strand	ATG	TCC	CAT
template strand	TAC	AGG	GTA

mutant

coding strand	ATG	**TCA**	CAT
template strand	TAC	**AGT**	GTA

First Letter ↓	Second Letter: U	C	A	G	Third Letter ↓
U	phenylalanine	serine	tyrosine	cysteine	U
	phenylalanine	serine	tyrosine	cysteine	C
	leucine	serine	STOP	STOP	A
	leucine	serine	STOP	tryptophan	G
C	leucine	proline	histidine	arginine	U
	leucine	proline	histidine	arginine	C
	leucine	proline	glutamine	arginine	A
	leucine	proline	glutamine	arginine	G
A	isoleucine	threonine	asparagine	serine	U
	isoleucine	threonine	asparagine	serine	C
	isoleucine	threonine	lysine	arginine	A
	methionine and START	threonine	lysine	arginine	G
G	valine	alanine	aspartate	glycine	U
	valine	alanine	aspartate	glycine	C
	valine	alanine	glutamate	glycine	A
	valine	alanine	glutamate	glycine	G

The Genetic Code

a. **Identify** the type of mutation.

b. **Determine** the amino acid sequence that will be produced by the wild type coding strand after transcription and translation.

c. **Describe** the effect that the mutation in the mutant strand will have on the structure of the protein for which the DNA codes.

d. **Describe** the effect that this mutation will have on the function of the protein for which the DNA codes.

6. The graph below shows changes in the average beak depth of a seed-eating ground finch on a Galápagos island. When the island experienced seasons of drought (1975 to 1978), plants favored by the finches produced fruits with hard shells. When the island experienced seasons of abundant rain (1985 to 1990), plentiful and smaller fruits were produced, and these fruits had softer shells than those produced during seasons of drought.

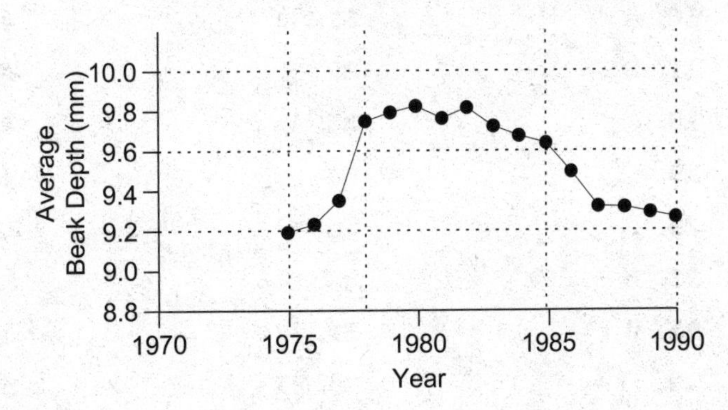

a. **Describe** how the beak depth of the finch changed during seasons of abundant rain.

b. **Explain** why the beak changed in response to seasons of abundant rain.

c. **Describe** how and **explain** why the beak changed in response to seasons of drought.

d. **Make a claim** as to how the population size changed (increased, decreased, or remained the same) during the seasons of drought. **Justify** your claim.

Answer Key for Practice Exam 2

Section I: Multiple-Choice Questions

1. C
2. C
3. B
4. B
5. C
6. B
7. B
8. D
9. A
10. C
11. B
12. C
13. B
14. D
15. D
16. A
17. B
18. C
19. C
20. B
21. A
22. B
23. D
24. C
25. A
26. C
27. A
28. B
29. D
30. D
31. D
32. B
33. D
34. A
35. C
36. A
37. B
38. D
39. A
40. D
41. D
42. B
43. C
44. A
45. C
46. C
47. D
48. C
49. B
50. A
51. D
52. A
53. B
54. C
55. C
56. A
57. C
58. A
59. D
60. B

Scoring Your Practice Exam

Section I: Multiple-Choice Questions

TOTAL number of multiple-choice questions out of 60 you answered correctly	

Section II: Free-Response Questions

Each long free-response question can be worth 8 to 10 points, depending upon the question and the exam.

Each short free-response question is worth 4 points.

Score for free-response question 1 (0–10 points)	
Score for free-response question 2 (0–10 points)	
Score for free-response question 3 (0–4 points)	
Score for free-response question 4 (0–4 points)	
Score for free-response question 5 (0–4 points)	
Score for free-response question 6 (0–4 points)	
TOTAL for Section II (0–36)	

Combined Score (Section I + Section II)

Total for Section I (from above)		× 0.83 =	
Total for Section II (from above)		× 1.39 =	
TOTAL combined score		=	

AP Grade Estimate	
60–100	5
50–59	4
41–49	3
33–40	2
0–32	1

Answers and Explanations for Practice Exam 2

Section I: Multiple-Choice Questions

1. **C.** The beak depth for *G. fortis* varied from 3.8 to 7.8 mm (a range of 4 mm) on Santa Cruz, the greatest of all the occurrences. *G. fuliginosa* varied from 2.7 to 4.3 (1.6 mm range) on Los Hermanos and 2.6 to 4 (1.4 mm range) on Santa Cruz. *G. fortis* varied from 3.5 to 5.5 (2 mm range) on Daphne Major.

2. **C.** The average beak depth increased by 0.8 mm (4.4 to 5.2 mm) for *G. fortis*. The average beak depth decreased by 0.2 mm for *G. fuliginosa* (3.5 to 3.3 mm). Therefore, the greatest average change in beak depth on Santa Cruz after 22 years is the 0.8 mm increase in *G. fortis*.

3. **B.** The beak of a bird is primarily used for feeding. When *G. fortis* arrived on Santa Cruz, its beak depth range overlapped that of *G. fuliginosa*. Competition for food ensued, and natural selection favored those beak depths for which seed sizes were most available. Known as character displacement, this evolutionary response minimizes competition.

4. **B.** Because the solution in the beaker turned blue, you know that the IKI diffused through the bag, into the solution in the beaker, and reacted with the starch. Since the contents of the bag did not turn blue, you know that the starch did not diffuse from the beaker into the bag. Thus, you know the bag is permeable to IKI and not permeable to starch. Since no test for glucose was reported, no conclusion can be drawn for the bag's permeability to glucose.

5. **C.** Signaling molecules are small molecules that bind to larger receptor proteins embedded in the plasma membrane. In general, the binding of the signaling molecule to the receptor protein induces a change in the shape of the receptor protein. Although you do not need to know this to answer this question, the receptor protein here is a receptor tyrosine kinase (RTK) that, when activated by the signaling molecule, causes two RTKs to form a pair (dimer). Autophosphorylation follows, allowing the subsequent phosphorylation of relay proteins.

6. **B.** Activation of this receptor protein (a receptor tyrosine kinase, or RTK) induces autophosphorylation, which, in turn, phosphorylates relay proteins. The relay protein then exchanges a GTP for the GDP on a nearby G protein, which, in this case, is the Ras protein.

7. **B.** In a signaling cascade, the first activated protein kinase is able to activate multiple copies of the second protein kinase. Then, each copy of the second protein kinase is able to activate multiple copies of the third. Because each protein kinase functions repeatedly until it is dephosphorylated, each step of the cascade magnifies the response.

8. **D.** The purpose of a signaling cascade (or any signal transduction pathway) is to elicit a cellular response. In general, these cellular responses include the activation of cytoplasmic enzymatic activity, the activation of a second messenger, or the activation or inhibition of transcription factors that regulate transcription. The product of the signaling cascade shown in the figure is an activated protein that enters the nucleus. Since the figure designates a mitogen-activated protein signaling cascade, the cellular response is to stimulate cell division. Cell division can be initiated by activating transcription factors that regulate transcription of DNA. The products of transcription are mRNAs that are translated to produce proteins that stimulate cell division.

9. **A.** As described in the opening description of the question set, the figure shows a mitogen-activated protein (MAP) signaling cascade whose products (mitogens) stimulate cell division. A permanently activated Ras protein would produce a constant supply of mitogens, which would result in uncontrolled growth and cell division. Uncontrolled cell growth and division are the defining characteristics of cancer.

10. **C.** The RNA genome of a retrovirus contains the enzyme reverse transcriptase. In process I in the figure, reverse transcriptase makes a DNA complement of the viral RNA molecule. In process II in the figure, reverse transcriptase makes a second DNA molecule to complete a viral double-stranded DNA (dsDNA) molecule.

11. B. Once the viral double-stranded DNA incorporates into the host DNA (as a provirus), the viral genes are transcribed by RNA polymerase (in the same manner as the genes of the host).

12. C. The assembly of HIV requires two copies of the viral RNA (with the reverse transcriptase enzyme), viral proteins for the protein coat (capsid), and glycoproteins for the viral envelope. Viral mRNA is translated to produce the proteins; some of those proteins enter the endoplasmic reticulum, where they are modified into glycoproteins. The glycoproteins are transported to the plasma membrane by vesicles, where they merge with the virus as it leaves the cell.

13. B. The viral genes exist in the dsDNA generated in the cytoplasm. The dsDNA enters the nucleus and merges with the host DNA, where it gets transcribed into mRNA along with the host DNA.

14. D. The RNA of RNA viruses is copied by an RNA polymerase or reverse transcriptase. Neither enzyme has a proofreading function or the ability to repair copying errors. As a result, RNA viruses mutate much faster than DNA viruses.

15. D. The outgroup species is chosen so that it is the most primitive of the taxa being studied. Since the outgroup lacks all three of the recorded traits, it branches from the tree first. Species III possesses only trait X, while species I and II both possess traits X and Y. Therefore, species I and II, but not species III, are descendants of a common ancestor possessing trait Y. Finally, species I and species II differ by the absence/presence of trait Z. Two ways of representing this tree are shown below.

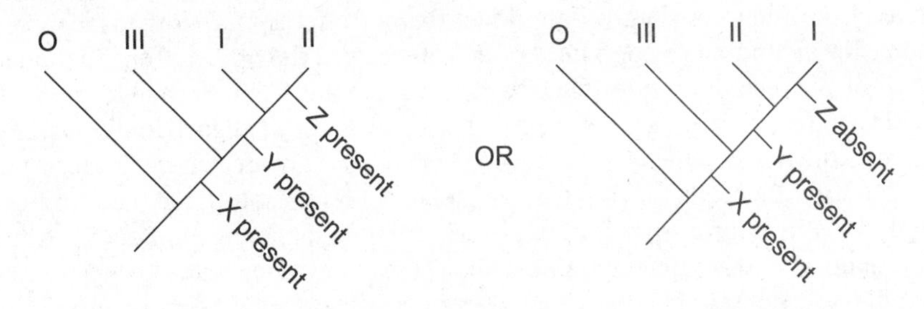

16. A. Since species I and II share the most traits, they are the most closely related.

17. B. The reaction is in equilibrium because the concentrations of reactants and product are such that the rate of the forward reaction (formation of product C) equals that of the reverse reaction (formation of A and B from C), and the *net* production of C (or A and B) is zero. By decreasing C, the forward reaction will again exceed that of the reverse reaction, producing more C, until equilibrium is reached again. Adding more enzyme would only help if there were a considerable excess of reactants (essentially, not enough enzyme to go around) and if the reaction were not already in equilibrium. Increasing the temperature could increase the rate of reactions (in *both* directions), but if the reaction is in equilibrium, the net production of product will still be zero.

18. C. It is important that cells have a large surface area relative to their volume in order to maximize the ability of cells to import necessary nutrients and export wastes. The greater the surface area is, the greater the area for carrying out these processes. A large cell may not have enough surface area to accommodate the transport needs of the cell. Another limitation to cell size is the genome-to-volume ratio. The genome of a cell (the cell's genetic material or chromosomes) remains fixed in size and fixed in its ability to control the activity of the cell (by producing RNA, which in turn produces proteins). The genome may not be able to accommodate the protein (and enzyme) needs of a large cell.

19. C. Divergence of species is indicated by horizontal changes. All of the horizontal changes and, thus, all of the divergence, occur in the area indicated by period IV.

20. B. Area III represents a period of time with little or no evolutionary change maintained by stabilizing selection.

21. A. Adaptive radiation can occur when a single species is introduced into an unoccupied area with many available niches. Rapid evolution of many species occurs as the available niches are exploited. Once all the niches are filled, evolutionary rates decline.

22. **B.** Each plot shows the accumulation of oxygen consumption as time passes. The slopes of the plots represent the *rates* of oxygen consumption (changes in O_2 with time, expressed as $ml \cdot g^{-1} \cdot min^{-1}$, or $ml/g \cdot min$). Because the slopes are constant (nearly straight lines), the respiratory rates are constant but differ for the two temperatures. Oxygen production is recorded in ml/g, which means the amount of oxygen produced, in milliliters, is divided by the weight, in grams, of the hamster. As a result, oxygen production by hamsters with different sizes is averaged over their weights. Although these hamsters may, indeed, have different respiratory rates (due to size or even physiological differences), these differences are not detectable in the reported data.

23. **D.** In this experiment, time is an independent variable and respiratory rate is the dependent variable. Temperature is also an independent variable, but this variable was assigned two values (10°C and 25°C) for the purpose of examining how temperature affects respiratory rate over time. But within each chamber, the temperature must be constant (either 10°C or 25°C). Only time and respiratory rate are allowed to vary. Since respiratory rate is indirectly measured by changes in the quantity of oxygen consumed, the amount of oxygen in each chamber is allowed to change. Note that KOH was added to the animal chambers to absorb any CO_2 produced, thus maintaining a constant amount of CO_2 in the chambers.

24. **C.** A single tree can support large numbers of insects (represented by the second tier of the pyramid), which can, in turn, support many birds (or other insects) that eat the insects. One or two large predators, such as hawks, may occupy the fourth tier.

25. **A.** A deletion (or an addition) of a nucleotide in a DNA strand that codes for mRNA produces a frameshift mutation. As a result, reading from left to right downstream from the mutation, the last nucleotide in every codon will become the second, the second nucleotide will become the first, and the first nucleotide will become the last nucleotide of the preceding codon. Such an arrangement is likely to change many of the amino acids in the sequence (depending upon where in the sequence the frameshift begins) and, thus, considerably affect the final sequence of the polypeptide. Answer choice D may have no effect at all because a change in the third position of a codon will often code for the same amino acid. (This results from the "wobble" of the third position of the tRNA anticodon.) Answer choice B will result in a missing amino acid, while answer choice C will change one amino acid to a different amino acid. These changes may alter the effectiveness of the polypeptide, but not as severely as changing many amino acids, as would occur in answer choice A. The inherited disorder sickle-cell anemia is caused by the replacement of one amino acid by another in two chains of the hemoglobin protein, severely reducing the effectiveness of hemoglobin in carrying oxygen. However, a frameshift in the mRNA coding for hemoglobin would certainly make it entirely ineffective.

26. **C.** Mitochondria and chloroplasts do not function independently from the eukaryotic host cell. Some of the ancestral mitochondria and chloroplast genes are now located in the genome of the host cell. Thus, respiration and photosynthesis cannot occur without the manufacture of necessary enzymes by the host genome.

27. **A.** The graph shows that the cuvette with the higher concentration of substrate (succinate) has a higher rate of respiration. The vertical axis of the graph records the percent transmittance of the solution; the greater the transmittance, the more reduced DPIP in the cuvette, and the greater amount of respiration. Since the *y*-axis represents respiration and the *x*-axis represents time, the respiratory *rate* is the slope of a plotted line. (Note that *rate* means a change in a variable with time, or $\frac{\Delta y}{\Delta t}$, graphically represented by the slope of a plotted line.)

28. **B.** Pyruvate, the product of glycolysis, is the initial substrate for the Krebs cycle. All of the steps of the Krebs cycle that normally reduce NAD^+ to NADH can also reduce DPIP if NAD^+ and pyruvate are continuously provided. However, only succinate is provided, so only succinate can contribute electrons (normally for the reduction of FAD to $FADH_2$).

29. **D.** The cDNA lacks the introns found in the original DNA segment because it is copied (with reverse transcriptase) from the mRNA after it has undergone splicing. Because it contains only the exons, the cDNA produced from the spliced mRNA is the segment with the fewest nucleotides. The number of nucleotides in the introns of a primary RNA transcript varies dramatically, from hundreds to hundreds of thousands of nucleotides. The intron nucleotides are removed by snRNPs during RNA splicing (processing of RNA while still in the nucleus), thus reducing the length of the final RNA considerably. A DNA segment with a point mutation (a single nucleotide replaced, added, or deleted) will differ from the original DNA segment by, at most, only one nucleotide. The primary RNA transcript, containing both introns and exons, will contain the same number of nucleotides as the original DNA segment.

30. **D.** Although disease, competition, and carrying capacity each could be responsible for the decline between 18 and 30 months, no data to support these hypotheses are provided. The only answer choice that is supported by the provided data is that the rate of growth is zero at 18 months and 30 months.

31. **D.** The bands and centromere of the chromosome segment in box 2 of the orangutan are in reverse order (inverted) compared to the bands and centromere of the chromosome segment in box 2 for the human, chimpanzee, and gorilla. Similarly, the segment of the chromosome in box 3 for the orangutan is inverted compared to the other species.

32. **B.** Vesicles that form by phagocytosis usually merge with a lysosome, and their contents are digested by the hydrolytic enzymes inside the lysosome.

33. **D.** All available niches on the mainland were long ago filled by species that, over millions of years, became highly specialized for their niches. Whereas the woodpecker finch is well adapted for seeking insects on the Galápagos Islands, various species of woodpeckers are even more specialized for the same purpose on the mainland. Had an actual woodpecker found its way to the Galápagos Islands before the adaptive radiation of finches, evolution of the woodpecker finch would have been unlikely.

34. **A.** The change in the nitrogen growth rate, approximately 0.21, was greatest of the four nutrients. Or to put it another way, growth would be limited the most if nitrogen were absent. Note that adding phosphorus had little to no effect on growth rate.

35. **C.** If only the father expresses the trait of an X-linked dominant allele, all daughters and no sons will receive the X-chromosome and the trait. His sons receive his Y chromosome and cannot inherit the X-linked trait from him. All the remaining inheritance patterns are possible (including X-linked recessive, a choice not given), as illustrated in the labeled pedigrees below. In each pedigree below, all possible genotypes, based on sex and inheritance pattern, are shown; those genotypes that do not fit the pedigree are crossed out. You should work out Punnett squares to confirm each pedigree.

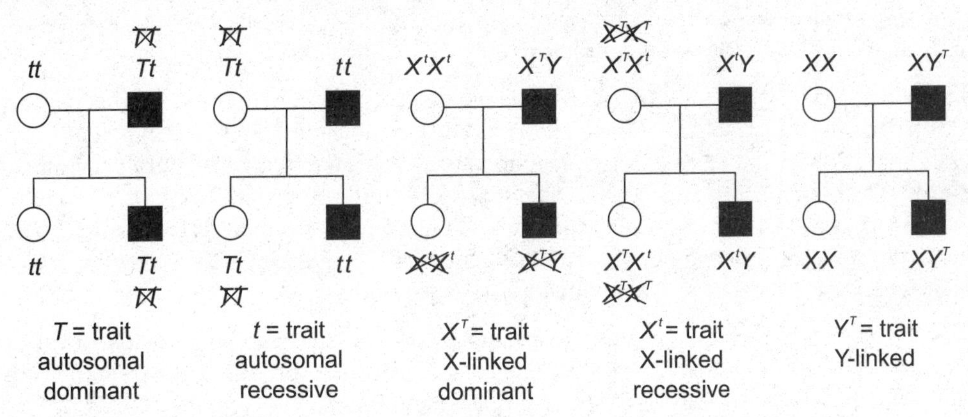

36. **A.** Called evaporative cooling, the water left behind cools because of the energy used (heat of vaporization) to convert the evaporating water molecules from the liquid to the gaseous state.

37. **B.** If the proton channels associated with ATP synthase are blocked, H^+ cannot move from the intermembrane space (the compartment between the inner and outer membranes) across the cristae membranes into the matrix. The expectation, then, is that H^+ concentration increases in the intermembrane space.

38. **D.** The more often a phase appears, the longer it takes for that phase to occur. According to the data, metaphase appeared the fewest number of times; so, compared to the other phases, metaphase took the least amount of time to occur. Interphase took the most time, but not longer than all other phases combined (according to the data presented).

39. **A.** Because lysosomes are sites of chemical breakdown (catabolic processes), the abnormal accumulation of substances within them results from the absence or malfunction of a lysosomal enzyme. These are generally called lysosomal storage diseases and include diseases where various undigested materials, including carbohydrates and lipids, accumulate due to the absence of a correctly functioning lysosomal enzyme. Since cytosol glycogen and blood glucose levels are normal, plasma membrane transport is not the source of the problems associated with the disease.

40. **D.** Allopatric speciation occurs when a geographic barrier creates reproductive isolation between two parts of what was once originally a single population without reproductive barriers. Once gene flow stops between the two new populations (reproductive isolation), changes in the gene pools of each population can occur independently (by mutation, genetic drift, natural selection, nonrandom mating, or gene flow from a third population). Changes in the gene pool (evolution) of the populations may lead to behavioral, physiological, or structural differences that prevent individuals of the two populations from mating or producing fertile offspring (reproductive barriers). Once these barriers are established, the two populations represent two species.

41. **D.** Pure water, with no solutes, has a water potential of zero. The presence of solutes in water or in the water of the potato cells decreases the water potential (makes it more negative). The potato, with central vacuoles filled with water containing solutes, has a negative water potential.

42. **B.** Water moves across a selectively permeable membrane from a region of higher water potential to a region of lower water potential. Because pure water has a water potential of zero and the potato has a water potential of less than zero, water moves into the potato across the plasma membrane. Plasmolysis would occur only if water moved in the opposite direction, out of the potato cells. When plasmolysis occurs, water leaves the vacuoles, the vacuoles collapse, cell turgor drops, and the plasma membrane shrinks and pulls away from the cell wall.

43. **C.** Chlorophylls *a* and *b* have absorption peaks in the long wavelength area (between 650 and 750 nm). Of the two, chlorophyll *a* has the greatest absorption at the longest wavelength of light.

44. **A.** The action spectrum shown at the top of the graph is a plot of photosynthetic rate against light wavelengths absorbed by all pigments. The highest rate of photosynthetic activity occurs at about 450 nm.

45. **C.** Light that is not absorbed is reflected. The wavelengths of light that are not absorbed in photosynthesis occur mostly in the 525 to 575 nm region. This can be seen most clearly by examining the plot for the photosynthetic rate (the action spectrum). Not coincidentally, the color of light in the 525 to 575 nm region is green, the color we see when we look at a leaf.

46. **C.** The reaction $NAD^+ + H^+ + 2e^- \rightarrow NADH$ occurs in glycolysis as energy from pyruvate is used to attach the electrons and protons to NAD^+. The result is NADH, a molecule to be used in oxidative phosphorylation (where the reverse of this reaction occurs), if oxygen is present, to generate 3 ATPs. The reverse of this reaction occurs during anaerobic reactions. In the absence of oxygen, the Krebs cycle and oxidative phosphorylation cannot occur, and no NAD^+ is released from NADH. In order to at least continue with glycolysis, where a net of 2 ATPs is generated, the NADH is converted back to NAD^+ during alcohol or lactic acid fermentation, allowing glycolysis to continue.

47. **D.** This polypeptide is drawn to show only its primary structure, that is, a linear display of its amino acids. The secondary, tertiary, and quaternary structures of polypeptides (proteins) illustrate the complex three-dimensional shapes that are essential for their proper functioning.

48. **C.** This is a drawing of a phospholipid. The phosphate group on the left forms a hydrophilic head. The two long hydrocarbon chains extending to the right are hydrophobic tails. Hydrophilic ("water loving") means that a molecule (or group of atoms) has a strong affinity for or is capable of mixing with or dissolving in water. Hydrophobic ("water fearing") means the molecule resists mixing with water.

49. **B.** This molecule, an α-glucose polysaccharide, is starch or glycogen, both energy storage molecules. Starch is found in plant cells, while glycogen is found in animal cells. In contrast, cellulose (not shown), a β-glucose polysaccharide, functions as a structural element in the cell walls of plants.

50. **A.** Plants with an *r*-selected life history are opportunistic, pioneer species that reproduce and mature quickly and produce many offspring. These species are usually the first species to invade a habitat when succession begins, when a previously uninhabited region becomes available, or after a climax community has been destroyed.

51. **D.** Hardwood trees are in greatest abundance in the climax stages of succession for a temperate forest community typical of the eastern United States.

52. **A.** Reproductive isolation is the initial requirement for the evolution of two species from one species. Mutations, genetic drift, and gene flow can cause changes that result in a population evolving into a new species, but *two* species will be created from one species only if the population is divided into two groups that are reproductively isolated.

53. **B.** Since the trait is X-linked, the cross is $X^RX^r \times X^rY$. Note that there is no allele on the Y chromosome because an X-linked gene occurs only on the X chromosome. The Punnett square below shows that the ratio for the different traits among the offspring is 1:1:1:1, or 1 each of red-eyed female (X^RX^r), white-eyed female (X^rX^r), red-eyed male (X^RY), and white-eyed male (X^rY).

	X^r	Y
X^R	X^RX^r	X^RY
X^r	X^rX^r	X^rY

54. **C.** Since the expected ratio for the genotypes is 1:1:1:1, or 25% each, the expected number of each type of fly is 0.25 × 100, or 25. Using the formula provided on the Equations and Formulas pages for determining χ^2 gives the following:

$$\chi^2 = \sum \frac{(O-E)^2}{E} =$$

$$\frac{(30-25)^2}{25} + \frac{(20-25)^2}{25} + \frac{(20-25)^2}{25} + \frac{(30-25)^2}{25} = \frac{25}{25} + \frac{25}{25} + \frac{25}{25} + \frac{25}{25} = 4$$

55. **C.** The degrees of freedom (*df*) are the number of observed phenotypic categories minus 1, or 4 – 1 = 3.

56. **A.** For a *df* of 3, the critical value for χ^2 is 7.81 ($p = 0.05$, the generally accepted probability for accepting the null hypothesis). Because the value for χ^2 for the data obtained in this question ($\chi^2 = 4$) is less than the critical value ($\chi^2 = 7.81$), the null hypothesis is accepted. The interpretation of these results is that there is more than a 5% chance that the difference between expected and observed values can be explained by chance alone. To help understand what this means, consider the possible extremes: If there were a 100% chance that the difference between expected and observed data could be explained by chance, then the results would be very accurate. That is, the expected and observed results don't differ at all. In contrast, if there were a 0% chance that the difference between expected and observed data could be explained by chance, then the data could only be explained by some variable in the experiment and the hypothesis is incorrect (and must be rejected). If this were the case, answer choices C or D, or some other explanation, could explain the results.

57. **C.** Comparing plates I and II indicates that both kinds of bacteria can grow on regular agar (with no tetracycline).

58. **A.** Comparing plates III and IV indicates that normal bacteria die in the presence of tetracycline but that at least some of the treated bacteria survive. This shows that the treated bacteria have successfully absorbed the plasmid and have expressed the plasmid gene for antibiotic resistance.

59. **D.** Not all of the treated bacteria were successfully transformed. Some of the bacteria did not absorb the plasmid, while others may have absorbed the plasmid but were unable to express the antibiotic-resistant gene.

60. **B.** Some untreated bacteria are resistant to tetracycline. Antibiotic resistance in untreated bacteria may have been acquired by mutation. Another explanation could be that the untreated bacteria or the agar plate on which they were growing was contaminated with bacteria from the treated-bacteria batch.

Section II: Free-Response Questions

Scoring Standards for the Free-Response Questions

To score your answers, award points to your response using the standards given below. For each item listed below that matches the content and vocabulary of a statement or explanation in your response, add the indicated number of points to your score (to the maximum allowed for each section). Your score for each question will range from 0 to 10 points.

Words appearing in parentheses in answers represent alternative wording.

Question 1 (10 points maximum)

a. (3 points maximum)

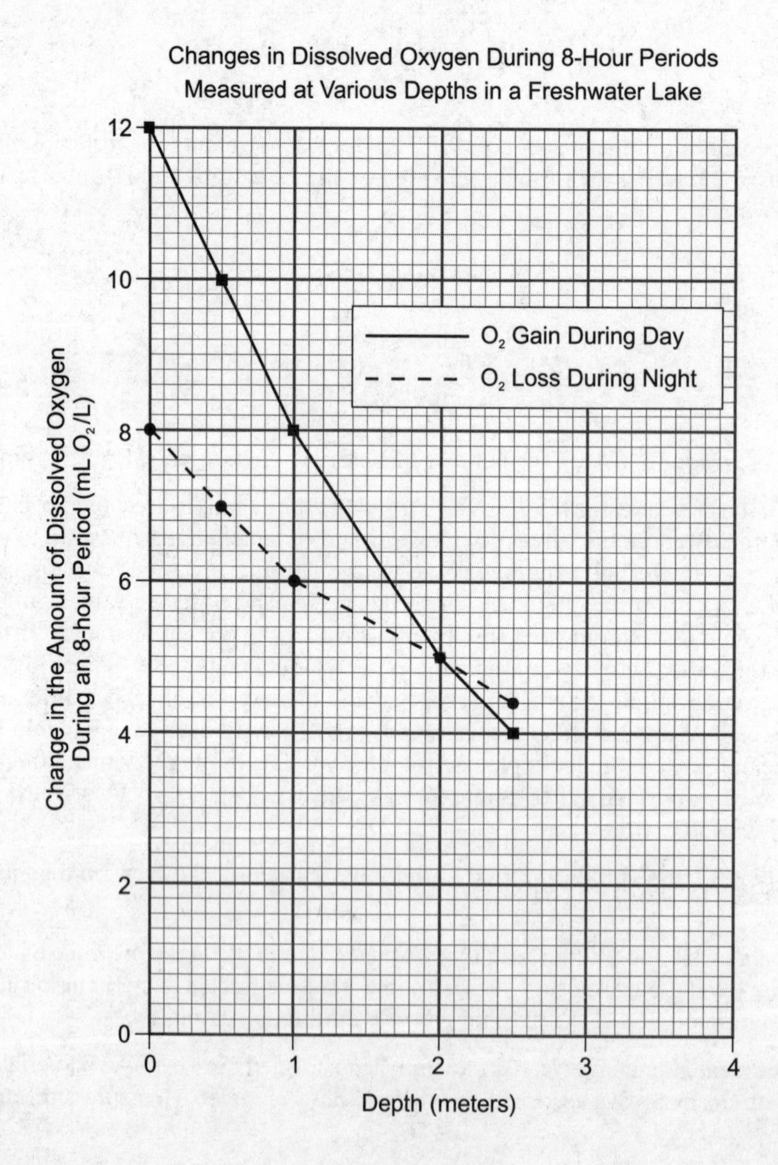

1 pt: Dissolved oxygen concentration (dependent variable) is displayed on the *y*-axis, and depth (independent variable) is displayed on the *x*-axis.

1 pt: Axes are properly labeled and scaled.

1 pt: Both plots are correctly drawn.

1 pt: Each plot is identified with a label or in a legend.

1 pt: A title is given to the graph.

b. (3 points maximum)

1 pt: Temperature: Oxygen solubility in water decreases as water temperature increases. As a result, the oxygen concentrations in the lake for both day and night would be lower if water temperatures were higher.

1 pt: Season or time of year: As the seasons change, temperatures change. During winter, colder atmospheric temperatures would chill the lake water and result in higher concentrations of oxygen in the water. Oxygen solubility in water increases as water temperature decreases.

1 pt: Altitude: The solubility of a gas in a liquid is directly proportional to the pressure of that gas above the liquid. At a higher elevation, where the atmospheric pressure is lower (as is the partial pressure of oxygen), less oxygen would be measured in the lake.

1 pt: Latitude: Higher latitudes would result in lower water temperatures. If the lake in the northern hemisphere were closer to the north pole, average atmospheric and water temperatures would be lower and, as a result, oxygen concentrations in the lake would be higher. Oxygen solubility in water increases as water temperature decreases.

c. (2 points maximum)

1 pt: During the day, photosynthesis in plants produces O_2.

1 pt: During the day and night, respiration consumes O_2.

1 pt: Respiration during the night occurs in both plants and animals.

d. (2 points maximum)

1 pt: At lower depths, the attenuation of light results in reduced photosynthetic activity (or, at lower depths, fewer plants are able to carry out photosynthesis because of reduced levels of light).

1 pt: At lower depths, less oxygen is available for animals to carry out respiration.

1 pt: Less food (plants and animals) is available to animals at lower depths because most plants live closer to the surface.

Question 2 (10 points maximum)

a. (2 points maximum)

(1) 1 pt: In the beginning, the population is experiencing exponential growth (or J-shape growth).

1 pt: In the beginning, resources are readily available.

1 pt: In the beginning, competition and predation (or disease) is minimal.

(2) 1 pt: When the population approaches its peak size, competition, predation, or disease increase and resources become more limited.

b. (2 points maximum)

1 pt: The population size at the end of the first and second years is 100 and 300 individuals, respectively, so the change is 200 individuals.

1 pt: Rate of population increase $\frac{300-100}{100}=2$.

c. (3 points maximum)

(1) 1 pt: The population is fluctuating (or fluctuating about a carrying capacity that would be predicted from a logistic model for population growth).

(2) 1 pt: Resources, living conditions, or competition may be varying from season to season.

(3) 1 pt: The life-history strategy for this population most closely resembles a *K*-selection strategy.

1 pt: In the *K*-selection strategy, individuals have multiple births over a long lifetime, producing few offspring per pregnancy, with offspring requiring considerable parental care (all relative to species with *r*-selection).

(4) 1 pt: Possible examples include snowshoe hares, lynx, elephants, and humans.

d. (3 points maximum)

(1) 1 pt: The population is experiencing a crash (or a catastrophic event).

(2) 1 pt: Resources may have been depleted, disease may have overwhelmed the population, or environmental conditions may have deteriorated or otherwise changed.

(3) 1 pt: Under certain conditions (see above), any population can experience a population crash.

1 pt: Species with an *r*-selection life-history strategy experience rapid growth, often followed by rapid decline. Such species are pioneer or opportunistic species that enter a habitat and reproduce quickly, producing many offspring that require little or no parental care, mature quickly, and have short life spans.

(4) 1 pt: Possible examples include insects, grasses, and weeds.

Question 3 (4 points maximum)

a. (1 point maximum)

1 pt: Hydrogen bonding between the amino acids creates helices.

1 pt: Hydrogen bonding between amino acids creates pleated sheets.

1 pt: Amino acids with hydrophobic R groups clump toward the inside of the protein.

1 pt: Amino acids with hydrophilic R groups clump toward the outside of the protein.

1 pt: Disulfide bonds between certain amino acids create bends or folds in the protein.

b. (1 point maximum)

1 pt: A substrate binds to an active site in the enzyme. This causes a change in the enzyme such that it facilitates a chemical reaction in the substrate.

c. (1 point maximum)

1 pt: Nonpolar (hydrophobic) regions of amino acids will reorient so that they face the nonpolar molecules of the solvent.

1 pt: When immersed in a nonpolar solvent, the structure of an enzyme will change.

1 pt: When the structure of an enzyme changes, it may no longer function.

d. (1 point maximum)

1 pt: When a protein is heated, as when it is cooked, the protein structure is disrupted (denatured).

1 pt: When immersed in a strong acid, the structure of a protein will be disrupted (denatured), much like that which results from the application of heat (cooking).

1 pt: Denatured proteins are often insoluble and coagulate (form solids); (denatured fish proteins change from a translucent, gel-like texture to a solid, opaque texture).

Question 4 (4 points maximum)

a. (1 point maximum)

1 pt: A mutation was most likely the origin of the first appearance of deafness.

b. (1 point maximum)

25%, 0.25, or ¼ and a Punnett square similar to the one below. In this example, *D* represents the allele for normal hearing and *d* represents the recessive allele for deafness. Individuals who are *DD* are normal hearing, *Dd* are carriers with normal hearing, and *dd* are deaf. The Punnett square shows that the probability that a cross of two carriers, *Dd* × *Dd*, have a 25% chance of having a deaf child.

	D	d
D	DD	Dd
d	Dd	dd

c. (1 point maximum)

1 pt: The founder effect describes a population of migrating individuals whose allele frequencies are, by chance, not the same as that of its population of origin.

d. (1 point maximum)

1 pt: Gene flow describes the movement of individuals between populations resulting in the removal of alleles from a population when they leave (emigration) or the introduction of alleles when they enter (immigration).

Question 5 (4 points maximum)

a. (1 point maximum)

1 pt: The mutation is a point mutation (a single nucleotide replacement).

b. (1 point maximum)

1 pt: methionine-serine-histidine

c. (1 point maximum for any one of the following.)

1 pt: There will be no effect on protein structure. (There are no amino acid differences between the wild type and the mutant.)

1 pt: The wild type and mutant DNAs code for the same three amino acids.

d. (1 point maximum)

1 pt: There will be no effect on protein function.

1 pt: It is a silent mutation.

Question 6 (4 points maximum)

a. (1 point maximum)

1 pt: During seasons of abundant rain, the average beak depth decreased.

b. (1 point maximum)

1 pt: For fruits that are small and soft, smaller beaks are more efficient at obtaining seeds than larger beaks.

1 pt: Natural selection favored finches with smaller beaks, as smaller beaks are more proficient at obtaining smaller seeds.

c. (1 point maximum)

1 pt: During seasons of drought, average beak depth increased. Natural selection favored larger and stronger beaks, as they were more proficient at obtaining seeds from fruit with harder shells.

d. (1 point maximum)

1 pt: The population size would decrease during periods of drought because fewer fruits would be available and finches with smaller beaks would not be able to easily access the seeds within the hard-shelled fruits.